Lecture Notes in Chemistry 58

B. O. Roos (Ed.)

Lecture Notes in Quantum Chemistry

European Summer School in Quantum Chemistry

Springer-Verlag Berlin Heidelberg GmbH

Editor

Björn O. Roos
Department of Theoretical Chemistry
Chemical Centre
P. O. Box 124
University of Lund
S-22100 Lund

ISBN 978-3-540-55371-7 ISBN 978-3-642-58150-2 (eBook)
DOI 10.1007/978-3-642-58150-2

Originally published by Springer-Verlag Berlin Heidelberg New York in 1992

Typesetting: Camera ready by author

52/3140-5432 - Printed on acid-free paper

Introduction

The European Summer School in Quantum Chemistry (ESQC) has been arranged on two occasions: the first in August 1989 and then again in August 1991. A third school is planned to take place in August 1993. The total number of participants in the two schools was about 150.

The aim of ESQC was described in the first announcement:

Computational chemistry has today reached a status where available methods and program systems are also used by scientists who are not specialists in the field. We also see increasing interest from the chemical and pharmaceutical industry to apply modern quantum chemical methods in their research. As a result there is a growing need to acquire the background knowledge necessary for a skilful use of these methods. This summer school will attempt to assist in distributing such knowledge.
The emphasis will be more on understanding than on the technical aspects of the methods, and much time will be devoted to discussion of different electronic structure problems and the choice of appropriate methods for their solution. The course will consist of both lectures and exercises.

The teachers of the school produced a set of lecture notes which were used as the basic teaching material. The set was distributed to the students at the start of the school. There has also been a considerable demand for these notes outside the summer schools. This is the reason why we are now publishing part of them in the present form. It should be emphasized, though, that the material in this book does not cover all the topics presented at the school. Such topics as: integrals and integral derivatives, SCF theory, many body perturbation theory, and intermolecular forces, have for various reasons not been included. Some of this material will be published separately. Most of the exercise material has also been omitted.

In spite of this incompleteness, we believe that the book contains enough material to be useful as a textbook for students and researchers interested in learning more about modern methodology in quantum chemistry.

Lund in February 1992

Björn Roos

Contents

Mathematical Tools in Quantum Chemistry

Per-Åke Malmqvist
Theoretical Chemistry
Chemical Centre
University of Lund

1. Important Objects of Quantum Chemistry

Introduction

A mathematical object is defined by a loose analogy to objects in the syntax of ordinary languages. The objects are those parts of a formula that are combined to form other objects, substituted for other objects, or mapped onto other objects, while the "verbs" are designators of how to combine, or which mapping to apply. Obviously, this classification depends on context, upon the usage of the formula, but also on the interpreters attitude: One man's operations are another man's objects.

This becomes particularly clear to the student of Quantum Mechanics. His previous experience was probably that functions specified an action on some specified number to produce another number. This notion was quickly replaced by regarding (wave-)functions as objects, to be acted upon by operators to produce other functions. After a while it becomes clear that it is often much more profitable to regard the operators as objects, to be combined by composition and commutation rules and perhaps mapped onto each other by functions. Any of these levels of abstraction can be found also in computer programs.

We will not try to give a definite description or classification of mathematical objects here. This section should be regarded merely as a collection of useful facts and nomenclature. We will cover the most common terms regarding continuous spaces in general and vector spaces, operators and matrices. We will not touch upon spinors, nor on tensors.

Spaces in general, linear and non-linear.

The set of possible values of a type of objects is called a "space". Thus, we already know some of the simplest spaces. In particular, the set of real numbers is called $\mathcal{R}$, and the set of complex numbers is called $\mathcal{C}$. Since "space" is such a general word, we will confine ourselves to some of the continuous spaces. We shall define here only some classification and nomenclature of major importance. As a first classification ground, consider the *size* of the space.

We have first the simplest multidimensional spaces, namely the set of possible *n-tuples of real numbers,*, called $\mathcal{R}^n$. Similarly, with complex numbers, we get $\mathcal{C}^n$. Their *dimensionality* are n and 2n, respectively.

Next in complexity are the more general n-dimensional continuous spaces. Such a space can be cut up into a finite set of pieces, each of which can be mapped onto $\mathcal{R}^n$by a *continuous and invertible* mapping. If we need no cutting, then the space is topologically equivalent to $\mathcal{R}^n$; other n-dimensional spaces form other equivalence classes, depending upon how many cuts we need, and how the cut edges were connected. As an example, consider a spherical surface, or a torus. As a 3-dimensional example, consider the set of all possible rotations around a fixed point, which can be mapped, *e.g.*, on three Euler angles. However, we will not concern ourselves with topological classification here.

Next, we have infinite-dimensional spaces. The simplest examples are the *set of infinite sequences of real numbers,* $\mathcal{R}^\infty$, and similarly $\mathcal{C}^\infty$ (but in a sense, they are actually the same space). We also have spaces which, although not identical to $\mathcal{R}^\infty$, can be continuously and invertibly mapped onto $\mathcal{R}^\infty$, at least in some neighborhood around any given point.

All these spaces have a reasonably simple structure. In terms of size, there are spaces which are more difficult to handle, *i.e.*, those which are not representable in $\mathcal{R}^\infty$, but we will have no need to consider them. The spaces we have dealt with here will be called n-dimensional and infinite-dimensional spaces, without attempting any further classification.

Apart from size, there are classifications which are based on properties of the elements: Is some generic type of operation or function defined in this space?

Specifically, is there a distance between any two elements? If so, this is a *metric* or *normed* space. A distance is defined as a function with the following properties:

$$\begin{aligned}
&x, y \in S \quad \Rightarrow \quad \exists \quad \mathrm{dist}(x,y) \in R \\
&\mathrm{dist}(x,y) > 0 \quad \Leftrightarrow \quad x \neq y \\
&\mathrm{dist}(x,x) = 0 \\
&\mathrm{dist}(x,y) = \mathrm{dist}(y,x) \\
&\mathrm{dist}(x,y) \leq \mathrm{dist}(x,z) + \mathrm{dist}(z,y)
\end{aligned} \tag{1.1}$$

If there is a metric, we may define convergence. Given a sequence of elements $\{x_i\} \in S$, is there an element towards which the series converges? It turns out that a crucial question is this: assume that we know that the sequence is a so-called Cauchy sequence, *i.e.*,

$$\lim_{\substack{n\to\infty \\ m\to\infty}} \mathrm{dist}(x_m, x_n) = 0$$

Does this imply that there is an element $x \in S$ such that

$$\lim_{n\to\infty} \mathrm{dist}(x_n, x) = 0 \quad ?$$

If so, the space is called a *Banach space*. The spaces we are interested in will in general be Banach spaces.

Is it possible to apply linear operations to the space? These are defined as *scaling* an element by a real or complex number, and addition of two elements to obtain a third. The following rules must be satisfied: Let $\alpha, \beta, \ldots$ be scalars (real or complex) and $x, y.. \in S$. Then

$$\begin{aligned} \alpha x + \beta y &\in S \\ \alpha x + \alpha y &= \alpha(x+y) \\ \alpha(\beta x) &= (\alpha\beta)x \\ \alpha x + \beta x &= (\alpha+\beta)x \end{aligned} \tag{1.2}$$

Such a space is a *linear space* which is also called a *vector space*. The elements are called *vectors*. If the scalars are restricted to be real, it is a real vector space; otherwise, a complex vector space.

If the vector space is *normed*, we must change the distance definition somewhat. First of all, it must be possible to write it in the form $\mathrm{dist}(x,y) = ||x-y||$, *i.e.*, it must be invariant to translation. Furthermore, we can remove the symmetry condition and replace it with the condition of positive homogeneity. In full:

$$\begin{aligned} x \in S \quad &\Rightarrow \quad \exists \quad ||x|| \in R \\ ||x|| > 0 \quad &\Leftrightarrow \quad x \neq 0 \\ ||x|| = 0 \quad &\Leftrightarrow \quad x = 0 \\ ||\alpha x|| &= |\alpha| \cdot ||x|| \\ ||x+y|| &\leq ||x|| + ||y|| \end{aligned} \tag{1.3}$$

Finally, in most vector spaces, a scalar product is defined. This is written $\langle x|y\rangle$ and has the properties

$$\begin{aligned} x, y \in S \quad &\Rightarrow \quad \langle x|y\rangle \in C (or\ R) \\ \langle x|y\rangle &= \langle y|x\rangle^* \\ \langle x|x\rangle > 0 \quad &\Leftrightarrow \quad x \neq 0 \\ \langle x|x\rangle = 0 \quad &\Leftrightarrow \quad x = 0 \\ \langle x|\alpha y + \beta z\rangle &= \alpha\langle x|y\rangle + \beta\langle x|z\rangle \end{aligned} \tag{1.4}$$

This is called a *scalar product space*. Note that a scalar product automatically has a norm defined, namely $|x| = \sqrt{\langle x|x\rangle}$.

If a Cauchy sequence converges, using the distance defined by $|x_n - x_m|$, in a scalar product space, then it is a *Hilbert* space.

Some additional important properties of a scalar product are: Cauchy's inequality

$$|\langle x|y\rangle| \leq |x| \cdot |y| \tag{1.5}$$

The Parallellogram Identity:

$$|x+y|^2 + |x-y|^2 = 2|x|^2 + 2|y|^2 \tag{1.6}$$

The norm polarization

$$|x+y|^2 - |x-y|^2 = 4\langle x|y\rangle \tag{1.7}$$

Matrices and vector representations.

The spaces we have defined could (at least piecewise) be mapped onto $\mathcal{R}^n$, and this means precisely the same as saying that their elements $x, y, .. \in S$ can be *represented by* arrays r,s,.. in $\mathcal{R}^n$. This means that any operations to be carried out that involves elements $x, y, .. \in S$, can be replaced by some equivalent operation involving their mappings, which are the arrays $\mathbf{r} = f(x)$, $\mathbf{s} = f(y)$ etc. However, spaces in general tend to be complicated, and the details of each representation differs.

By contrast, any n-dimensional *vector* space has a generic way, by means of a *basis set,* to be mapped onto $\mathcal{R}^n$, such that the basic operations becomes simply the corresponding element-by-element scalar operations on the arrays. Furthermore, the basis set can always be choosen such that the scalar product assumes the standard form

$$\langle \mathbf{r}|\mathbf{s}\rangle = \langle (r_1 \ldots r_n)|(s_1 \ldots s_n)\rangle = \sum_{i=1}^{n} r_i^* s_i \tag{1.8}$$

For this reason, the n-dimensional vector space $\mathcal{R}^n$with this scalar product is referred to as the *standard representation* of n-dimensional vector spaces.

The following terminology is important: The set $\Omega = \{x_1 \ldots x_k\}$ of vectors $x_i \in S$ is *linearly dependent,* iff there exists a set of scalars $\alpha_1 \ldots \alpha_k$, not all zero, such that $\alpha_1 x_1 + \cdots + \alpha_k x_k = 0$. If this is not possible, then the vectors are *linearly independent.* A vector x_i for which $\alpha_i \neq 0$ is one of the linearly dependent vectors. The set of vectors defines a vector subspace S_1 of S, called $\mathrm{span}(\Omega)$, which consists of all possible vectors $x = \alpha_1 x_1 + \cdots + \alpha_k x_k$. This definition also provides a mapping from the array $(\alpha_1, \ldots, \alpha_k) \in \mathbf{R}^k$ to the vector space $\mathrm{span}(\Omega)$. If Ω is a linearly independent set, then the dimension of S_1 is k, and then the vectors constitutes a *basis set* in S_1. If it is linearly dependent, then there is a subset $\Omega_1 \in \Omega$ of size $k_1 = \mathrm{card}(\Omega_1)$ which *is* linearly independent and spans the same space. Then k_1 is the dimension of S_1.

Furthermore, the basis set is easily transformed to yield a new set $\{b_i\}$ with the property that $\langle b_i|b_j\rangle = \delta_{ij}$ Such a set is called *orthonormal.* The standard procedure is the *Gram-Schmidt ON-algorithm:*

$$\begin{aligned} v_k &= x_k - \sum_{i=1}^{k-1} b_i \langle b_i|x_k\rangle \\ b_k &= v_k/|v_k| \end{aligned} \tag{1.10}$$

This procedure depends upon $|v_k| \neq 0$ in each step, which is guaranteed when we start from linearly independent vectors. That it generates orthonormal vectors is shown by recursion, and by the hermiticity (or symmetry) of $\langle|\rangle$.

So there is always a basis set representation of a vector space, and this representation maps operations in the expected way. Addition of two vectors is represented as

$$\begin{gathered} x = \sum r_i b_i, \quad y = \sum s_i b_i, \quad z = \sum t_i b_i, \\ \text{and } x = y + z \\ \Leftrightarrow \\ r_i = s_i + t_i \end{gathered} \tag{1.11}$$

and so on.

Furthermore, if the basis set is orthonormal, then

$$\langle x|y\rangle = \langle \sum_i r_i b_i | \sum_j s_j b_j\rangle = \sum_{ij} r_i^* s_j \langle b_i|b_j\rangle = \sum_{ij} r_i^* s_j \delta_{ij} = \sum_{ij} r_i^* s_i \tag{1.12}$$

One concludes that all n-dimensional vector spaces are in fact the same abstract space, with the familiar standard representation $\mathcal{R}^n$. Extension to the 'enumerably infinite' spaces with representation $\mathcal{R}^\infty$ is straightforward.

Linear operators and matrices.

In a vector space S, a *linear operator* $\hat{A}$ has the properties

$$\begin{gathered} x \in S \quad \Rightarrow \quad (\hat{A}x) \in S \\ \hat{A}(\alpha x + y) = \alpha(\hat{A}x) + \hat{A}y \end{gathered} \tag{1.13}$$

Using an ON-basis we get the corresponding representation

$$\begin{gathered} x = \sum_i r_i b_i \quad , \quad y = \sum_i s_i b_i \quad , \quad x = \hat{A}y \\ \Leftrightarrow \\ r_i = \langle b_i|x\rangle = \langle b_i|\hat{A}y\rangle = \langle b_i|\hat{A}\sum_j s_j b_j\rangle = \sum_j \langle b_i|\hat{A}b_j\rangle s_j = \sum_j A_{ij} s_j \end{gathered} \tag{1.14}$$

This is simply a *matrix product,* if we interprete **r** and **s** as *column* $n \times 1$ *matrices*. **A** is the *representation* of $\hat{A}$ in the basis set $\{b_i\}$. Thus, it becomes expedient to change the notation somewhat: The linear operators in $\mathcal{R}^n$ are $n \times n$ matrices, while the vectors are *column* $n \times 1$ matrices, rather than rows as we have shown them before.

A valuable concept in matrix algebra is the *hermitean conjugate* of a matrix: If **A** is an $n \times m$ matrix, then $\mathbf{A}^\dagger$ is an $m \times n$ matrix with elements

$$(A^\dagger)_{ij} = A_{ji}^*$$

For example, the scalar product can then be written simply as

$$\sum r_i^* s_i = \mathbf{r}^\dagger \mathbf{s}$$

The question arises: Can objects be defined in S which are represented by the $1 \times n$ *row* matrix $\mathbf{r}^\dagger$? Well, not quite in S itself. It turns out that to any vector x in S there corresponds a so-called *bra functional,* written as $\langle x|$, with the defining property

$$\langle x|y \equiv \langle x|y\rangle \text{ for any } y \in S$$

The set of bra functionals form in themselves a vector space similar to S, and called the *dual* to S. The elements are also called *bra vectors*. In order to get a simple and consistent notation, we also change the notation of the usual vectors: Where we used to write simply x, we now write $|x\rangle$, which is called a *ket* vector. The names are due to Dirac and refers to the common expression $\langle x|\hat{C}y\rangle$, now written as $\langle x|\hat{C}|y\rangle$ and called a 'bracket', *i.e.*, $\langle bra|c|ket\rangle$. An important point is that this notation allows operators in the form of $|x\rangle\langle y|$ to be inserted into formualae. Of particular importance is the expression of the basis set representation on the form $x = \sum b_i \langle b_i|x\rangle$, which we now write as

$$|x\rangle = \sum_i |b_i\rangle\langle b_i|x\rangle$$

for any $|x\rangle$, *i.e.*,

$$\hat{1} \equiv \sum_i |b_i\rangle\langle b_i| \tag{1.15}$$

which is called the *resolution of the identity.* Left multiplying with $\hat{A}$ yields

$$\hat{A} \equiv \sum_i |\hat{A}b_i\rangle\langle b_i|$$

Left multiplying with the identity resolution then gives

$$\hat{A} \equiv \sum_i |b_i\rangle\langle b_i|\hat{A}|b_j\rangle\langle b_j| = \sum_{ij} A_{ij}|b_i\rangle\langle b_j| \tag{1.16}$$

The above holds true in an ON-basis. What about a general basis set? For any basis set $\{|c_i\rangle\}$, define the *metric matrix,* or *overlap matrix,* as

$$\mathbf{S} \quad : \quad S_{ij} = \langle c_i|c_j\rangle$$

Left-multiply $|x\rangle = \sum_j r_j|c_j\rangle$ with $\langle c_i|$:

$$\langle c_i|x\rangle = \sum_j \langle c_i|c_j\rangle r_j = \sum_j S_{ij}r_j$$

$\Rightarrow$

$$r_i = \sum_j (S^{-1})_{ij}\langle c_j|x\rangle$$

Also, we get

$$|x\rangle = \sum_i r_i c_i = \sum_{ij} |c_i\rangle(S^{-1})_{ij}\langle c_j|x\rangle$$

i.e.,

$$\hat{1} = \sum_{ij} |c_i\rangle(S^{-1})_{ij}\langle c_j| \tag{1.17}$$

in a general basis. Sandwiching $\hat{A}$ between identity resolutions gives

$$\hat{A} \equiv \sum_{ijkl} |c_i\rangle(S^{-1})_{ij}\langle c_j|\hat{A}|c_k\rangle(S^{-1})_{kl}\langle c_l| \tag{1.18}$$

These expressions show that a non-ON basis is more difficult to handle: there are repeated inserts of the inverse S matrix, and double as many summation indices. However, there is a simple way to improve things: Define the *dual basis* $\{|\bar{c}_i\rangle\}$ as

$$|\bar{c}_i\rangle = \sum_j (S^{-1})^*_{ji}|c_j\rangle \tag{1.19}$$

$\Leftrightarrow$

$$\langle\bar{c}_i| = \sum_j (S^{-1})_{ij}\langle c_j|$$

We now get

$$
\begin{aligned}
r_i &= \langle \tilde{c}_i | x \rangle \\
\hat{1} &= \sum_i |c_i\rangle\langle \tilde{c}_i| \\
\hat{A} &= \sum_{ij} |c_i\rangle\langle \tilde{c}_i|\hat{A}|c_j\rangle\langle \tilde{c}_j| \\
&= \sum_{ij} A_{ij}|c_i\rangle\langle \tilde{c}_j|
\end{aligned}
\tag{1.20}
$$

etc. The conclusion is that if an ON-basis cannot be used, at least use a dual basis. Note also that the ON-basis is now merely a special case where the basis and its dual happen to be identical.

We will also need to shift occasionally from one basis to another. This can be done by a matrix multiply, so in the representation as an array it looks precisely as if an operator had been applied.

$$
|x\rangle = \sum_i r_i|b_i\rangle = \sum_j s_j|c_j\rangle
$$

$\Rightarrow$

$$
r_i = \sum_j \langle b_i|c_j\rangle s_j = \sum_j T_{ij} s_j \tag{1.21}
$$

(if we assume b_i to be an orthonormal basis.)

Finally, we define an important concept: hermitean conjugation of an operator. To every operator $\hat{A}$ there is a hermitean conjugate $\hat{A}^\dagger$, with the property

$$
\langle x|\hat{A}y\rangle \quad \equiv \quad \langle \hat{A}^\dagger x|y\rangle \tag{1.22}
$$

for all x and y. It turns out that this definition corresponds exactly to the hermitean conjugate of the representation matrix provided that an ON-basis is used. This implies that concepts which involve hermitean conjugation can be carried over to corresponding concepts for matrices, if an ON-basis is used.

Important types of matrices and operators.

We can now classify different types of operators as follows:

$$
\begin{aligned}
\hat{X} \text{ is } \textit{hermitean} &\quad\Leftrightarrow\quad \hat{X}^\dagger = \hat{X} \\
\textit{antihermitean} &\quad\Leftrightarrow\quad \hat{X}^\dagger = -\hat{X} \\
\textit{nonhermitean} &\quad\Leftrightarrow\quad \hat{X}^\dagger \neq \hat{X} \\
\textit{unitary} &\quad\Leftrightarrow\quad \hat{X}^\dagger\hat{X} = \hat{1} \\
\textit{idempotent} &\quad\Leftrightarrow\quad \hat{X}^2 = \hat{X}
\end{aligned}
$$

For matrices, we can use precisely the corresponding definition, but we also add:

$$
\begin{aligned}
\mathbf{X} \text{ is } \textit{orthogonal} &\quad\Leftrightarrow\quad \mathbf{X}^T\mathbf{X} = \mathbf{1} \\
\textit{diagonal} &\quad\Leftrightarrow\quad \mathbf{X} = \mathrm{dg}(d_1 \ldots d_n) \quad\Leftrightarrow\quad X_{ij} = \delta_{ij} d_j \\
\textit{upper triangular} &\quad\Leftrightarrow\quad X_{ij} = 0 \text{ if } i > j \\
\textit{lower triangular} &\quad\Leftrightarrow\quad X_{ij} = 0 \text{ if } i < j
\end{aligned}
$$

One can also define orthogonal *operators*, if a complex-conjugation operator has been defined on the vector space. The last three classifications are basis-set specific and cannot be meaningfully applied to general operators.

An important concept is that of *spectral properties* of matrices. A matrix can be *diagonalized* if there is a basis in which the matrix is diagonal. If so, there exists a basis set transformation, in this context called *similarity transformation*, of the form

$$\mathbf{T}^{-1}\mathbf{X}\mathbf{T} = \mathrm{dg}(d_1 \ldots d_n) = \mathbf{D}$$

or equivalently

$$\mathbf{X}=\mathbf{T}\mathbf{D}\mathbf{T}^{-1} \tag{1.23}$$

The latter form is then called the *spectral resolution* of **X**; the values $d_1 \ldots d_n$ are the *eigenvalues*, and each column $\mathbf{t}_i$ is an *eigenvector*, corresponding to the eigenvalue d_i:

$$\mathbf{X}\mathbf{t}_i = d_i\mathbf{t}_i \tag{1.24}$$

Similarly, for operators we define the set of eigenvalues and eigenvectors

$$\hat{X}|v_i\rangle = \lambda_i|v_i\rangle \tag{1.25}$$

and, if the eigenvectors form a basis set, we get the spectral resolution

$$\hat{X} = \sum_i \lambda_i|v_i\rangle\langle\tilde{v}_i| \tag{1.26}$$

We can now list some of the most important properties of the various types of operators and matrices: Any *hermitean, antihermitean, unitary,* or *idempotent* operator has a spectral resolution where the eigenvectors form an ON-basis, so that

$$\hat{X} = \sum_i \lambda_i|v_i\rangle\langle v_i| \tag{1.27}$$

Furthermore, all eigenvalues of a hermitean operator are *real*, of an antihermitean operator, *pure imaginary*, and for a unitary operator they *lie on the unit circle* $|\lambda_i| = 1$. The idempotent operator is hermitean, and has only eigenvalues 0 and 1. Similarly for the corresponding matrices, and we can then add that the diagonalizing matrix **T** is unitary. However, the idempotent matrix is an exception: It still has eigenvalues 0 or 1, but is diagonalized by a unitary matrix only if it is also a hermitean matrix, which is not always true. Also, we add that a real orthogonal matrix is also a real unitary matrix, and in that case the eigenvalues form mirror pairs. This means that any eigenvalue with imaginary part $\Im(\lambda_i) \neq 0$ has a companion $\lambda_j = \lambda_i^*$.

Finally, using the eigenvalues there are some further subdivision possible: If the product of eigenvalues of a unitary matrix or operator is equal to +1, it is called a *special unitary* (SU) matrix or operator. Similar for real orthogonal matrices, where the only possible choice is +1 or -1: the former case is called *special orthogonal* (SO) matrices. For a matrix, this product equals the determinant of the matrix. For both matrices and operators, the sum of eigenvalues is called the *trace* of the matrix or operator. This equals the sum of the diagonal elements of a matrix representation.

Operations and functions in an operator space.

The linear operators on a vector space forms themselves a vector space, called *operator space*. In this context, the original vector space is called the *carrier* space for the operators. The operator space is sometimes normed, but usually not. Since operator products are defined, we have here a vector space where a product of vectors to give a vector is defined. Such a vector space is also called a *linear algebra*. Operations and functions can be defined in the operator space; thus we can define *superoperators* for which the operator space is the carrier space. The hierarchy is not usually driven any further. Functions are usually named in analogy to their analytical counterparts. To be specific, assume that $\hat{A}$ has a spectral resolution

$$\hat{A} = \sum_i \lambda_i |v_i\rangle\langle v_i| \tag{1.28}$$

Then the analytic function $f(z)$ is extended to the operator space by the definition

$$f(\hat{A}) = \sum_i f(\lambda_i)|v_i\rangle\langle v_i| \tag{1.29}$$

Some of the most important applications are:

- The *resolvent* $\hat{R}(z)$ is defined, for some particular operator $\hat{A}$, as $\hat{R}(z) = (z\hat{1} - \hat{A})^{-1}$. This analytic function of the parameter-dependent operator $z\hat{1} - \hat{A}$ has the following property (assuming again the existence of a spectral resolution):

$$\oint_C \hat{R}(z)\,dz \quad = \quad 2\pi i\,\hat{P}(C) \tag{1.30}$$

  where C is a simple closed curve, traversed in counterclockwise direction, and $\hat{P}(C)$ is the projection operator for precisely those eigenvectors, whose eigenvalue lies inside the curve. The resolvent, when regarded as a function of z, has a pole where z coincides with any eigenvalue. A *reduced* resolvent has had one or more of its poles subtracted away.

- The *exponential* function

$$\hat{A} = \exp\hat{B} \tag{1.31}$$

  can be written as a convergent Taylor series for any bounded operator $\hat{B}$. For an *unbounded* operator, it can still usually be defined as the solution of a differential equation

$$\begin{aligned} \hat{A}(0) &= \hat{1} \\ \frac{\partial}{\partial\lambda}\hat{A}(\lambda) &= \hat{B}\hat{A}(\lambda) \end{aligned} \tag{1.32}$$

  which is then taken to define

$$\hat{A}(\lambda) = \exp(\lambda\hat{B})$$

  An important case is the exponential of an *antihermitean* operator, which can be shown to be an SU operator:

$$\hat{U} = \exp(\hat{X}) \quad , \quad \hat{X} = -\hat{X}^\dagger \tag{1.33}$$

  If the operators are matrices, then X is bounded, so the exponential always exists. This is used to obtain the so-called *exponential parametrization* of the space of

SU(n) matrices. One then needs the inverse. However, it is easily shown that the above equation is fulfilled for infinitely many matrices X. An inverse can still be defined by restricting the size of X, and a common choice is to demand that all the eigenvalues of X should fall on the imaginary axis between $-\pi i$ and $+\pi i$, with exclusion of the former end point.

Returning to the more general equation, $\hat{A} = \exp \hat{B}$ can be inverted if $\hat{A}$ is non-singular and has a spectral resolution. The eigenvalues of B must be restricted to the strip

$$-i\pi < \Im(b_i) \leq i\pi \tag{1.34}$$

Taylor expansion of the inverse, $\hat{B} = \log \hat{A}$, in powers of $\hat{A} - \hat{1}$, is convergent if all eigenvalues of $\hat{A}$ lie within the unit circle $|a_i - 1| < 1$. The resulting B is then limited by

$$\begin{aligned} |\Im b_i| &< \frac{\pi}{2} \\ \Re b_i &< \frac{1}{2} \log(2 + 2\cos(2\Im b_i)) \end{aligned} \tag{1.35}$$

Exercises for Chapter 1.

(1) From the postulates (1.2) which define a vector space, one can deduce the existence of a zero vector 0, which is unchanged by scaling, and which gives $x + 0 = 0 + x = x$ when added to any vector x. Is this zero vector unique, or could there be several zeroes with this property? Is there any other vector, whose existence follows from the postulates? (Hint: see if the space consisting of the single element 0 is a vector space.)

(2) Assume D to be an arbitrary domain: A finite set of elements, or a continuous space, or whatever. The elements are called $a, b \ldots$. Then assume there is a vector space V with elements x,y.... Prove that the set of all possible functions $D \mapsto V$,*i.e.*,$\{f : x = f(a)\}$, in itself constitutes a vector space, if we define scaling and vector add as:

$$\begin{aligned} f = \alpha g \qquad &\Leftrightarrow f(a) = \alpha\, g(a), \quad \forall a \in S \\ f = g + h \qquad &\Leftrightarrow f(a) = g(a) + h(a), \quad \forall a \in S \end{aligned}$$

Can you calculate the dimension of this space?

(Important special cases: If V=$\mathcal{R}$ or $\mathcal{C}$, and if D is some connected region of $\mathcal{R}$ or $\mathcal{C}$, then the set of *continuous* functions is called C(D). Examples: C($\mathcal{R}$)=The set of all continuous functions of a real variable. Also C($[0, 2\pi]$) = The set of continous functions on the interval $[0, 2\pi]$.)

(3) Consider the subset of functions $f \in C(D)$, where D is an interval in $\mathcal{R}$, and $\int_D |f(x)|\, dx < \infty$. Prove that this space (which is called $\mathcal{L}_1(D)$), is a metric vector space.

(4) Consider the subset of functions $f \in C(D)$, where D is an interval in $\mathcal{R}$, and $\int_D |f(x)|^2\, dx < \infty$. Prove that this space (which is called $\mathcal{L}_2(D)$), is a scalar product space, if we define the scalar product

$$\langle f|g \rangle = \int_D f(x)^* g(x)\, dx.$$

(Note: the troublesome part here is to prove that the integral converges. Leave out that part of the proof.)

(5) Prove that $\mathcal{L}_1(D)$ can *not* be a scalar product space. (Hint: Assume that the Parallellogram Identity (1.6) is true, or prove it to yourself. Then select two functions f and g which do not fulfill (1.6))

(6) Assume that $\mathbf{X}$ is an antihermitean matrix, $\mathbf{X}$=-$\mathbf{X}^\dagger$. Prove that $\mathbf{U} = \exp(\mathbf{X})$ is a unitary matrix.

(7) Assume $\mathbf{X} = \theta(\mathbf{u}\mathbf{v}^\dagger - \mathbf{v}\mathbf{u}^\dagger)$ where $|u| = |v| = 1$, $\mathbf{u}^\dagger\mathbf{v} = 0$. Prove that

$$\exp(\mathbf{X}) = \mathbf{Q} + \cos\theta\,\mathbf{P} + \frac{\sin\theta}{\theta}\mathbf{X}$$

where

$$\mathbf{P} = \mathbf{u}\mathbf{u}^\dagger + \mathbf{v}\mathbf{v}^\dagger,$$

$$\mathbf{Q} = \mathbf{1} - \mathbf{P}$$

(Hint: Evaluate a couple of powers of $\mathbf{X}$ and see what happens.)

2. Fundamentals of Convergent Procedures

The O(x) pseudo-function.

In most articles related to approximation theory, or to asymptotic convergence rates, a pseudo-function O(x) is prominently used. We call it a pseudo-function, since it does not identify a specific function, but rather a norm of behaviour to which other functions are compared. Specifically, the following expressions are used:

$$\begin{aligned} f(x) \sim O(g(x)) \text{ means } & \lim_{g(x)\to 0} \frac{f(x)}{g(x)} = A \neq 0 \\ f(x) < O(g(x)) \text{ means } & \lim_{g(x)\to 0} \frac{f(x)}{g(x)} = 0 \\ f(x) > O(g(x)) \text{ means } & \lim_{g(x)\to 0} \frac{g(x)}{f(x)} = 0 \end{aligned} \tag{2.1}$$

The symbol $\sim$ ('similar to') can also be used to compare any two functions, in which case one can say

$$f(x) \sim g(x) \text{ means } \lim \frac{f(x)}{g(x)} = A \neq 0 \tag{2.2}$$

Here, the type of limit ($\to 0$ or $\to \infty$) must be specified if it is not clear from context. If comparison is to the pseudo-function O, then usually the argument tells us which limit to take.

Occasionally, the O(x) function is used also with *large* arguments. In computability theory, for instance, one can see statements such as "Algorithm XX multiplies two n-digit numbers in a time $\sim O(n \log n \log\log n)$". However, it will be clear from context which meaning is currently attached to the O(x) function, and in this paper we will use it only to compare small quantities.

One also frequently writes that $|f(x)| \lesssim g(x)$, meaning that $|f(x)|$ is *asymptotically bounded* by some $A \cdot g(x)$:

$$\limsup_{x\to\infty} \frac{|f(x)|}{g(x)} < A \tag{2.3}$$

Finally, the term "exponentially decreasing" is frequently used. This does *not* mean that $\exists \gamma_0 : f(x) \sim e^{-\gamma_0 x}$. Instead, it means

$$\begin{cases} |f(x)| < O(e^{-\gamma x}) & \text{if } \gamma < \gamma_0 \\ |f(x)| > O(e^{-\gamma x}) & \text{if } \gamma > \gamma_0 \end{cases} \tag{2.4}$$

The Taylor series.

The Taylor series expansion of a function $f(x)$ around the point x_0 is defined as a power series

$$a_0 + a_1(x - x_0) + a_2(x - x_0)^2 + a_3(x - x_0)^3 + \cdots$$

with partial sums

$$s_n(x) = \sum_{i=0}^{n} a_i (x - x_0)^i \tag{2.5}$$

which fulfill

$$f(x) - s_n(x) \lesssim O(x^{n+1})$$

if such a series exists. We quote the following well-known facts:

- If $\limsup_{n\to\infty} \sqrt[n]{|a_n|} = \rho$ exists, then the function $f^{(T)}(x) = \lim_{n\to\infty} s_n(x)$ exists within the circle $|x - x_0| < 1/\rho$ if $\rho > 0$; else, it exists for all x. $R = 1/\rho$ is called the *convergence radius*; the latter case is indicated by saying $R = \infty$.
- Else, the sequence $\sqrt[n]{|a_n|}$ is unbounded. In that case, $s_n(x)$ diverges everywhere except for $x = x_0$, and one says that the convergence radius is $R = 0$.

Now assume a non-zero convergence radius. One can then prove that:

- Inside the circle, $f^{(T)}(x)$ is an analytic function.
- If $f(x)$ is analytic inside the circle, then $f^{(T)}(x) = f(x)$ inside the circle.
- If $f(x)$ is analytic in x_0, then the coefficients are given by

$$a_n = \frac{1}{n!}\frac{d^n f}{dx^n}\Big|_{x=x_0} \tag{2.6}$$

- R is the largest possible radius such that $f^{(T)}$ is analytic within the circle, *i.e.* if R is finite, then there is at least one non-analytic point on the circumference of the circle.

There is a common belief that a convergent Taylor series also means that the series converges to the intended function. This is usually true in most areas of applied mathematics, but there are some *very common counterexamples* in some applications of perturbation theory. One such example is the energy of a system perturbed by an external field F, under circumstances which allow *field ionization* for all non-zero F. Here, the unperturbed energy gets immersed in an infinitely deep continuum as soon as $F \neq 0$ The tunneling probability has the property of being limited by $O(F^N)$ for any arbitrarily large N. The energy *can* be expressed as a function of F, but only if complex energies are allowed, corresponding to finite life time. The energy then has the form

$$E(F) = (E_0 + E_1 F + E_2 F^2 + \cdots) \quad + \quad \Xi(F), \tag{2.7}$$

where Ξ is some complex function with Taylor expansion $\Xi^{(T)}(F) \equiv 0$. The first part of the r.h.s. *may* have a finite convergence radius, which then allows definition of dipole mnoment, polarizability, etc. in the form of energy derivatives, eventhough this at first may seem to be prohibited by the possibility of field ionization.

Another common example arises in the expansion of the interaction energy of two atoms as a function of the inverse interatomic distance $1/R_{AB}$. This turns out to be a Taylor series which differ from the correct energy by a real contribution with zero Taylor expansion. The difference may be attributed to so-called exchange repulsion. In this case, it is nowadays known that the Taylor series has a zero convergence radius, so that the energy expression constitutes an example of an *asymptotic series* (to be defined in a moment) which is *non-convergent* for all R_{AB}.

Asymptotic series.

Sometimes, one is not so much concerned with the pointwise convergence of a series; one merely wants each partial sum to be asymptotically better than the last. Such an *asymptotic series* is almost always an *inverse power series*, and it is then defined as follows:

$$f(x) \sim a_0 + a_1 x^{-1} + a_2 x^{-2} + a_3 x^{-3} + \cdots \tag{2.8}$$

with partial sums

$$s_n(x) = \sum_{i=0}^{n} a_i x^{-i}$$

iff

$$f(x) - s_n(x) \lesssim O(x^{-(i+1)})$$

There is no suggestion that $\lim_{n\to\infty} s_n(x) = f(x)$. By comparison to what is known about Taylor series, by defining $g(z) = f(1/z)$ we immediately find:

- If g(z) is analytic in the origin, then the asymptotic series converges towards f(x) for all $|x| > 1/R$, where R is the convergence radius of the Taylor series for $g(z)$. If $R = 0$, it diverges for all x.
- Else, it either does not converge, or it converges towards a function which differs from f(x). From the definition, it then follows that the difference is limited by all inverse powers of x. The function $e^{-|x|}$ provides an explicit example of such a function.

A famous example of a *non-convergent* asymptotic series is the *Stirling formula* for the factorial function:

$$\log(z!) = \frac{1}{2}\log 2\pi + (z + \frac{1}{2})\log z - z + \frac{1}{12z} - \frac{1}{360z^3} + \cdots \tag{2.9}$$

As it stands, the series diverges for any z. It is usually used together with the recursion $z! = (z+1)!/(z+1)$, which is first employed to push the argument arbitrarily far into the asymptotic region, to provide any requested accuracy of the Stirling formula.

Note, however, that the common belief, that an asymptotic series is always non-convergent, is wrong.

Application of divergent series.

It is not true that a divergent series cannot be applied in computations. The key to the matter is the *analyticity* of virtually all functions derived in applied mathematics, and the essential *uniqueness* of analytical functions. With the latter, we refer to the fact that if two analytic functions are identical in any open disk, they are in fact identical in every point where they have both been defined – with the exception that we may have to select branches of multivalued functions. Thus, as long as some functional dependencies on a variable remain in the terms, it does not matter if some reshuffling of the terms changes the convergence region of a series, as long as these convergence regions overlap somewhere. The point is to make the reshuffling in such a way, that the resulting convergence region covers our intended argument.

There are many different ways to sum a divergent series. The two simplest ways are: either, to add and subtract a function with a known series, with the same type of divergence, or similarly to multiply and divide with such a series expansion. This operation is intended to *remove* the non-analytic point(s) closest to the origin, thereby leaving a rest with larger convergence circle. A slightly less easy way is to make a linear combination of low powers of the expanded function. The trick is then to find such a combination which in itself has a larger convergence circle, and the result is thereby reexpressed as the solution of an algebraic equation where the coefficients can be evaluated as summable series. The latter trick is valuable when the poor convergence is known to arise from branch points of some very large and complicated algebraic equation, such as an eigenvalue problem. The straightforward deformation of the convergence region can be more or less difficult, but can be exemplified: Consider the lower eigenvalue $E(\lambda)$ of the perturbed two-by-two matrix

$$\begin{pmatrix} 0 & \lambda \\ \lambda & 1 \end{pmatrix}$$

which has the Taylor expansion

$$E(\lambda) = -\lambda^2 + \frac{2}{2}\lambda^4 - \frac{2\cdot 6}{2\cdot 3}\lambda^6 + \frac{2\cdot 6\cdot 10}{2\cdot 3\cdot 4}\lambda^8 + \cdots \tag{2.10}$$

This series cannot be used to evaluate E(1), since the convergence radius is only $\frac{1}{2}$. This is due to two branch points, at $\lambda = \pm\frac{1}{2}i$. The simple replacement of λ with μ, which we define as

$$\mu = \frac{2\lambda^2}{1+2\lambda^2} \quad ; \quad \lambda^2 = \frac{\mu}{2(1-\mu)}$$

can be substituted into the terms of the series, and the terms themselves expanded in powers of μ. The results are (after some boring algebra)

$$E(\mu) = -\frac{1}{2}\mu - \frac{1}{4}\mu^2 - \frac{1}{4}\mu^3 - \frac{3}{16}\mu^4 - \frac{3}{16}\mu^5 - \cdots \tag{2.11}$$

The new series has a convergence radius of 1; the value of μ to insert (corresponding to $\lambda = 1$) is $\mu = \frac{2}{3}$, so we have been able to transform the divergent series into a convergent one. This was done by a transformation, whereby the original branch points were transformed into a single branch point located at $\mu = -1$. We also introduced a pole at $\mu = +1$. The resulting convergence circle covers the intended argument, $\mu = \frac{2}{3}$.

There exists also *convergence acceleration* schemes, which act rather like black-box algorithms, where one can enter the partial sums of a series in one column, perform a sequence of 'difference and quotient operations' following a given prescription, and end up with a rapidly converging column of values. This type of scheme generally works fine for the intended purpose, and especially if the non-analyticities responsible for the poor convergence are regular poles rather than branch points. The application of such methods to the partial sums of a divergent series is not recommended.

Iteration, recursion and defect-correction.

Very often, a sequence of quantities $\{Q_i\}$ have a relation between neighboring elements, which can be exploited by repeatedly performing the same operations to successively generate one quantity after another. This procedure is called either *iteration* (from Lat. *iterum*= again) or *recursion* (Lat. *recurrere*= run back). The procedure may terminate, or it may be formally infinite. The only distinction between the two words seems to be that recursion is used to describe a procedure where each of the quantities are considered as potentially interesting, while iteration is used when only a final result (such as the convergence limit) is of interest.

Let us investigate some iterative scheme

$$x^{(k+1)} = F(x^{(k)}) \tag{2.12}$$

as an attempt to find a *fixpoint* x^f of F(x):

$$x^f = F(x^f) \tag{2.13}$$

Let us define an error $e^{(k)}$ by

$$x^{(k)} = x^f + e^{(k)}$$

We immediately get

$$\begin{aligned} e^{(k+1)} &= x^{(k+1)} - x^f = F(x^f + e^{(k)}) - F(x^f) \\ &= F'(x^f)e^{(k)} + \frac{1}{2}F''(x^f)e^{(k)^2} + \cdots \\ &= Ae^{(k)} + Be^{(k)^2} + R(e^{(k)}) \\ &\text{where } R(e) \lesssim O(e^3) \end{aligned} \tag{2.14}$$

Some conclusions can be drawn immediately:

- If $0 < |A| < 1$, there will be some disk around $x = x^f$ within which the procedure will converge with an exponentially decreasing error. If F(x) was unsuitable, then $|A| > 1$ and the procedure will initially diverge, and can never converge towards the intended root.
- If $A = F'(x^f) = 0$, there will be some disk within which e can be mapped onto some new variable z, such that

$$e \mapsto z : z = e + O(e^2)$$
$$e^{(k+1)} = Be^{(k)^2} + R(e^{(k)})$$
$$\Rightarrow$$
$$z^{(k+1)} = Bz^{(k)^2}$$

i.e., the mapping is intended to remove the residual term so the exact solution can be found:

$$z^{(k)} = \frac{1}{B}(Bz^{(0)})^{(2^k)}$$

Within this disk, we get then

$$e^{(k)} \sim O(z^{(k)}) \sim O((Bz^{(0)})^{(2^k)}),$$

i.e.,

$$e^{(k)} \sim O(const.^{(2^k)}) \tag{2.15}$$

Such convergence is called *quadratic*, and is obviously very fast if $const \approx Be^{(0)}$ is appreciably smaller than 1. The convergence region is given by $|z| < |1/B|$, and this can be used as a rough estimate also for the convergence region of x, provided that it is contained in the region where $e \approx z$ is valid.

A full investigation of the convergence properties should ideally subdivide the x space into domains where the procedure diverges, domains where it ultimately converges towards some root, and in particular one convergent domain which is mapped onto itself by the iteration procedure, and which contains the desired solution. This domain, if it exists, is called ***the convergence region,*** and within it, we want to know how fast the convergence is.

Such a full investigation is virtually impossible except in some very simple cases, and is even then usually very difficult. In particular for quadratically convergent methods, the convergence region is usually bounded by a *fractal* instead of a regular curve. Out of necessity, the convergence properties studied are usually some necessary criterion on f and c ***near*** the desired solution, and the influence of that criterion on the asymptotic error.

A particular case is the iterative solution of some equation $f(x) = 0$, which is usually done with some variant of the so-called *defect-correction* methods. Assume that we want to find a solution to the equation

$$f(x) = 0 \tag{2.16}$$

close to some given point x_0. It is then solved by iterating the formulae

$$\begin{cases} r^{(k)} = f(x^{(k)}) \\ x^{(k+1)} = c^{(k)}(x^{(k)}, r^{(k)}) \end{cases} \tag{2.17}$$

where $r^{(k)}$ is the *defect,* or *residue,* and $c^{(k)}$ is the *correction function.* The latter may, as indicated, be different in each step. This may be due to an adaptive modification of the correction function, based upon the result of previous iterations, or it may be due to a switching between use of different correction functions. However, the superscript of $c^{(k)}$ will now be dropped for simplicity.

We must require that

$$x = c(x, 0) \tag{2.18}$$

which is ***necessary*** (but not sufficient) for the $x^{(k)}$ to converge to a solution of (2.16). The reverse requirement

$$x = c(x, y) \quad \Rightarrow \quad y = 0 \tag{2.19}$$

is highly desirable, since this is a ***sufficient*** condition for a converged x to be a solution to (2.16).

In this formulation, let us assume that $f(x)$ can be written as

$$f(x) = f_1 e + f_2 e^2 + R_1(e),$$

where

$$R_1(e) \lesssim O(e^3),$$

and similarly

$$c(e, y) = c_{01} y + c_{11} e y + c_{02} y^2 + R_2(e, y),$$

where

$$R_2(e, y) \lesssim O(e^2 y) + O(e y^2) + O(y^3).$$

Direct substitution into (2.17) gives

$$e^{(k+1)} = Ae^{(k)} + Be^{(k)^2} + R(e^{(k)})$$

where

$$\begin{aligned} A &= c_{01}f_1 \\ B &= c_{11}f_1 + c_{01}f_2 + c_{02}f_1^2 \\ R(e) &\lesssim O(e^3) \end{aligned} \tag{2.20}$$

We will not carry this analysis further, since in writing the correction function on the form $c(x, y)$ we have not explicitly included the dependence on derivatives of f. However, we note that this section provides reason for a classification of defect correction schemes by a new concept, called convergence order. A procedure which guarantees the asymptotic inequalities (within some disk)

$$\begin{aligned} |e^{(k+1)}| &\lesssim |e^{(k)}|^\gamma \quad , \quad \gamma \leq \gamma_0 \\ |e^{(k+1)}| &\gtrsim |e^{(k)}|^\gamma \quad , \quad \gamma \geq \gamma_0 \end{aligned} \tag{2.21}$$

is said to have *convergence order* γ_0. We note that $\gamma_0 > 1$ *guarantees* fast convergence for starting point $x^{(0)}$ sufficiently close to the intended root.

A very important method is the *Newton-Raphson* procedure:

$$x^{(k+1)} = x^{(k)} - f(x)/f'(x) \tag{2.22}$$

Series expansion gives

$$e^{(k+1)} = \frac{f_2}{f_1}e^{(k)^2} - 3\frac{f_2^2 - f_1f_3}{f_1^2}e^{(k)^3} + \cdots, \tag{2.23}$$

so it has second-order (=quadratic) convergence, and unless higher-order derivatives of f are large, the convergence radius can be estimated as $|f_1/f_2|$. In this case, note that the correction function was given the form

$$c(x, y) = x - d(y)$$

This is a common form, but by no means the only one. Often, the corrected x is instead given by some nonlinear procedure. As an example, an SCF procedure can be implemented by applying corrections to the Fock matrix, which is diagonalized to provide new occupied orbitals. The variable x is best represented (in SCF) by a density matrix, which then depends in a highly non-linear way upon the corrections to the Fock matrix.

Finally, we note that if it is impractical to calculate the derivative, it can be replaced by a difference approximation, where the *two* last iterations are used:

$$x^{(k+1)} = x^{(k)} - f(x^{(k)})\frac{x^{(k)} - x^{(k-1)}}{f(x^{(k)}) - f(x^{(k-1)})} \tag{2.24}$$

This method is called *regula falsi,* and it has *superlinear* convergence, which means that $\gamma_0 > 1$. In this one-dimensional case, $\gamma_0 = \frac{1+\sqrt{5}}{2} \approx 1.618$.

We will return to the analysis of iterative equation solving in Chapter 4, but then we will extend it to many dimensions.

Exercises for Chapter 2.

(8) When $n \to \infty$, is $ne^{-n} \sim O(e^{-n})$? – Is ne^{-n} an exponentially decreasing function?

(9) Find the Taylor series for $\log(x)$ expanded around $x_0 = 1$. Determine the convergence radius. Explain which feature of the function is responsible for the limited convergence radius. Then show that a better way of calculating the logarithm is provided by

$$\log(x) = 2((\frac{x-1}{x+1}) + \frac{1}{3}(\frac{x-1}{x+1})^3 + \frac{1}{5}(\frac{x-1}{x+1})^5 + \cdots)$$

which is valid for $\Re(x) > 0$. This is an example of a deformation of the convergence region by a simple variable substitution.

(10) Is it possible to find an asymptotic series $\sum_i a_i x^{-i}$ for any of the following functions? If so, is the series convergent? If it is convergent, what is its sum?

(a) $f(x) = 1/(x\log(x))$

(b) $f(x) = 1/(x + x^2 + e^{-x})$

(11) Consider a sequence $s_n, n = 1, 2, \ldots$, which is slowly converging with

$$s_n = s_\infty + A/n + B/n^2 + C/n^3 + D/n^4 + \cdots$$

Calculate the sequences $a_n, b_n, c_n, \ldots$, as

$$a_n = s_{2n} + (s_{2n} - s_n)/1$$
$$b_n = a_{2n} + (a_{2n} - a_n)/3$$
$$c_n = b_{2n} + (b_{2n} - b_n)/7$$

etc, where the denominators are $2^1 - 1, 2^2 - 1, 2^3 - 1, \ldots$, etc., and show that each of the sequences are converging towards s_∞, and that each new sequence has faster asymptotic convergence than the previous one. This procedure is called *Richardson*-extrapolation and is often used for its simplicity.

(12) Suppose you want to solve the equation $x = e^{-x}$ on a pocket calculator. Suggest a simple method, and verify that it works by working out its asymptotic convergence properties: What is the convergence order? If this is a first-order procedure, what is the convergence rate $|A| \stackrel{def}{=} \lim_{k\to\infty} \left|\frac{e^{(k+1)}}{e^{(k)}}\right|$? Show that the NR method, applied as $f(x) = x - e^{-x} = 0$, takes the form

$$x^{(k+1)} = (1 + x^{(k)})/(1 + e^{x^{(k)}})$$

Try a few iterations and see how it performes.

3. Current Methods for Eigensystem Problems

The numerical solution of large matrix eigenvalue problems is one of the most frequently encountered computation tasks of a standard, yet nontrivial nature. Consequently, there is great theoretical and practical interest in the invention, modification and assessment of methods to cope with this problem. This chapter will describe some well-established methods, currently of widespread use in quantum-chemistry programs.

In general, the problem is to find either a few, or all, of the possible solutions of the equation

$$\mathbf{A}\mathbf{v}^K = \lambda^K \mathbf{S}\mathbf{v}^K \tag{3.1}$$

where **A** and **S** are given $n \times n$ matrices, **S** is non-singular, and the set of scalar *eigenvalues* $\{\lambda^K\}_{K=1}^n$ and *eigenvectors* $\{\mathbf{v}^K\}_{K=1}^n$ are often jointly referred to as the *eigensystem* $\{(\lambda^K, \mathbf{v}^K)\}_{K=1}^n$ of **A** using the *metric* **S**. Following Laplace, the eq. (3.1) is often called a *secular equation.*

We will confine ourselves to the case when **A** and **S** are real symmetric matrices and **S** is positive definite, since this simplifies the problem considerably, while excluding more troublesome cases which may be of more interest to specialists.

We will further consider two sets of problems. On the one hand, there are those problems where we require *all* possible solutions, and where general methods can conveniently cope with matrices up to a size of n $\approx$ 100 on student-size machines, or up to n $\approx$ 1000 on larger machines. On the other hand, there are cases where we require only *some* solutions of a very large problem. In the latter case, the limiting factors are mainly the number of operations needed to form the products of **A** and **S** with a vector, and the number of parameters which must be stored to represent the matrices **A** and **S**. In large CI problems, **S** is a unit matrix, and only a small fraction (typically of the order 1%) of the **A** elements are non-zero. Such problems can be routinely dealt with at most research labs for dimensions up to n $\approx$ 500000. Much larger problems can be handled at some places. As an extreme example, a couple of years ago a CI calculation was performed using slightly more than *1 billion* determinant functions. The size limits can be expected to grow fast in the near future.

Diagonalization Methods

We will start out by assuming that $\mathbf{S} = \mathbf{1}$ and leave the more general case until later. When the entire eigensystem is required, the eigenvectors are collected as columns into a matrix **U**, and the eigenvalues λ as diagonal elements into a diagonal matrix **D**. The equation then takes the form

$$\mathbf{AU} = \mathbf{UD} \tag{3.1.1}$$

It is easily shown that the eigenvectors can be assumed to form an orthonormal set of real vectors. This implies that

$$\mathbf{U}^T\mathbf{U} = \mathbf{U}\mathbf{U}^T = \mathbf{1} \tag{3.1.2}$$

Such a matrix is called a *real orthogonal matrix*. Multiplying $\mathbf{U}^T$ with eq. (3.1.1) gives

$$\mathbf{U}^T\mathbf{A}\mathbf{U} = \mathbf{D} \tag{3.1.3}$$

The solving of this equation is referred to as a *diagonalization* of **A**. Multiplying eq. (3.1.1) with $\mathbf{U}^T$ gives

$$\mathbf{A} = \mathbf{U}\mathbf{D}\mathbf{U}^T \tag{3.1.4}$$

which is called the *spectral resolution* of **A**. Diagonalization methods work by modifying **A** by a succession of *orthogonal transformations* of the type (3.1.3), until the result can be accepted as approximating a diagonal matrix.

The Jacobi Method.

The Jacobi method is traditional (first presented 1846) and simple. More efficient general methods were not found until 1954 (W. Givens, Oak Ridge), and some special features make it the first choice for many applications still today. In order to describe the method in more detail, consider the problem of finding a solution **A**,**V** to the problem

$$\begin{aligned} \mathbf{A} &= \mathbf{U}^T\mathbf{A}^{(0)}\mathbf{U} \\ \mathbf{V} &= \mathbf{V}^{(0)}\mathbf{U} \end{aligned} \tag{3.1.5}$$

where $\mathbf{A}^{(0)}$ is a given real symmetric $n \times n$ matrix and $\mathbf{V}^{(0)}$ is some $n \times m$ matrix given as input, while **A** is the resulting diagonal matrix and **V** is a required output of some kind. Note that **U** is not explicitly used. As we shall see, it is the product of a large number of simple, unitary matrices. Obvious choices for $\mathbf{V}^{(0)}$ are: Leave out $\mathbf{V}^{(0)}$ altogether ($m = 0$) if only eigenvalues are required, or, let $\mathbf{V}^{(0)} = \mathbf{1}$ to obtain the eigenvectors. A common choice is to let $\mathbf{V}^{(0)}$ contain the MO-coefficients of an SCF calculation, while $\mathbf{A}^{(0)}$ contains a Fock matrix in MO-basis. The transformed MO-coefficients are obtained directly as the solution **V** without transformations, and yet one has the advantage of an input Fock matrix which is already nearly diagonal in later iterations (the "prediagonalized" SCF scheme).

The Jacobi method works by successively transforming the matrices **A** and **V** 'in place', in a way which ensures that the non-diagonality measure

$$\tau^2(\mathbf{A}) = \sum_{p>q} A_{pq}^2 \tag{3.1.6}$$

decreases and ultimately converges (quadratically!) towards 0.

In the iteration step k, either the procedure has converged, or else we can find some index pair $i < j$ for which $A_{ij}^{(k)} \neq 0$. We can then calculate

$$\begin{aligned} u &= (A_{jj}^{(k)} - A_{ii}^{(k)})/2A_{ij}^{(k)} \\ t &= sign(u)/(|u| + \sqrt{1+u^2}) \\ c &= 1/\sqrt{1+t^2} \\ s &= ct \end{aligned} \tag{3.1.7}$$

We then define a *plane rotation* matrix $\mathbf{R}^{(k)}$ with elements

$$\begin{array}{ll} R_{ii}^{(k)} = c & R_{ij}^{(k)} = s \\ R_{ji}^{(k)} = -s & R_{jj}^{(k)} = c \end{array} \tag{3.1.8}$$

while all the other elements $R_{pq}^{(k)} = \delta_{pq}$. The matrix $\mathbf{R}^{(k)}$ is an SO-matrix, and so will any product of such matrices be.

A plane rotation is then applied to $\mathbf{A}$ and to $\mathbf{V}$ as

$$\begin{aligned} \mathbf{A}^{(k+1)} &= \mathbf{R}^{(k)T}\mathbf{A}^{(k)}\mathbf{R}^{(k)} \\ \mathbf{V}^{(k+1)} &= \mathbf{V}^{(k)}\mathbf{R}^{(k)} \end{aligned} \tag{3.1.10}$$

It is now fairly easy to show that

$$A_{ij}^{(k+1)} = 0 \tag{3.1.11}$$

and

$$\tau^2(\mathbf{A}^{(k+1)}) = \tau^2(\mathbf{A}^{(k)}) - A_{ij}^{(k)2} \tag{3.1.12}$$

The latter fact alone is enough to ensure convergence, provided that any non-zero non-diagonal element is ultimately selected to be zeroed. On convergence, we will have found a solution of eq. (3.1.5):

$$\begin{aligned} \mathbf{A} &= \lim_{k\to\infty} \mathbf{A}^{(k)} \\ \mathbf{V} &= \lim_{k\to\infty} \mathbf{V}^{(k)} \end{aligned} \tag{3.1.13}$$

where $\mathbf{U}$ was never calculated, but is given by

$$\mathbf{U} = \lim_{k\to\infty} \mathbf{R}^{(k)}\mathbf{R}^{(k-1)}\ldots\mathbf{R}^{(0)}$$

Superficially, eqs. (3.1.11–12) seem to imply that diagonalization could be achieved in $n(n-1)/2$ steps. However, every time a rotation is used to set a specific A_{ij} to zero, previously zeroed elements in the same row and column will be destroyed. This observation also hints at the proper selection of which element to eliminate in each step: It is wasteful to spend operations on a rotation to eliminate a very small element which will anyway later get destroyed, so we should eliminate one of the largest elements. The best choice is usually the largest element; however, for any but the smallest matrices, it takes more time to search for the largest element than to perform the rotation. Common practice is to simply loop over all elements in a predetermined order, but to skip the rotation for elements which fall below a certain threshold. This threshold is then successively lowered before each such sweep. Its value is based on the non-diagonality measure (3.1.6) (due to *v. Neumann*) or some other convenient measure.

The Givens-Householder Method.

As we saw above, the Jacobi method is prevented from reaching convergence in a finite number of steps, because each rotation reintroduces non-zero elements. However, A can be transformed to *tridiagonal* form in a finite number ($(n-1)(n-2)/2$) of rotations. Since finding the eigenvalues of a tridiagonal matrix is a very rapid procedure, this reduction was suggested by Givens as a suitable first step of a diagonalization procedure.

Later, an improved tridiagonalization was found by Householder. His method uses $(n-2)$ orthogonal transformations, each with a transformation matrix of the form

$$\mathbf{H}^{(k)} = \mathbf{1} - 2\mathbf{w}^{(k)}\mathbf{w}^{(k)T} \tag{3.1.14}$$

where $\mathbf{w}^{(k)}$ is a unit vector, which is chosen such that the k-th row and column is tridiagonalized, while the previously introduced zeroes are untouched. For a large matrix, this requires about $\frac{2}{3}n^3$ operations — half as much as the Givens method, and one-third as much as one sweep of the Jacobi method.

After tridiagonalization, there are a number of methods in use to find the eigenvalues. They are all very fast. The currently most popular methods are variations of Wilkinsons QL (also called QR) algorithm. This uses about $9n^2$ operations to transform a tridiagonal matrix to a new tridiagonal matrix with smaller non-diagonal elements. The method converges quadratically to produce a diagonal matrix.

After the eigenvalues have been obtained, one usually wants also the eigenvectors. Again, there are several methods, all with an operation count proportional to n^3. A convenient choice is to use the QL algorithm again. (When a non-degenerate eigenvalue is accurately known, one step of the QL algorithm will produce the corresponding eigenvector.)

Standard library routines are usually delivered either as three separate parts (tridiagonalize, eigenvalues, eigenvectors) or else as two subroutines (obtain only eigenvalues, or obtain eigenvalues *and* eigenvectors). If there are problems in dealing with degenerate matrices, this is because of an inappropriate choice of the last, eigenvector solution, part, and one may consider replacing it.

Using a non-unit metric.

We now reintroduce the metric S, and assume it to be positive definite. We can then find a new set of basis vectors $\{\mathbf{t}_i\}$, which are orthonormal under the metric S:

$$\mathbf{t}_i^T\mathbf{S}\mathbf{t}_j = \delta_{ij} \tag{3.1.15}$$

Collecting the basis vectors into a matrix **T**, which can then be used to transform any vectors and matrices, eq. (3.1.15) takes the form

$$\mathbf{T}^T\mathbf{S}\mathbf{T} = \mathbf{1} \tag{3.1.16}$$

leading to

$$\mathbf{T}\mathbf{T}^T = \mathbf{S}^{-1} \tag{3.1.17}$$

The simple Gram-Schmidt ON-process could have been applied to any initial basis set to find a set of solutions to (3.1.15). The equations above are all equivalent. Any two solutions to these equations are related by some orthogonal matrix **V** :

$$\mathbf{T} = \mathbf{T}'\mathbf{V} \tag{3.1.18}$$

which follows directly from (3.1.17), and this freedom of choice corresponds to some of the freedom we have in selecting the initial basis vectors before applying the Gram-Schmidt process. The natural choice is to start with the standard unit vectors in order, and the array **T** will then be triangular. In fact, we can then write

$$\begin{aligned} \mathbf{L}^T\mathbf{L} &= \mathbf{S} \\ \mathbf{T}^{(GS)} &= \mathbf{L}^{-1} \end{aligned} \tag{3.1.19}$$

where **L** is a lower-triangular matrix. The first of these equations is called the *Choleski decomposition* of **S**. We see that the famous Choleski decomposition is nothing but a Gram-Schmidt-procedure in disguise. The point is that there will be, at most laboratories, very fast and accurate standard library routines available, not only for the Choleski decomposition but also for the subsequent inversion of the triangular matrix **L**.

There are two other common choices of **T**. They both start by considering the diagonalization of the metric **S**

$$\mathbf{W}^T\mathbf{S}\mathbf{W} = \mathbf{D} \tag{3.1.20}$$

Now sandwich this equation between $\mathbf{D}^{-1/2}$ (which is simply the diagonal matrix with elements $D_{ii}^{-1/2}$):

$$\mathbf{D}^{-1/2}\mathbf{W}^T\mathbf{S}\mathbf{W}\mathbf{D}^{-1/2} = \mathbf{1} \tag{3.1.21}$$

which shows that eq. (3.1.16) is trivially fulfilled if we define **T** as

$$\mathbf{T}^{(C)} = \mathbf{W}\mathbf{D}^{-1/2} \tag{3.1.22}$$

which is called the *canonical orthogonalization.* Finally, we also have the possibility

$$\mathbf{T}^{(L)} = \mathbf{T}^{(C)}\mathbf{W}^T = \mathbf{W}\mathbf{D}^{-1/2}\mathbf{W}^T \tag{3.1.23}$$

which is also a solution (by virtue of eq. (3.1.18)) and is called *Lowdin's* orthonormalization, or *symmetric* orthonormalization (since $\mathbf{T} = \mathbf{S}^{-1/2}$ is symmetric).

With $\mathbf{T}^{(GS)}$ and $\mathbf{T}^{(C)}$, one has the opportunity of removing from consideration any vector subspace containing *near-null* vectors, *i.e.*, such vectors **v** that $\mathbf{S}\mathbf{v} \approx 0$. (There may be some difficulty in practice, if a library subroutine was used instead of hand-coding, in the case of $\mathbf{T}^{(GS)}$).

After having obtained the matrix **T**, we proceed by transforming the original equation:

$$\mathbf{A}\mathbf{v}^k = \lambda^k\mathbf{S}\mathbf{v}^k \tag{3.1, \textit{repeated}}$$

is solved by calculating the transformed matrix

$$\mathbf{A}' = \mathbf{T}^T\mathbf{A}\mathbf{T} \tag{3.1.24}$$

then solving the standard diagonalization problem

$$\mathbf{A}'\mathbf{w}^k = \lambda^k\mathbf{w}^k \tag{3.1.25}$$

and finally transforming the vectors:

$$\mathbf{v}^k = \mathbf{T}\mathbf{w}^k \tag{3.1.26}$$

Large Matrix Eigenproblems.

The diagonalization of a large CI Hamiltonian matrix (n = 500000) would require 3000 Gbytes of memory and would take 30 years on a modern supercomputer. The result would be rather uninteresting, however. It is easily shown that (by a number of criteria) the spectrum of the CI hamiltonian matrix is qualitatively different from that of the hamiltonian operator, with the exception of only a few low states for which basis set and configurations have been properly selected. Thus, we want methods which use computer resourses only in the evaluation of a few selected eigenvalues and -vectors.

At the same time, a lot of memory space can be saved. Quite apart from the space needed for the vectors (2000 Gbytes), the space set aside for the matrix elements is needed only if elements of the matrix are to be manipulated by transformations. However, if a defect-correction method is employed to solve (3.1) as it stands, then all we need is the ability to form, for a given vector $\mathbf{v}$, the products $\mathbf{Av}$ and $\mathbf{Sv}$. In the CI case, $\mathbf{S} = \mathbf{1}$ and also the multiplication with $\mathbf{A}$ can be made using the electrostatic integral list without calculating any elements A_{ij}. The reduction in storage requirement is dramatic: Even if only non-zero elements of A were stored, we would need (assuming 1 % non-zero elements) 10 Gbytes of values and an additional 10 Gbytes indices. By contrast, the integral list will typically be of the order 10 Mbytes in size. This approach, to form products $\mathbf{Av}$ directly from some basic set of data rather than to calulate matrix elements, is in this context called *direct CI methods.*

For large matrix eigenproblems, there are essentially two groups of methods used today. Let us write the problem in terms of a defect:

$$(\mathbf{A}-\lambda\mathbf{S})\mathbf{v} = \mathbf{r} \tag{3.2.1}$$

and the goal is to manipulate λ and $\mathbf{v}$ so, that the defect $\mathbf{r}$, the *residual vector,* decreases towards zero.

The first group of methods then manipulate a very small subset of vector elements v_i at a time, and a direct method continually updates the affected elements r_j. Such methods are collectively known as *relaxation* methods, and they are primarily used in situations where, for each elements v_i to be changed, the set of affected r_j, and the matrix elements A_{ij}, are immediately known. This applies in particular to *difference approximations* and also to the so-called *Finite Element Method* for obtaining *tabular* descriptions of the wave function, *i.e.* a list of values of the wave function at a set of electron positions. (For some reason, such a description is commonly referred to as a "numerical" solution to the Schrodinger equation). Relaxation methods have also been applied to the CI problem in the past, but due to their slow convergence they have been replaced by analytical methods. Even for "numerical" problems, the relaxation methods are slowly yielding to "analytical" methods.

The second group of methods use the residual vector $\mathbf{r}$, in each step of an iterative process, to form a new member of a set of basis vectors. A new $\mathbf{v}$ is formed as a linear combination of basis vectors, such that the resulting residual vector is orthogonal to the currently employed expansion space. If it is non-zero, one can always use it to form a new vector, linearly independent of the current basis set. Thus, the procedure will not stop until eq. (3.2.1) is accurately satisfied; on the other hand, it is guaranteed to terminate when the basis set spans the entire vector space. Although the latter point is uninteresting from a practical point of view, it means that these methods have guaranteed convergence to a solution in a finite number of steps. Note that it also means that a classification of the procedure in terms of convergence order is impossible.

In this orientation, we will briefly describe the most common methods of the second group only. This group can be classified as *subspace iteration* methods.

The Lanczos Method.

In its original form, the Lanczos algorithm was intended as a tridiagonalization method. When used as such, the iterations are continued until termination. The algorithm is most easily described by assuming the tridiagonal result, and by specifying that the basis set should be orthonormal:

$$\mathbf{B}^T\mathbf{A}\mathbf{B} = \begin{pmatrix} d_1 & t_1 & . & . & . \\ t_1 & d_2 & t_2 & . & . \\ . & t_2 & d_3 & t_3 & . \\ . & . & t_3 & d_4 & . \\ . & . & . & . & . \end{pmatrix} = \mathbf{T} \tag{3.2.2}$$

where $\mathbf{B}^T\mathbf{B} = \mathbf{1}$. Then

$$\mathbf{A}\mathbf{B} = \mathbf{B}\mathbf{T} \tag{3.2.3}$$

or

$$\mathbf{A}\mathbf{b}_i = t_{i-1}\mathbf{b}_{i-1} + d_i\mathbf{b}_i + t_i\mathbf{b}_{i+1} \tag{3.2.4}$$

This leads immediately to the recursion formula

$$\begin{aligned} \mathbf{b}_i &= \mathbf{x}_i/x_i \\ \sigma_i &= \mathbf{A}\mathbf{b}_i \\ \mathbf{d}_i &= \sigma_i^T\mathbf{b}_i \\ \mathbf{x}_{i+1} &= \sigma_i - d_i\mathbf{b}_i - t_{i-1}\mathbf{b}_{i-1} \\ t_i &= x_{i+1} \end{aligned} \tag{3.2.5}$$

(Terminate if $t_i = 0$, else repeat)

The recursion starts with an arbitrary vector $\mathbf{x}_1$, and the term containing t_{i-1} is ignored the first time. When the process terminates, the generated basis functions span the same space as the smallest set of eigenvectors which have a non-zero projection on the starting vector. The specification 'smallest' means that only one representative of each degenerate set of eigenvectors will appear. These basis functions are precisely the Gram-Schmidt orthogonalization of the vectors $\mathbf{x}_1, \mathbf{A}\mathbf{x}_1, \mathbf{A}^2\mathbf{x}_1 \ldots$, which is called a *Krylov* sequence. The distinctive quality of Lanczos method is that this is achieved without any need to save the basis vectors. However, the method in this form is unstable – the basis vectors generated will gradually become less orthogonal to the first vectors. One may put in an explicit orthogonalization, but then all vectors must be stored.

When used for eigenvalue problems, the process is started with the best possible guess to the eigenvector, and the tridiagonal matrix in each step is diagonalized. The procedure can be terminated when the coefficient in the last row of the interesting eigenvector of $\mathbf{T}$ is close to zero. This will occur in a few iterations if the set of eigenvectors with non-zero projection on $\mathbf{x}_1$ have *clustered* eigenvalues, *i.e.* if the set of *different values* of the eigenvalues is small.

However, a CI hamiltonian matrix have eigenvalues spread out in a very wide range, and this reduces the value of the Lanczos algorithm for CI purposes. Attempts have been made to *precondition* the Lanczos procedure, such that the iterations are governed by a clustered matrix while at the same time a stable subspace of the original matrix is constructed. The success of such a scheme has yet to be demonstrated.

The Davidson Method.

The original Davidson's method is easily extended in various ways, and it is just as easy to give a fairly general description first and specialize later.

First, consider any specific step of the iteration process, and assume that at present we are using N orthonormal basis vectors $\mathbf{b}_i$, collected as before in $\mathbf{B}$. Then form the *small Davidson matrix* $\tilde{\mathbf{A}}$ as

$$\tilde{\mathbf{A}} = \mathbf{B}^T\mathbf{A}\mathbf{B} = \mathbf{B}^T\mathbf{\Sigma} = \mathbf{\Sigma}^T\mathbf{B} \tag{3.2.6}$$

The columns of $\mathbf{\Sigma} = \mathbf{A}\mathbf{B}$ are the *sigma vectors* $\sigma_i = \mathbf{A}\mathbf{b}_i$ and they are stored, together with the vectors $\mathbf{b}_i$, on disk. In each iteration, most of $\tilde{\mathbf{A}}$ is already calculated, and only the new entries have to be added. The $\tilde{\mathbf{A}}$ matrix is then diagonalized

$$\tilde{\mathbf{A}}\mathbf{U} = \mathbf{U}\mathbf{D} \tag{3.2.7}$$

and the interesting columns of $\mathbf{U}$, the *selected roots,* are collected in a matrix $\mathbf{U}^S$ (very often a single column):

$$\tilde{\mathbf{A}}\mathbf{U}^S = \mathbf{U}^S\mathbf{D}^S \tag{3.2.8}$$

We can now form the *residue vectors* as columns in $\mathbf{R}$:

$$\mathbf{R} = \mathbf{\Sigma}\mathbf{U}^S - \mathbf{B}\mathbf{U}^S\mathbf{D}^S \tag{3.2.9}$$

If $\mathbf{R} = 0$ we have converged; else we can proceed with the preconditioning step: Assume that the M non-zero columns of $\mathbf{R}$ are $\mathbf{r}_1 \ldots \mathbf{r}_M$, and determine a set of new vectors as

$$\mathbf{q}_i = \mathbf{X}_i\mathbf{r}_i \tag{3.2.10}$$

Here, $\mathbf{X}_i$ is a *preconditioner,* which should, ideally, be as close as possible to $(\mathbf{A} - \lambda_i)^{-1}$, where λ_i is the true eigenvalue approximated by D^S_{ii}. Finally, the vectors $\mathbf{q}_1 \ldots \mathbf{q}_M$ are orthonormalized to the old basis set, and to themselves, and the result is the new basis vectors $\mathbf{b}_{N+1} \ldots \mathbf{b}_{N+M}$. The sigma vectors are calculated, new elements are added to $\tilde{\mathbf{A}}$, etc.

The simplest implementation is to add only *one* new basis vector at a time, and to use a *diagonal preconditioner*

$$q_j = \frac{r_j}{A_{jj} - D^S} \tag{3.2.11}$$

This defines the *standard Davidson* method. A very common improvement is the *Davidson-Liu* method, which uses several vectors at a time, and a different, diagonal, preconditioner for each root:

$$q_{ji} = \frac{r_{ji}}{A_{jj} - D^S_{ii}} \tag{3.2.12}$$

This variant is especially valuable in near-degenerate cases and when several roots are required.

Furthermore, the preconditioner can be improved: First separate out a small subset of preselected configurations, and diagonalize $\mathbf{A}$ in this subspace. Split up

$$\mathbf{r}_i = \mathbf{r}'_i + \mathbf{r}''_i$$

where the prime denotes the small subspace (typically about 100 configurations), and then

$$q_{ji} = \sum_K \frac{u_j^K(\mathbf{u}^{K^T}\mathbf{r}'_i)}{\mu^K - D_{ii}^S} + \frac{r''_{ji}}{A_{jj} - D_{ii}^S} \tag{3.2.13}$$

where $\{\mu^K, \mathbf{u}^K\}$ is the eigensystem of the submatrix of $\mathbf{A}$ projected on the preselected space. This improves convergence very much when there are low-lying near-degeneracy. No particular name has been coined for this *specific* procedure, but the term *generalized Davidson* has been suggested as a generic term to denote use of a improved preconditioner.

It should be noted that a seemingly *worse* approximate preconditioner sometimes *improves* convergence. This happens *e.g.* when a Slater determinant basis is used, and the reason is that the diagonal approximation breaks the spin symmetry. If the preconditioner is replaced by

$$q_{ji} = \frac{r_{ji}}{A_{jj}^0 - \Delta} \tag{3.2.14}$$

where A_{jj}^0 is the average of A_{jj} over all determinants which belong to the same configuration, one will get a faster convergence.

In recent years, it has been found that use of increasingly accurate approximations to the inverse hamiltonian only gives a limited improvement in convergence speed, unless its application is somewhat altered: instead of solving approximately the equation $(\mathbf{A} - \lambda)\mathbf{q}_i = \mathbf{r}_i$, one must add to the right-hand side a scale factor times the current approximative wave function. The preconditioned update is then frequently so good that it no longer makes sense to store a large number of vectors on disk, and it is then advantageous to store only a few of the most recent eigenvector approximations, with their corresponding sigma vectors, and to overwrite the oldest approximation with the newest update in round-robin fashion. In this case, the basis set consists of the latest eigenvector approximations in themselves, which means that the small eigenproblem (3.2.8) will now be of the general type, involving an overlap matrix.

Finally, it should be noted that, with minor modification, the standard Davidson procedure is able to produce systematically the terms of the Moller-Plesset perturbation series to any order, provided that a full CI space is used. This has been used to study the convergence properties of the perturbation series.

Bibliography.

The methods described here are described in the following original articles:

W. Givens, ORNL-1574, Oak Ridge National Laboratory (1954).

A. S. Householder, J. Soc. Ind. Appl. Math. 6,189(1958).

J. H. Wilkinson, "The Algebraic Eigenvalue Problem",
Oxford University Press 1965.

C. Lanczos, J. Res. Natl. Bur. Stand. 45,255(1950).

E. R. Davidson, J.Comput. Phys. 17,87(1975)

B. Liu, in "Numerical Algorithms in Chemistry: Algebraic Methods",
c. Moler and I. Shavitt (eds.)
LBL-8158 Lawrence Berkeley Laboratory 1978.

A good general reference is:

B. N. Parlett, "The Symmetric Eigenvalue Problem",
Prentice-Hall, N. J. 1980.

Exercises for Chapter 3.

(13) The 9×9 matrix $A_{ij} = 0.1 + 0.9\,\delta_{ij}$ has one eigenvalue close to 2. Use Davidson's method to find the eigenvalue and the corresponding eigenvector. Note: In this case, the preconditioner (3.2.11) will be singular. Common practice is that whenever the preconditioning formula will require a divide by 0, it is replaced by 1.

Hint: Is there some "direct CI"-like way of computing the sigma vectors without having to write it all up on paper?

4. Iterative Methods for Equation Systems

Iterative Solution of Non-Linear Equation Systems

We have already had a brief introduction to the iterative solutions of equations on the form f(x)=0. We implicitly assumed that both f and x were complex- or real-valued. However, most of it carries over directly to multi-dimensional cases. Some new nomenclature has to be presented, and in practice the computational complexity and the convergence problems may be more severe, but the basics are precisely the same. We will assume here that the problem has been mapped onto $\mathcal{R}^n$. None of the problems concerning the parametrization, *i.e.* how to select a proper mapping of x onto $\mathcal{R}^n$, will be considered here; our problem will be assumed to have the already parametrized form $\mathbf{f}(\mathbf{x})=0$, with both $\mathbf{x}$ and $\mathbf{f}$ in $\mathcal{R}^n$. We begin by studying some general iterative procedure

$$\mathbf{x}^{(k+1)} = \mathbf{F}(\mathbf{x}^{(k)}) \tag{4.1}$$

We want to find the condition for this procedure to converge to some specific *fixpoint* $\mathbf{x}^f$ of $\mathbf{F}$, *i.e.* a point such that $\mathbf{x}^f = \mathbf{F}(\mathbf{x}^f)$. Let $\mathbf{x}^{(k)} = \mathbf{x}^f + \mathbf{e}^{(k)}$. This gives

$$\mathbf{e}^{(k+1)} = \mathbf{x}^{(k+1)} - \mathbf{x}^f = \mathbf{F}(\mathbf{x}^f + \mathbf{e}^{(k)}) - \mathbf{F}(\mathbf{x}^f) \tag{4.2}$$

Assume $\mathbf{F}$ differentiable. In component form, the above equation gives

$$\begin{aligned} e_i^{(k+1)} &= (F_i(\mathbf{x}^f) + \sum_j \frac{\partial F_i(\mathbf{x}^f)}{\partial x_j} e_j^{(k)} + O(|\mathbf{e}^{(k)}|^2)) - F_i(\mathbf{x}^f) = \\ &= \sum_j \frac{\partial F_i(\mathbf{x}^f)}{\partial x_j} e_j^{(k)} + O(|\mathbf{e}^{(k)}|^2) \end{aligned} \tag{4.3}$$

In order to get an estimate of the right-hand side, we introduce the following two concepts: First of all, a matrix $\mathbf{J}$ called the Jacobian of $\mathbf{F}$:

$$\mathbf{J} \quad : \quad J_{ij}(\mathbf{x}) = \frac{\partial F_i(\mathbf{x})}{\partial x_j} \tag{4.4}$$

Then, a matrix norm which we write

$$\rho(\mathbf{J}) = \sup_{\mathbf{d} \neq 0} \frac{|\mathbf{J}\mathbf{d}|}{|\mathbf{d}|} \quad , \quad \mathbf{d} \in R^n \tag{4.5}$$

If $\mathbf{J}$ has a spectral resolution, then $\rho(\mathbf{J})$ is the magnitude of the largest eigenvalue and is called the *spectral radius*. We get

$$|\mathbf{e}^{(k+1)}| \lesssim \rho(\mathbf{J}(\mathbf{x}^f)) \times |\mathbf{e}^{(k)}| \tag{4.6}$$

This shows that the spectral radius of the Jacobian of $\mathbf{F}(\mathbf{x}^f)$ must be smaller than 1 for convergence, and that it should be as small as possible to ensure fast convergence.

Now we want to apply this to some simple defect-correction schemes. Simplest is of course

$$\mathbf{F}(\mathbf{x}) = \mathbf{x} + \mathbf{A}\mathbf{f}(\mathbf{x}) \tag{4.7}$$

$\Rightarrow$

$$\mathbf{J}(\mathbf{x}^f) = \mathbf{1} + \mathbf{A}\mathbf{j}(\mathbf{x}^f)$$

where we use j for the Jacobian of f, and A is some matrix. Since ρ defined a matrix norm, we conclude that fast convergence implies that A should be as close as possible to $-\mathbf{j}^{-1}(\mathbf{x}^f)$ This is unknown, of course, until convergence. We may expect, however, that using a variable $\mathbf{A}(\mathbf{x}) = -\mathbf{j}^{-1}(\mathbf{x})$ will result in fast convergence. We than have to take into account derivatives of A. In component form, we get

$$F_i(\mathbf{x}) = x_i + \sum_k A_{ik}(\mathbf{x}) f_k(\mathbf{x}) \tag{4.8}$$

$\Rightarrow$

$$J_{ij}(\mathbf{x}) = \delta_{ij} + \sum \frac{\partial A_{ik}(\mathbf{x})}{\partial x_j} f_k(\mathbf{x}) + A_{ik}(\mathbf{x}) j_{kj}(\mathbf{x}) \tag{4.9}$$

But at $\mathbf{x}^f$, we have all $f_k(\mathbf{x}^f) = 0$. This shows that

$$\mathbf{J}(\mathbf{x}^f) = \mathbf{1} + \mathbf{A}(\mathbf{x}^f)\mathbf{j}(\mathbf{x}^f) \tag{4.10}$$

also when $\mathbf{A}(\mathbf{x})$ is variable. Thus, if the Jacobian of $\mathbf{f}(\mathbf{x})$ is known, we can use the multidimensional Newton-Raphson (NR-) method:

$$\mathbf{x}^{(k+1)} = \mathbf{x}^{(k)} - \mathbf{j}^{-1}(\mathbf{x}^{(k)})\mathbf{f}(\mathbf{x}^{(k)}) \tag{4.11}$$

It seems that we have to invert a matrix in each iteration, which takes roughly $\frac{4}{3}n^3$ operations. However, rewrite as a linear equation system in a correction $\mathbf{d}$:

$$\begin{aligned} \mathbf{j}(\mathbf{x}^{(k)})\mathbf{d}^{(k)} &= -\mathbf{f}(\mathbf{x}^{(k)}) \\ \mathbf{x}^{(k+1)} &= \mathbf{x}^{(k)} + \mathbf{d}^{(k)} \end{aligned} \tag{4.12}$$

It may appear as if this is no great improvement, since finding a solution to a linear equation system with direct methods requires about $\frac{2}{3}n^3$ operations, about half as many as the inversion. However, the solution of the linear equation system can be accomplished by iterative methods where, in each step, some product $\mathbf{j}\mathbf{v}$ is formed. Superficially, this cuts down the number of operations, but still requires the Jacobian to be computed and stored. However, for a very large class of important problems, such a product can be efficiently computed without the need of precalculating or storing the Jacobian.

Before entering into a description of the linear equation solvers, we will have a look at the type of problems where large equation systems usually occur. It turns out that special properties of the Jacobian can be useful.

A Digression: Why Symmetric Jacobians?

In Quantum Chemistry, equation systems arise primarily from optimization problems. For simplicity, let us call equations of the type $\mathbf{f}(\mathbf{x}) = 0$ *conventional* equation systems, while those of the type $f(x) = \lambda v(x)$ are called *quasi-secular* equations. As we shall see shortly, the former arise from optimization problems with no constrains on the parameters. The quasi-secular equation is obtained when the optimization is done on a subset of the parameter space. Let us assume that some differentiable function $Q(\mathbf{x})$ is to be minimized. We then define the following functions of $\mathbf{x}$:

- The *gradient* of Q(x) is a vector $\mathbf{g} \in \mathcal{R}^n$ with the components

$$g_i(\mathbf{x}) = \frac{\partial Q(\mathbf{x})}{\partial x_i} \tag{4.13}$$

- The *Hessian* of Q(x) is an $n \times n$ matrix $\mathbf{H}$ with components

$$H_{ij} = \frac{\partial^2 Q(\mathbf{x})}{\partial x_i \partial x_j} \tag{4.14}$$

Around the point x, the Taylor expansion of Q is

$$Q(\mathbf{x}+\mathbf{d}) = Q(\mathbf{x}) + \mathbf{g}^T(\mathbf{x}) \cdot \mathbf{d} + \frac{1}{2}\mathbf{d}^T\mathbf{H}(\mathbf{x})\mathbf{d} + O(|\mathbf{d}|^3) \tag{4.15}$$

The conditions for x to be at least a local minimum are that

- x is a *stationary* point, *i.e.* $\mathbf{g}(\mathbf{x})=0$.
- The Hessian is *positive definite, i.e.*,

$$\mathbf{d} \neq 0 \quad \Rightarrow \quad \mathbf{d}^T\mathbf{H}(\mathbf{x})\mathbf{d} > 0$$

Now, since the Hessian is the second derivative matrix, it is real and symmetric, and therefore hermitian. Thus, all its eigenvalues are real, and it is positive definite if all its eigenvalues are positive. We find that minimization amounts to finding a solution to $\mathbf{g}(\mathbf{x})=0$ in a region where the Hessian is positive definite. Convergence properties of iterative methods to solve this equation have earlier been studied in terms of the Jacobian. We now find that for this type of problems the Jacobian is in fact a Hessian matrix.

Now assume that some constraint $c(\mathbf{x}) = 0$ is added to the problem. It can be shown that the stationarity condition now takes the form

$$g_i(\mathbf{x}) = \lambda \frac{\partial c(\mathbf{x})}{\partial x_i}, \tag{4.16}$$

where the unknown λ is called the *Lagrange multiplier*; it's value will be determined by the condition $c(\mathbf{x}) = 0$.
(An intuitively appealing way of regarding this condition is that the gradient should be 'projected' onto the hypersurface of allowed variations of x, and that this projection should be zero. This projection is obtained by subtracting $\lambda\frac{\partial c(\mathbf{x})}{\partial x_i}$, which is simply a vector of unknown length along the normal to the surface $c(\mathbf{x}) = 0$.)

This equation has the form of a quasi-secular equation. In case Q and c have the forms

$$Q(\mathbf{x}) = \frac{1}{2}\mathbf{x}^\mathbf{T}\mathbf{A}\mathbf{x}$$

$$c(\mathbf{x}) = \frac{1}{2}\mathbf{x}^\mathbf{T}\mathbf{S}\mathbf{x} - const. = 0$$

then we can drop the prefix 'quasi-'.

Conventional equation systems and quasi-secular equations are usually interrelated, not only because similar techniques can be used to solve them, but as we have seen, also because an underlying common problem may have been formulated either way.

The Microiterations: Linear Equation Solvers.

After this digression, let us return to the linear equation system, repeated here for convenience:

$$\mathbf{j}(\mathbf{x}^{(k)})\mathbf{d}^{(k)} = -\mathbf{f}(\mathbf{x}^{(k)})$$

The iteration counter k and the argument $\mathbf{x}^{(k)}$ refers to the *macroiterations* made in the Newton-Raphson procedure, and they are obviously constant within the context of this section. Let us drop them for convenience. Also, let us explicitly assume that the Jacobian is in fact a positive definite Hessian, and that $\mathbf{f}(\mathbf{x}^{(k)})$ is a gradient. The equation to be solved is thus rewritten in the form

$$\mathbf{Hd} = -\mathbf{g} \tag{4.17}$$

Now, we intend to solve this equation by a sequence of *microiterations*, so let us introduce a residue vector and a microiteration counter l:

$$\mathbf{r}^{(l)} = \mathbf{Hd}^{(l)} + \mathbf{g} \tag{4.18}$$

The iterations are assumed to proceed by making $\mathbf{r}^{(l)}$ smaller and smaller until the equation can be regarded as solved by some accuracy criterion.

First, we have an abundancy of *relaxation* methods. Recall from chapter 3 that such a method is characterized by the subsequent variation of only a few parameters at a time, and that they demand an efficient bookkeeping of which entries of r to update, when the current subset of elements of d are varied. Of this class, the simplest methods in use are the Gauss-Seidel "family" of methods. Essentially only one element at a time gets updated. Let us simplify by using an algorithmic notation, where the iteration counter is dropped, and we use the replacement operator := instead of equalities:

Loop over elements i in some order;

Determine a change $t := -\lambda r_i / H_{ii}$

Apply the change to $d_i := d_i + t$

Update $\mathbf{r}$ *as* $r_j := r_j + H_{ij}t$

End of loop.

One such "sweep" through the elements is applied over and over again until the error has dropped low enough. Some remarks: in the basic Gauss-Seidel method, $\lambda = 1$, and this means that the correction brings r_i to zero. A slightly larger λ is used in practice, and then this is called an *overrelaxation* method. Furthermore, in order to use vector processors or polyprocessor machines, one may have to abandon the idea of applying all updates immediately. This last point, together with different schemes to utilize particular types of sparseness, has lead to a very large family of variations on this simple theme.

However, it is not a good method to apply to typical Quantum-Chemistry problems. It works best for matrices which have small off-diagonal values and which have a small *condition number*. This last number is a measure of the spread of eigenvalues:

$$\mathrm{cond}(\mathbf{H}) = \frac{\sup_i |\lambda_i|}{\inf_i |\lambda_i|} \tag{4.19}$$

None of this usually applies to Quantum Chemical problems.

Then, there is a class of *subspace iteration* methods. Just as in chapter 3, we can use a collection of basis vectors and solve the equation in terms of a linear combination. If the solution vector is still not accurate enough, this means that there is part of the residue vector which is orthogonal to all the previous basis vectors. After preconditioning, this part is added to the previous basis set. This method is simple to program, it can be made quite efficient by judicious choice of preconditioner, and it works with *general* matrices, not just hermitean or positive definite ones. However, if the matrix is in fact positive definite and hermitean, there is a very interesting twist to it: Just like in the Lanczos method, there is no need to store all the basis vectors. This is a very important point, of course. Also, unlike the Lanczos method, there is no problems in using a preconditioner. This type of methods are called *conjugate gradient* methods. There is a large variety of them. Because of their usefulness, we describe here a common form of these methods:

Suppose $\mathbf{A}$ is positive definite, symmetric, and a reasonable approximation to the inverse hessian $\mathbf{H}$. Select an arbitrary positive scale factor ρ_1 and set

$$\mathbf{r}_1 = -\mathbf{g}, \quad \mathbf{p}_1 = \rho_1 \mathbf{A}\mathbf{r}_1, \quad \mathbf{s}_1 = \mathbf{H}\mathbf{p}_1, \quad \gamma_1 = \mathbf{r}_1^T \mathbf{A}\mathbf{r}_1$$

Then iterate:

$$\delta_k = \mathbf{p}_k^T \mathbf{s}_k, \quad \eta_k = \rho_k \gamma_k, \quad \alpha_k = \frac{\eta_k}{\delta_k},$$

$$\mathbf{d}_{k+1} = \mathbf{d}_k + \alpha_k \mathbf{p}_k, \quad \mathbf{r}_{k+1} = \mathbf{r}_k - \alpha_k \mathbf{s}_k,$$

$$\gamma_{k+1} = \mathbf{r}_{k+1}^T \mathbf{A}\mathbf{r}_{k+1}, \quad \beta_k = \frac{\gamma_{k+1}}{\eta_k},$$

$$\mathbf{p}_{k+1} = \rho_{k+1}(\mathbf{A}\mathbf{r}_{k+1} + \beta_k \mathbf{p}_k), \quad \mathbf{s}_{k+1} = \mathbf{H}\mathbf{p}_{k+1}$$

where ρ_{k+1} is some positive scale factor. This is one of many algorithms described and discussed in detail in Hestenes' monograph (see below).

Finally, we shall describe an idea which at first seems odd: Instead of solving the linear equation problem, we solve a related eigensystem problem. It is easily verified that the equation to be solved can be written on the form

$$\begin{pmatrix} 0 & \mathbf{g}^T \\ \mathbf{g} & \mathbf{H} + \mu \mathbf{1} \end{pmatrix} \begin{pmatrix} 1 \\ \mathbf{d} \end{pmatrix} = \mu \begin{pmatrix} 1 \\ \mathbf{d} \end{pmatrix} \tag{4.20}$$

In this formula, μ stands for $\mathbf{g}^T\mathbf{d}$ This is certainly an eigensystem equation. However, we must add a small correction to $\mathbf{H}$, and moreover, this correction is not known in advance. It turns out that this correction can be left out, unless it is important that the linear equation is exactly solved. This is not necessary if the object is to find a good step for a macroiteration. Moreover, it turns out that, in such a context, the discrepancy introduced between this method and the exact NR-steps has the same asymptotic dependence as the error. Therefore, the method is *still* a second-order method with this modification, and there is no way to say *a priori* that this method is better or worse than the exact NR-iterations. This method is called the *augmented Hessian* (AH-)method. It is seen to be equivalent to a Newton-Raphson using a *shifted* hessian. This can be very advantageous, since this shift tends to keep the step down, and to keep the shifted hessian positive definite, when one is far from a solution. The size of the shift is automatically decided by the sizes of d and g. This feature is sometimes said to give the Augmented Hessian a larger convergence region than Newton-Raphson. However, bear in mind that the NR method is just as easily shifted, and is then exactly equivalent to AH. The difference is that the shift is determined automatically in AH.

Naturally, the success of this method, which would have seemed a bit implausible some years ago, reflects the advances made in eigensystem solution methods.

All the above methods can be applied using some *approximative* Hessian. Naturally, this will usually mean a lower convergence order. Of particular interest is a large variety of *update* methods, in which an approximate Hessian is improved by corrections based on the residue vectors. These methods can be regarded as a multi-dimensional analogue to regula falsi, and they are superlinear. However, the convergence order is smaller for larger systems, as follows:

n=1	$\gamma=1.618$
n=2	$\gamma=1.466$
n=24	$\gamma=1.1$
n=463	$\gamma=1.01$
n=1000	$\gamma=1.005$

In practice, the asymptotic convergence order in not so important. For a large system it takes so many iterations to reach the asymptotic behaviour, that convergence to machine accuracy has occurred much earlier. Of prime importance is instead to have a good enough approximation to the true Hessian.

Bibliographic notes.

The conjugate gradient method was discovered independently by Hestenes and Stiefel at about the same time. It was named in a joint paper. The original articles are:

M. R. Hestenes, J. Optim. Th. and Applic. **1**, 322-334 (1951).
M. R. Hestenes and E. Stiefel, J. Res. NBS **29**, 409-439 (1952).

A large number of different algorithms of this type are described in
M. R. Hestenes, *"Conjugate Gradient Methods in Optimization"*,
Application of Mathematics, vol 12,
Springer (New York) 1980.

Various update methods carry the names of Broyden, Davidon, Fletcher, Goldfarb, Powell, and Shanno, in different combinations. We recommend a comprehensive textbook, namely
R. Fletcher, *"Practical Methods of Optimization"*,
Wiley 1980.

A very good source to the use and properties of many methods can be found in the review article
J. Olsen, D.L.Yeager, and P. Jorgensen, *"Optimization and characterization of a Multiconfigurational Self-Consistent Field (MC-SCF) State"*,
in Adv. Chem. Phys. **54**,1 (1983).

The Method of Second Quantization

Jeppe Olsen

Theoretical Chemistry, Chemical Centre,

University of Lund, S-22100 Lund, Sweden

In the usual formalism of quantum mechanics, the first quantization formalism, observables are represented by operators and the wave functions are normal functions. In the method of second quantization, the wave functions are also expressed in terms of operators. The formalism starts with the introduction of an abstract vector space, the Fock space. The basis vectors of the Fock space are occupation number vectors, with each vector defined by a set of occupation numbers (0 or 1 for fermions). An occupation number vector represents a Slater determinant with each occupation number giving the occupation of given spin orbital. Creation and annihilation operators that respectively adds and removes electrons are then introduced. Representations of usual operators are expressed in terms of the very same operators.

The second quantization method is thus another representation of quantum mechanics providing alternative representations of wave functions and operators, and has, just like for example the interaction representation, its advantages and weaknesses. One of the advantages of the second quantization is that the antisymmetry of wave functions is ensured by proper definitions of the creation and annihilation operators. Another advantage is that the wave functions and operators are defined in terms of the same elementary operators so changes in wave functions can be reexpressed in terms of operators and changes of operators. The representation of operators in terms of annihilation and creation of electrons has also been instrumental in developing efficient algorithms for the calculation of the action of an operator on a wave function. The other contributions tothis book exemplifies the usefulness. The present chapter is about the major disadvantage of the method : it is yet another representation to learn!!

Second Quantization

1. Fock space and elementary operators

Let $\{\phi_i(\mathbf{r},\sigma), i = 1, m\}$ denote a basis of m orthonormal spin orbitals where $\mathbf{r}$ are the spatial coordinates and σ is the spin coordinate. A Slater determinant is an antisymmetric linear combination of one or more of these spin orbitals. The occupation of a given Slater determinant can be written as an occupation number vector, $|\mathbf{n}\rangle$, where n_i is one if spin orbital ϕ_i is occupied in the Slater determinant and n_i is zero if spin orbital ϕ_i is unoccupied.

$$|\mathbf{n}\rangle = |n_1, n_2,\cdots,n_m\rangle, \; n_i = 0,1 \quad \text{for } i=1,2,\cdots,m \tag{1.1}$$

The occupation number vectors are basis vectors in an m-dimensional abstract linear vector space, the Fock space, F(m). For a given spin orbital basis, there is a one-to-one mapping between a Slater determinant and an occupation number vector in the Fock space. The occupation number vectors are not Slater determinants; they do not have any spatial structure, they are just basis vectors in a linear vector space. Much of the terminology which is used for Slater determinants is, however, used for occupation number vectors. The sum $\sum_{i=1}^{m} n_i$ counts the number of electrons in the occupation number vector.

The Fock space F(m) is a sum of subspaces F(m,N) where each subspace F(m,N) contains all occupation number vectors that can be obtained by distributing N electrons into m spin orbitals. The subspace consisting of occupation number vectors with zero electrons contains a single vector, the true vacuum state,

$$|\text{vac}\rangle = |0_1, 0_2,\cdots,0_m\rangle \tag{1.2}$$

Approximations to an exact N electron wave function are expressed in terms of vectors in the Fock subspace F(m,N).

For an orthonormal basis of spin orbitals, we define the inner product between two occupation number vectors $|\mathbf{n}\rangle$ and $|\mathbf{k}\rangle$ as

$$\langle\mathbf{n}|\mathbf{k}\rangle = \delta(\mathbf{n},\mathbf{k}) = \prod_{i=1,m} \delta_{n_i k_i}. \tag{1.3}$$

This definition is consistent with the definition of overlap between two Slater-determinants having the same number of electrons. The overlap between Slater determinants having a different number of electrons is not defined. The extension to have a well-defined, but zero, overlap between two occupation number vectors with different numbers of electrons is a special feature of the Fock-space formulation that allows a unified description of systems with a different number of electrons. As a special case of Eq. (1.3), the vacuum state is defined to be normalized

$$\langle vac | vac \rangle = 1. \tag{1.4}$$

To bring the vector space to live, we introduce some elementary operators, creation operators, through the definitions

$$a_i^\dagger | n_1, n_2, \cdots, 0_i, \cdots, n_m \rangle = \Gamma(\underline{n})_i | n_1, n_2, \cdots, 1_i, \cdots, n_m \rangle \tag{1.5a}$$

$$a_i^\dagger | n_1, n_2, \cdots, 1_i, \cdots, n_m \rangle = 0, \tag{1.5b}$$

where

$$\Gamma(\mathbf{n})_i = -1^{\left(\Sigma_{j=1}^{i-1} n_j\right)} . \tag{1.6}$$

The phase factor $\Gamma(\mathbf{n})_i$ is introduced in order to endow the antisymmetry of many-electron wave functions in the Fock space, as we soon will see. The definition that a_i operating on an occupation number vector gives zero if spin orbital i is occupied ($n_i = 1$) describes the fact that a Slater-determinant with the same spin orbital occurring twice is zero. Products of creation operators operating on the vacuum state can be used to generate arbitrary occupation number vectors

$$| \mathbf{n} \rangle = \prod_{i=1}^{n} \left(a_i^\dagger\right)^{n_i} | vac \rangle , \tag{1.7}$$

All the reference to phase factors has been avoided by requiring that the creation operators be written in ascending order.

The properties of the creation operators can be deduced from the above definitions. Operating twice with $a_i^\dagger$ on an occupation vector gives

$$a_i^\dagger a_i^\dagger | n_1, n_2, \cdots, 0_i, \cdots, n_m \rangle = a_i^\dagger \Gamma(\mathbf{n})_i | n_1, n_2, \cdots, 1_i, \cdots, n_m \rangle = 0$$

$$a_i^\dagger a_i^\dagger | n_1, n_2, \cdots, 1_i, \cdots, n_m \rangle = 0. \tag{1.8}$$

The action of $a_i^\dagger \; a_i^\dagger$ on any vector is thus zero and $a_i^\dagger \; a_i^\dagger$ is the zero operator

$$a_i^\dagger a_i^\dagger = 0. \tag{1.9}$$

The action of two operators $a_i^\dagger$ and $a_j^\dagger$ on an occupation number vector with spin orbital i and j, i<j, unoccupied can be written in two ways

$$\begin{aligned} a_i^\dagger a_j^\dagger | n_1, \cdots, 0_i, \cdots, 0_j, \cdots, n_n \rangle &= a_i^\dagger \Gamma(\mathbf{n})_j | n_1, \cdots, 0_i, \cdots, 1_j, \cdots, n_n \rangle \\ &= \Gamma(\mathbf{n})_j \Gamma(\mathbf{n})_i | n_1, \cdots, 1_i, \cdots, 1_j, \cdots, n_n \rangle \end{aligned} \tag{1.10}$$

$$\begin{aligned} a_j^\dagger a_i^\dagger | n_1, \cdots, 0_i, \cdots, 0_j, \cdots, n_n \rangle &= a_j^\dagger \Gamma(\mathbf{n})_i | n_1, \cdots, 1_i, \cdots 0_j, \cdots, n_n \rangle \\ &= -\Gamma(\mathbf{n})_j \Gamma(\mathbf{n})_i | n_1, \cdots, 1_i, \cdots, 1_j, \cdots, n_n \rangle, \end{aligned}$$

where the $-\Gamma(\mathbf{n})_j$ factor in the last equation comes from the fact that $a_j^\dagger$ operate on $a_i^\dagger | \mathbf{n} \rangle$, and this string contains one more electron than $| \mathbf{n} \rangle$ before spin orbital j. For the case i<j we thus have

$$\left(a_i^\dagger a_j^\dagger + a_j^\dagger a_i^\dagger \right) | n_1, \cdots, 0_i, \cdots, 0_j, \cdots, n_n \rangle = 0. \tag{1.11}$$

Substitution of dummy variables in Eq. (1.11) shows that Eq. (1.11) also is valid for i>j. Relations analogous to Eq. (1.11) holds trivially for occupation number vectors where i or j or both are occupied. The action of $a_i^\dagger \; a_j^\dagger + a_j^\dagger \; a_i^\dagger$

on any vector is thus zero. The definition of the creation operators dictates therefore the anticommutation relation.

$$a_i^\dagger a_j^\dagger + a_j^\dagger a_i^\dagger = \left[a_i^\dagger, a_j^\dagger\right]_+ = 0 \tag{1.12}$$

The creation operators $a_i^\dagger$ are the hermitian adjoint of the operators a_i. The properties of a_i can be inferred from the above equations. From Eq. (1.12) the hermitian conjugated operators are seen to satisfy the anticommutation relation

$$a_i a_j + a_j a_i = [a_i, a_j]_+ = 0 \tag{1.13}$$

The action of the operators a_i on occupation number kets can be determined by taking the scalar product of the conjugates of the vectors of Eqs. (1.5) and an arbitrary vector $|\mathbf{k}\rangle$. Using Eq. (1.5a) we obtain

$$\langle n_1, \cdots, 0_i, \cdots, n_m | a_i | k_1, \cdots, k_i, \cdots, k_m \rangle = \Gamma(\mathbf{n})_i \tag{1.14a}$$

if

$$k_j = n_j + \delta_{ji}; \quad j = 1, m \tag{1.14b}$$

and zero if Eq. (1.14b) is not satisfied. From Eq. (1.5b) we obtain

$$\langle n_1, \cdots, 1_i, \cdots, n_m | a_i | k_1, \cdots, k_i, \cdots, k_m \rangle = 0. \tag{1.14.c}$$

Eq. (1.14) proves that

$$a_i | n_1, n_2, \cdots, 1_i, \cdots, n_m \rangle = \Gamma(\mathbf{n})_i | n_1, n_2, \cdots, 0_i, \cdots, n_m \rangle$$

$$a_i | n_1, n_2, \cdots, 0_i, \cdots, n_m \rangle = 0 \tag{1.15}$$

The action of operator a_i reduces n_i from 1 to 0 if spin orbital i is occupied and gives zero if spin orbital i is unoccupied. The operators a_i are for that reason called electron annihilation operators. A special case of Eq. (1.15) is

$$a_i \,|\, vac> = 0 \tag{1.16}$$

We will now derive the commutation relationship between the creation- and annihilation-operators. For $a_i^\dagger$ and a_i we obtain using Eqs. (1.5) and (1.15)

$$a_i^\dagger \, a_i \,|\, n_1, \cdots, 0_i, \cdots, n_m> = 0 \tag{1.17a}$$

$$a_i^\dagger \, a_i \,|\, n_1, \cdots, 1_i, \cdots, n_m> = \,|\, n_1, \cdots, 1_i, \cdots, n_m> \tag{1.17b}$$

$$a_i \, a_i^\dagger \,|\, n_1, \cdots, 0_i, \cdots, n_m> = \,|\, n_1, \cdots, 0_i, \cdots, n_m> \tag{1.17c}$$

$$a_i \, a_i^\dagger \,|\, n_1, \cdots, 1_i, \cdots, n_m> = 0. \tag{1.17d}$$

The operator $a_i^\dagger \, a_i + a_i \, a_i^\dagger$ is therefore equal to the unit operator

$$a_i^\dagger \, a_i + a_i \, a_i^\dagger = 1. \tag{1.18}$$

For i different from j we obtain

$$\left(a_i^\dagger \, a_j + a_j \, a_i^\dagger\right) |\, n_1, \cdots, 1_j, \cdots, 0_i, \cdots n_m> \; = 0. \tag{1.19}$$

For other occupation numbers in spin orbitals i and j the operator $a_i^\dagger a_j + a_j \, a_i^\dagger$ also produce zero vectors, and we therefore have the operator identity

$$a_i^\dagger \, a_j + a_j \, a_i^\dagger = 0 \text{ for } i \neq j \tag{1.20}$$

Combining Eq. (1.20) and Eq. (1.18) shows that for all values of i and j

$$a_i^\dagger \, a_j + a_j \, a_i^\dagger = \left[a_i^\dagger, \, a_j\right]_+ = \delta_{ij}. \tag{1.21}$$

The creation- and annihilation-operators can be combined to produce other types of operators. Operators conserving the number of electrons of the occupation number vectors must contain an equal number of annihilation- and creation-operators. The elementary excitation operators $a_i^\dagger a_j$ are the simplest number conserving operators and give when operating on an arbitrary occupation number vector (see Exercise 1.)

$$a_i^\dagger a_j |\mathbf{n}\rangle = \varepsilon_{ij}\, \Gamma(\mathbf{n})_i\, \Gamma(\mathbf{n})_j\, (1 - n_i + \delta_{ij})\, n_j |\mathbf{k}\rangle, \tag{1.22}$$

where

$$k_l = n_l; \;\; \text{if } l \neq i, j$$
$$k_i = 1$$
$$k_j = \delta_{ij}$$

and the ε-function is defined as

$$\varepsilon_{ij} = \begin{cases} 1 \text{ for } i \leq j \\ -1 \text{ for } i > j \end{cases}. \tag{1.23}$$

All occupation number vectors in F(m,N) can be obtained from an occupation number vector $|\mathbf{n}\rangle$ with N electrons by applying one or several elementary excitation operators on $|\mathbf{n}\rangle$. If a single excitation operator is applied we obtain a single excitation, if two excitation operators are involved, we obtain a double excitation, etc.

The diagonal operator $a_i^\dagger a_i$ are called spin-orbital number operators since

$$a_i^\dagger a_i |\mathbf{n}\rangle = n_i |\mathbf{n}\rangle \tag{1.24}$$

The spin-orbital number operators are hermitian

$$\left(a_i^\dagger a_i\right)^\dagger = a_i^\dagger a_i \tag{24a}$$

and they constitute a commuting set of operators,

$$\left[a_i^\dagger a_i, a_j^\dagger a_j\right] = a_i^\dagger a_i\, a_j^\dagger a_j - a_j^\dagger a_j\, a_i^\dagger a_i$$

$$= a_i^\dagger \left(\delta_{ij} - a_j^\dagger a_i\right) a_j - a_j^\dagger\, a_j a_i^\dagger a_i$$

$$= - a_j^\dagger a_i^\dagger a_j + \delta_j a_i^\dagger a_j - a_j^\dagger a_j a_i^\dagger a_i$$

$$= - a_j^\dagger \left(\delta_{ij} - a_j a_i^\dagger\right) a_i + \delta_{ij} a_i^\dagger a_j - a_j^\dagger a_j a_i^\dagger a_i$$

$$= \delta_{ij} a_i^\dagger a_j - \delta_{ij} a_j^\dagger a_i$$

$$= 0 \tag{24b}$$

The occupation number vectors are thus the common eigenvectors for the hermitian and commuting set of operators $\left(a_1^\dagger a_1, a_2^\dagger a_2, \cdots, a_m^\dagger a_m\right)$ and there is a one to one correspondence between an occupation vector and a set of eigenvalues for $\left(a_1^\dagger a_1, a_2^\dagger a_2, \cdots, a_m^\dagger a_m\right)$. This is consistent with the definition of the occupation number vectors as being an orthonormal basis for the Fock space.

The operator

$$\hat{N} = \sum_i a_i^\dagger a_i, \tag{1.25}$$

gives the total number of electrons in a given occupation number vector

$$\hat{N} | \mathbf{n} \rangle = \sum_i n_i | \mathbf{n} \rangle \tag{1.26}$$

and is for that reason called the number operator. In Table 1 we have summarized the fundamentals of the second quantization formalism derived in this section. In the next section we discuss the representation of other operators in second quantization.

Table 1. Fundamentals of second quantization

Basis vectors	$\|\mathbf{n}\rangle = \|n_1, n_2, \cdots, n_m\rangle$; $n_i = 0,1$ for $i = 1,2,\cdots,m$
Inner product	$\langle \mathbf{n} \| \mathbf{k} \rangle = \prod_{i=1}^{m} \delta_{n_i k_i}$
Creation operators	$a_i^\dagger \|n_1,\cdots,0_i,\cdots,n_m\rangle = \Gamma(\mathbf{n})_i \|n_1,\cdots,1_i,\cdots,n_m\rangle$
	$\Gamma(\mathbf{n})_i = -1^{(\Sigma_{j=1}^{i-1} n_j)}$
	$a_i^\dagger \|n_1,\cdots,1_i,\cdots,n_m\rangle = 0$
Annihilation operators	$a_i \|n_1,\cdots,1_i,\cdots,n_m\rangle = \Gamma(\mathbf{n})_i \|n_1,\cdots,0_i,\cdots,n_m\rangle$
	$a_i \|n_1,\cdots,0_i,\cdots,n_m\rangle = 0$
Anticommutation relations	$a_i^\dagger a_j + a_j a_i^\dagger = \delta_{ij}$
	$a_i^\dagger a_j^\dagger + a_j^\dagger a_i^\dagger = 0$
	$a_i a_j + a_j a_i = 0$
Number operators	$a_i^\dagger a_i \|\mathbf{n}\rangle = n_i \|\mathbf{n}\rangle$
	$\Sigma_i a_i^\dagger a_i \|\mathbf{n}\rangle = \Sigma_i n_i \|\mathbf{n}\rangle$
Vacuum state	$\langle vac \| vac \rangle = 1$
	$a_i \|vac\rangle = 0$

2. Representation of operators in second quantization

The occupation number vectors are basis vectors in an abstract linear vector space and specify thus only the occupation of the spin orbitals. The occupation number vectors contain no reference to the basis set. The reference to the basis set is built into the operators in the second quantization formalism. Observables are described by expectation values of operators and must be independent of the representation given to the operators and states. The matrix elements of a first quantization operator between two Slater determinants must therefore equal its counterpart of the second quantization formulation. For a given basis set the operators in the Fock space can thus be determined by requiring that the matrix elements between two occupation number vectors of the second quantization operator, must equal the matrix elements between the corresponding two Slater determinants of the corresponding first quantization operators. Operators that are considered in first quantization like the kinetic energy and the coulomb repulsion conserve the number of electrons. In the Fock space these operators must be represented as linear combinations of multipla of the $a_i^\dagger a_j$ operators, that is each term must contain an equal number of creation and annihilation operators. The actual form of the operators depends on whether the first quantized operator is a one-electron operator or a two-electron operator. We treat these two cases separately below.

In the first quantization formalism, one-electron operators are written as

$$f^c = \sum_{i=1,N} f^c(\mathbf{r}_i, \sigma_i), \tag{2.1}$$

where the summation is over the electrons of the system, $\underline{r}_i$ is the spatial coordinate and σ_i the spin coordinate of electron i. The index c indicates that we are working in the coordinate representation. The operator in Eq. (2.1) which refers to each electron independently have only non-vanishing matrix elements between two Slater determinants, if the Slater determinants differ at most in the occupation of a single electron. The second quantization analogue of f^c must therefore have the structure

$$\hat{f} = \sum_{rs} f_{rs}\, a_r^\dagger\, a_s, \tag{2.2}$$

since $a_r^\dagger a_s$ moves just one electron in an occupation number vector. The summation in Eq. (2.2) is over all spin orbitals in order to obtain the largest possible flexibility in moving one electron. The order $a_r^\dagger a_s$ is chosen to ensure that the result of $\hat{f}$ operating on the vacuum state vanishes. The constants f_{rs} can be identified by calculating the matrix elements of $\hat{f}$ between two occupation number vectors. We consider three different cases

1: diagonal elements

$$\langle \mathbf{n} | \hat{f} | \mathbf{n} \rangle = \sum_r n_r f_{rr} \tag{2.3a}$$

2: $|\mathbf{n}_1\rangle$ and $|\mathbf{n}_2\rangle$ differ in one set of occupation numbers

$$|\mathbf{n}_1\rangle = |n_1, \cdots, 0_i, \cdots, 1_j, \cdots, n_m\rangle$$

$$|\mathbf{n}_2\rangle = |n_1, \cdots, 1_i, \cdots, 0_j, \cdots, n_m\rangle$$

$$\langle \mathbf{n}_2 | \hat{f} | \mathbf{n}_1 \rangle = \Gamma(\mathbf{n}_1)_j \, \Gamma(\mathbf{n}_2)_i \, f_{ij} \tag{2.3b}$$

3: $|\mathbf{n}_1\rangle$ and $|\mathbf{n}_2\rangle$ differ in more than one set of occupation numbers

$$\langle \mathbf{n}_2 | \hat{f} | \mathbf{n}_1 \rangle = 0 \tag{2.3c}$$

The matrix elements of Eqs. (2.3) equal the Slater-Condon matrix elements between Slater determinants if we choose

$$f_{rs} = \int d\mathbf{r}\, d\sigma\, \phi_r^*(\mathbf{r},\sigma)\, f^c(\mathbf{r},\sigma)\, \phi_r(\mathbf{r},\sigma) \tag{2.4}$$

The recipe for constructing a second quantization representation of a one-electron operator is thus to use Eq. (2.2) with the integrals of Eq. (2.4).

In the first quantization formalism, a two-electron operator like the electronic repulsion operator is given as

$$g^c = \frac{1}{2} \sum_{i \neq j} g^c(\mathbf{r}_i, \sigma_i, \mathbf{r}_j, \sigma_j). \tag{2.5}$$

Other two-electron operators are the mass-polarization and the spin-orbit coupling operator. A two-electron operator gives non-vanishing matrix elements between two Slater determinants if the determinants contain at least two electrons and if they differ in the occupation of at most two pairs of electrons. The second quantization representation of a two-electron operator must thus have the structure

$$\hat{g} = \frac{1}{2} \sum_{ijkl} g_{ijkl}\, a_i^\dagger\, a_k^\dagger\, a_l\, a_j. \tag{2.6}$$

The annihilation operators are written to the right of the creation operators to ensure that g operating on an occupation number vector with less than two electrons vanishes. Using that the annihilation operators anticommute and that the creation operators anticommute it is easy to show that the parameters g_{ijkl} can be chosen in a symmetric fashion

$$g_{ijkl} = g_{klij} \tag{2.7}$$

The factors g_{ijkl} can be identified by calculating the matrix elements of $\hat{g}$, $\langle \mathbf{n} | \hat{g} | \mathbf{m} \rangle$, between two occupation number vectors and requiring that the obtained expressions should be equal to the matrix elements between the corresponding Slater-determinants of the first quantization operators. We consider four different cases for these matrix elements

1: diagonal elements

$$\langle \mathbf{n} | g | \mathbf{n} \rangle = \frac{1}{2} \sum_{rs} n_r\, n_s\, (g_{rrss} - g_{rssr}) \tag{2.8a}$$

2: $|\mathbf{n}_1\rangle$ and $|\mathbf{n}_2\rangle$ differ in one set of occupation numbers

$$|\mathbf{n}_1\rangle = |n_1, \cdots, 0_i, \cdots, 1_j, \cdots, n_m\rangle$$

$$|\mathbf{n}_2\rangle = |n_1, \cdots, 1_i, \cdots, 0_j, \cdots, n_m\rangle$$

$$\langle \mathbf{n}_2 | \hat{g} | \mathbf{n}_1 \rangle = \Gamma(\mathbf{n}_1)_j\, \Gamma(\mathbf{n}_2)_i \sum_{r \neq ij} (g_{ijrr} - g_{irrj})(\mathbf{n}_1)_r \tag{2.8b}$$

3: $|\mathbf{n}_1\rangle$ and $|\mathbf{n}_2\rangle$ differ in two sets of occupation numbers

$$|\mathbf{n}_1\rangle = |n_1,\cdots,0_i,\cdots,0_j,\cdots,1_k,\cdots,1_l,\cdots,n_m\rangle$$

$$|\mathbf{n}_2\rangle = |n_1,\cdots,1_i,\cdots,1_j,\cdots,0_k,\cdots,0_l,\cdots,n_m\rangle \qquad \text{holds always if } i\leq j \text{ and } k\leq l$$

$$\langle\mathbf{n}_2|\hat{g}|\mathbf{n}_1\rangle = \Gamma(\mathbf{n}_1)_i\,\Gamma(\mathbf{n}_1)_j\,\Gamma(\mathbf{n}_2)_k\,\Gamma(\mathbf{n}_2)_l\,(g_{iljk} - g_{ikjl})$$

4: $|\mathbf{n}_1\rangle$ and $|\mathbf{n}_2\rangle$ differ in more than two sets of occupation numbers

$$\langle\mathbf{n}_2|\hat{g}|\mathbf{n}_1\rangle = 0. \tag{2.8d}$$

The matrix elements of Eqs. (2.8) equal the matrix elements obtained using the Slater-Condon rules for two-electron operators if we choose

$$g_{ijkl} = \int \mathbf{dr}\, d\sigma\, \mathbf{dr}'\, d\sigma'\, \phi_i^*(\mathbf{r},\sigma)\,\phi_j(\mathbf{r},\sigma)\, g^c(\mathbf{r}\sigma, \mathbf{r}'\sigma')\phi_k^*(\mathbf{r}',\sigma')\,\phi_l(\mathbf{r}',\sigma'). \tag{2.9}$$

The recipe for constructing a two-electron second quantization operator is thus given by Eqs. (2.6) and (2.9).

The second quantization representation of the electronic hamilton operator is thus

$$\hat{H} = \sum_{ij} h_{ij}\, a_i^\dagger a_j + \frac{1}{2}\sum_{ijkl} g_{ijkl}\, a_i^\dagger a_k^\dagger a_l a_j \tag{2.11}$$

where h_{ij} is the matrix element of the sum of the two one-electron operators, the electronic kinetic energy and the electron-nuclear attraction, and g_{ijkl} is the four index matrix element of the electronic-electronic repulsion.

Second quantization operators have been constructed for one- and two-electron operators such that the same matrix elements and thereby the same expectation values are obtained in first and second quantization. Since expectation values and matrix elements are the only observables, we have obtained another representation with the same physical contents. It is also seen that the phase factors in Eq. (1.b) were necessary to reproduce the Slater-Condon rules for matrix elements between Slater determinants.

Let O_1^c and O_2^c be two operators in first quantization

$$O_1^c = \sum_i O_1^c(r_i, \sigma_i) \tag{2.12}$$

$$O_2^c = \sum_i O_2^c(r_i, \sigma_i) \tag{2.13}$$

and let $\hat{O}_1$ and $\hat{O}_2$ be the corresponding second quantization representation of these operators. From the construction of the second quantization operators it is clear that the first quantization operator aO_1^c and bO_2^c where a and b are constants is represented by $a\hat{O}_1 + b\hat{O}_2$ in the Fock space. The relations for linear operators in a linear vector space also are valid, in particular

$$\hat{O}_1(\hat{O}_2\,\hat{O}_3) = (\hat{O}_1\,\hat{O}_2)\hat{O}_3 \tag{2.10}$$

$$(\hat{O}_1\,\hat{O}_2)^\dagger = (\hat{O}_2)^\dagger\,(\hat{O}_1)^\dagger$$

The product of the two first quantization operators $O_1^c\,O_2^c$ can be separated into a one-electron part and a two-electron part

$$O_1^c\,O_2^c = {}^1O_{12}^c + {}^2O_{12}^c \tag{2.14}$$

with

$${}^1O_{12}^c = \sum_i O_1^c(r_i, \sigma_i)\,O_2^C(r_i, \sigma_i) \tag{2.15}$$

$${}^2O_{12}^c = \frac{1}{2}\sum_{i\neq j}\left(O_1^c(r_i, \sigma_i)\,O_2^c(r_j, \sigma_j) + O_1^c(r_j, \sigma_j)\,O_2^c(r_i, \sigma_i)\right). \tag{2.16}$$

The two-electron operator is written so it is symmetric in the two particle indices. The second quantization representation, $\hat{O}_{12}$, of $O_1^c\,O_2^c$ is the sum of the second quantization representations of ${}^1O_{12}^c$ and ${}^2O_{12}^c$,

$$\hat{O}_{12} = \sum_{ij}\left({}^1O_{12}\right)_{ij}\,a_i^+a_j + \frac{1}{2}\sum_{ijkl}\left({}^2O_{12}\right)_{ijkl}\,a_i^+\,a_k^+\,a_l\,a_j \tag{2.17}$$

with

$$\left({}^{1}O_{12}\right)_{ij} = \int \mathbf{dr}\, d\sigma\, \phi_i^*(\mathbf{r},\sigma)\, O_1^c(\mathbf{r},\sigma)\; O_2^c(\mathbf{r},\sigma)\, \phi_j(\mathbf{r},\sigma) \tag{2.18}$$

$$\left({}^{2}O_{12}\right)_{ijkl} = \iint \mathbf{dr}\, d\sigma\, \mathbf{dr}' d\sigma'\; \phi_i^*(\mathbf{r},\sigma)\, \phi_k^*(\mathbf{r}',\sigma')$$

$$\left(O_1^c(\mathbf{r},s)\, O_2^c(\mathbf{r}',\sigma') + O_1^C(\mathbf{r}',\sigma')\, O_2^C(\mathbf{r},\sigma)\right)$$

$$\phi_l(\mathbf{r}',\sigma')\, \phi_j(\mathbf{r},\sigma)$$

$$= (O_1)_{ij}\,(O_2)_{kl} + (O_1)_{kl}\,(O_2)_{ij} \tag{2.19}$$

Inserting (2.19) into (2.17) allows us to rewrite

$$\hat{O}_{12} = \sum_{ij} (O_1\, O_2)_{ij}\; a_i^+ a_j$$

$$+ \sum_{ijkl} (O_1)_{ij}\,(O_2)_{kl}\left(a_i^+ a_j a_k^+ a_l - \delta_{jk}\; a_i^+ a_l\right)$$

$$= \hat{O}_1\, \hat{O}_2 + \sum_{ij}\left[(O_1 O_2)_{ij} - \sum_k (O_1)_{ik}\,(O_2)_{kj}\right] a_i^+ a_j \tag{2.20}$$

where $\hat{O}_1$ and $\hat{O}_2$ are the second quantization representations of O_1^c and O_2^c, respectively. The second quantization representation of $O_1^c O_2^c$ is thus in general not equal to $\hat{O}_1\, \hat{O}_2$.

In the limit of a full spin orbital basis, the second term in (2.20) can be shown to vanish. For a complete one-electron basis set the following relation holds,

$$\sum_k \phi_k(\mathbf{r},\sigma)\, \phi_k^+(\mathbf{r}',\sigma') = \delta(\mathbf{r}-\mathbf{r}')\, \delta_{\sigma\sigma'} \tag{2.21}$$

where $\delta(\mathbf{r}-\mathbf{r}')$ is the Dirac delta function and $\delta_{\sigma\sigma'}$ is the Kronecker delta function. The above relation allows us to rewrite

$$\left(O_1^c O_2^c\right)_{ij}$$

$$= \int d\mathbf{r}\, d\sigma\, \phi_i^*(\mathbf{r},\sigma)\, O_1^c(\mathbf{r},\sigma)\; O_2^c(\mathbf{r},\sigma)\, \phi_j(\mathbf{r},\sigma)$$

$$= \iint d\mathbf{r}\, d\mathbf{r}' d\sigma d\sigma' \phi_i^*(\mathbf{r},\sigma)\, O_1^c(\mathbf{r},\sigma)\; \delta(\mathbf{r}-\mathbf{r}')\, \delta_{\sigma\sigma'}\, O_2^c(\mathbf{r}',\sigma')\, \phi_j(\mathbf{r}',\sigma')$$

$$= \sum_k \int d\mathbf{r}\, d\sigma \phi_i^*(\mathbf{r},\sigma)\, O_1^c(\mathbf{r},\sigma)\; \phi_k(\mathbf{r},\sigma)$$

$$\int d\mathbf{r}'\, d\sigma' \phi_l^*(\mathbf{r}',\sigma')\, O_2^c(\mathbf{r}',\sigma')\; \phi_j(\mathbf{r}',\sigma')$$

$$= \sum_k \left(O_1\right)_{ik} \left(O_2\right)_{kj} \tag{2.22}$$

For a complete one-electron basis we have for the second quantization representation, $\hat{O}_{12}$, of $O_1^c O_2^c$,

$$\hat{O}_{12} = \hat{O}_1\, \hat{O}_2 \tag{2.23}$$

The above development indicates that commutation relationships that hold for first quantization operators do not necessarily hold for second quantization operators in a finite one-electron basis. Consider the canonical commutators

$$[r_k,\, p_l] = i\, \delta_{kl} \tag{2.24}$$

which holds exactly for first quantization operators. The second quantization representations of r_k and p_l are

$$\hat{r}_k = \sum_{ij} \left(r_k\right)_{ij}\, a_i^+\, a_j \tag{2.25}$$

$$\hat{p}_l = \sum_{mn} \left(p_l\right)_{mn}\, a_m^+\, a_n \tag{2.26}$$

and the commutator between the position and momentum operators become in the second quantization representation

$$[\hat{r}_k, \hat{p}_l] = \sum_{ijmn} (r_k)_{ij} (p_l)_{mn} a_i^+ a_j a_m^+ a_n$$

$$- \sum_{ijmn} (p_l)_{mn} (r_k)_{ij} a_m^+ a_n a_i^+ a_j$$

$$= \sum_{ijmn} (r_k)_{ij} (p_l)_{mn} [a_i^+ a_j, a_m^+ a_n] \tag{2.27}$$

The commutator between two elementary excitation operators is in next section shown to be

$$[a_i^+ a_j, a_m^+ a_n] = \delta_{jm} a_i^+ a_n - \delta_{in} a_m^+ a_j, \tag{2.28}$$

so (2.27) becomes

$$[\hat{r}_k, \hat{p}_l] = \sum_{ij} \left(\sum_m (r_k)_{im} (p_l)_{mj} - (p_l)_{im} (r_k)_{mj} \right) a_i^+ a_j \tag{2.29}$$

For a complete one-electron basis Eq. (2.22) gives

$$\sum_m (r_k)_{im} (p_l)_{mj} = (r_k p_l)_{ij} \tag{2.30}$$

$$\sum_m (p_l)_{im} (r_k)_{mj} = (p_l r_k)_{ij} \tag{2.31}$$

so for a complete basis set

$$[\hat{r}_k, \hat{p}_l] = \sum_{mn} ([r_k, p_l])_{mn} a_m^+ a_n$$

$$= i \sum_{mn} \delta_{kl} (1)_{mn} a_m^+ a_n$$

$$= i \delta_{kl} \sum_m a_m^+ a_m \tag{2.32}$$

For a complete one-electron basis, the canonical commutators become proportional to the number operators. For finite basis sets, the canonical commutator becomes a general one-body operator.

In Table II we summarize some of the characteristics of operators in first and second quantization

Table II. Characteristics for operators in first and second quantization.

First quantization	Second quantization
one-electron operator $\sum_i f(\mathbf{r}_i, \sigma_i)$	one-electron operator $\sum_{ij} f_{ij}\, a_i^\dagger a_j$
two-electron operator $\frac{1}{2}\sum_{i\neq j} g(\mathbf{r}_i, \sigma_i, \mathbf{r}_j\, \sigma_j)$	two-electron operator $\frac{1}{2}\sum_{ijkl} g_{ijkl}\, a_i^\dagger a_k^\dagger a_l\, a_j$
operators are independent of spin orbital basis	operators depend on spin orbital basis
operators depend on the number of electrons	operators are independent of the number of electrons
exact operators	projected operators

The dependence of the used orbital basis is opposite in first and second quantization. In first quantization, the Slater determinants depend on the orbital basis and the operators are independent of the orbital basis. In the second quantization formalism, the occupation number vectors are basis vectors in a linear vector space and contain no reference to the orbitals basis. The reference to the orbital basis is made in the operators. The fact that the second quantization operators are projections on the orbital basis means that a second quantization operator times an occupation number vector is a new vector in the Fock space. In first quantization an operator times a Slater determinant can normally not be expanded as a sum of Slater determinants. In first quantization we work directly with matrix elements. The second quantization formalism represents operators and wave functions in a symmetric way; both are expressed in terms of elementary operators. This can be used to write changes in the wave functions as changes in operators. The projected nature of the second quantization operators has several consequences, for example, relations that hold for exact operators like the commutation relationship between the coordinate operator and the momentum operator does not hold for the projected operators. Neither does the projected coordinate operator commute with the projected coulomb repulsion operator. These problems are fundamental problems that occur whenever a finite basis set is used. The same problems show up in first quantization but first at a level where matrix elements are actually evaluated.

3. Commutators and Anticommutators

In the manipulation of operators and matrix elements in second quantization the commutator

$$\left[\hat{A}, \hat{B}\right] = \hat{A}\hat{B} - \hat{B}\hat{A} \tag{3.1}$$

and the anticommutator

$$\left[\hat{A}, \hat{B}\right]_{+} = \hat{A}\hat{B} + \hat{B}\hat{A} \tag{3.2}$$

of two operators are often encountered. The elementary operators (creation- and annihilation-operators) satisfy the anticommutator relations (see Table 1). Commutators and anticommutators between strings of elementary operators can be simplified using these relations. The second quantization formalism deals extensively with manipulations of strings of elementary operators and it is important to have some training and a good strategy for the evaluation of commutators and anticommutators of such strings.

The following general relationships provide the framework for reducing complex commutators into sums of less complicated commutators

$$\left[\hat{A}, \hat{B}_1\hat{B}_2\right] = \left[\hat{A}, \hat{B}_1\right]\hat{B}_2 + \hat{B}_1\left[\hat{A}, \hat{B}_2\right] \tag{3.3a}$$

$$\left[\hat{A}, \hat{B}_1\hat{B}_2\cdots\hat{B}_n\right] = \sum_{k=0}^{n-1} \hat{B}_1\cdots\hat{B}_k\left[\hat{A}, \hat{B}_{k+1}\right]\hat{B}_{k+2}\cdots\hat{B}_n \tag{3.3b}$$

$$\left[\widehat{A}, \widehat{B}_1 \widehat{B}_2\right] = \left[\widehat{A}, \widehat{B}_1\right]_+ \widehat{B}_2 - \widehat{B}_1 \left[\widehat{A}, \widehat{B}_2\right]_+ \tag{3.3c}$$

$$\left[\widehat{A}, \widehat{B}_1 \widehat{B}_2 \cdots \widehat{B}_n\right] = \sum_{k=0}^{n-1} (-1)^k \widehat{B}_1 \cdots \widehat{B}_k \left[\widehat{A}, \widehat{B}_{k+1}\right]_+ \widehat{B}_{k+2} \cdots \widehat{B}_n \text{ (n even)} \tag{3.3d}$$

$$\begin{aligned} \left[\widehat{A}, \widehat{B}_1 \widehat{B}_2\right]_+ &= \left[\widehat{A}, \widehat{B}_1\right] \widehat{B}_2 + \widehat{B}_1 \left[\widehat{A}, \widehat{B}_2\right]_+ \\ &= \left[\widehat{A}, \widehat{B}_1\right]_+ \widehat{B}_2 - \widehat{B}_1 \left[\widehat{A}, \widehat{B}_2\right] \end{aligned} \tag{3.3e}$$

(See Exercise 2). Note the terms in the summations in Eqs. (3.3b) and (3.3d) where $k = 0$ and $k = n-1$ contain reference to $\widehat{B}_1$ and $\widehat{B}_n$ respectively only in the commutator.

The Jacobi identity gives a relation between double commutators

$$\left[\widehat{A}, \left[\widehat{B}, \widehat{C}\right]\right] + \left[\widehat{C}, \left[\widehat{A}, \widehat{B}\right]\right] + \left[\widehat{B}, \left[\widehat{C}, \widehat{A}\right]\right] = 0 \tag{3.4}$$

The proof of the above identities are given in Exercise 2.

The rewriting of commutators and anticommutators is guided by the simple rule that the particle rank of the operator should be reduced. The particle rank of an operator consisting of a string of p creation and q annihilation operators is $\frac{1}{2}(p+q)$. A reduction in the particle rank by one can be obtained in two different cases. 1) if one of the two strings of operators has an integer particle rank then the commutator reduces the particle rank by one. 2) if both the two strings of operators have a non integer particle rank then the anticommutator reduces the particle rank by one. In Exercise 4 it is proven that the particle rank is reduced by one if the two strings of operators have an integer particle rank. The proof in Exercise 4 can straightforwardly be generalized to cover the other cases considered above. The reduction of the particle rank is important as it can be done prior to and independently of the evaluation of matrix elements. Matrix element evaluations are simplified when the particle rank is lowered of the operators. The anticommutator relation of the creation and annihilation operators is a simple example of the above rule. Further examples are provided below. Let us consider the simplest nontrivial commutator,

$$\left[a_i^\dagger, a_j^\dagger a_k\right] \tag{3.5}$$

If we move one of the two operators in $a_j^\dagger a_k$ outside the commutator according to one of the relations in Eq. (3.3), we are left with the commutator or anticommutator of two elementary operators, i.e. two spin strings containing an odd number of elementary operators. In order to reduce the number of operators involved, it is the anticommutator that must be employed. We therefore rewrite Eq. (3.5) as

$$\left[a_i^\dagger, a_j^\dagger a_k\right] = \left[a_i^\dagger, a_j^\dagger\right]_+ a_k - a_j^\dagger \left[a_i^\dagger, a_k\right]_+ \tag{3.6}$$

and the anticommutation relations (Eqs. (1.12), (1.21)) can now be invoked to give

$$\left[a_i^\dagger, a_j^\dagger a_k\right] = -\delta_{ik}\, a_j^\dagger. \tag{3.7}$$

In a similar fashion we can show

$$\left[a_i, a_j^\dagger a_k\right] = \delta_{ij}\, a_k \tag{3.8}$$

More complicated commutator relations can be obtained by reexpressing these in terms of Eqs. (3.7) and (3.8). The commutator between two spin orbital excitations becomes using Eq. (3.3)

$$\left[a_i^\dagger a_j, a_k^\dagger a_l\right] = \left[a_i^\dagger, a_k^\dagger a_l\right] a_j + a_i^\dagger \left[a_j, a_k^\dagger a_l\right] \tag{3.9}$$

In Eq. (3.9), the commutator is rewritten as a sum of two commutators instead of a sum of two anticommutators, since one of the operators on the right hand side contains two elementary operators. Inserting Eqs. (3.7) and (3.8) in Eq. (3.9) gives

$$\left[a_i^\dagger a_j, a_k^\dagger a_l\right] = -\delta_{il}\, a_k^\dagger a_j + \delta_{jk}\, a_i^\dagger a_l. \tag{3.10}$$

We can now proceed to evaluate more complicated commutators. We leave the following as exercises to the reader

$$\left[a_i^\dagger a_j, a_k^\dagger a_l a_m^\dagger a_n\right] = \delta_{jk} a_i^\dagger a_l a_m^\dagger a_n - \delta_{il} a_k^\dagger a_j a_m^\dagger a_n$$

$$+ \delta_{jm} a_k^\dagger a_l a_i^\dagger a_n - \delta_{in} a_k^\dagger a_l a_m^\dagger a_j \qquad (3.11)$$

Expressions are often encountered containing several commutators. The following double commutator is easily evaluated using Eq. (3.10)

$$\left[a_i^\dagger a_j, \left[a_k^\dagger a_l, a_m^\dagger a_n\right]\right] = \delta_{lm} \delta_{jk} a_i^\dagger a_n - \delta_{lm} \delta_{in} a_k^\dagger a_j$$

$$- \delta_{kn} \delta_{jm} a_i^\dagger a_l + \delta_{kn} \delta_{il} a_m^\dagger a_j \qquad (3.12)$$

It is seen that the particle rank is reduced by two as a result of the double commutator.

4. Rotation of orbitals and Exponential Mappings

In the preceding sections, the occupation vectors were defined by the occupation of the basis orbitals ϕ. In many cases it is necessary to study occupation number vectors where the occupation numbers refer to a set of orbitals $\tilde{\phi}$, that can be obtained from ϕ by a unitary transformation. This is, for example, the case when optimizing the orbitals for a single or a multiconfiguration state. The unitary transformation of the orbitals is obtained by introducing operators that carry out orbital transformations when working on the occupation number vectors. We will use the theory of exponential mapping to develop operators that parameterizes the orbital rotations such that i) all sets of orthonormal orbitals can be reached, ii) only orthonormal sets can be reached and iii) the parameters are independent variables.

Any set of rotated orbitals, $\tilde{\phi}$ can be obtained from ϕ by a transformation with a unitary matrix $\mathbf{X}$,

$$\tilde{\phi} = \phi \, \mathbf{X}. \qquad (4.1)$$

Our task is therefore to find a parametrization of the unitary matrices. The elements of $\mathbf{X}$ cannot be used as independent parameters, since the unitary relation

$$\mathbf{XX}^{\dagger} = 1, \tag{4.2}$$

couples the elements of **X**. Instead, a parametrization can be obtained by writing the unitary matrices in terms of an exponential mapping. The exponential of a matrix **A** is defined as

$$\exp(\mathbf{A}) = \sum_{n=0}^{\infty} \frac{1}{n!} \mathbf{A}^n, \tag{4.3}$$

with A^0 defined as the unit matrix. The mathematically inclined reader can easily prove that the infinite summations of Eq. (4.3) converge for any choice of matrix in a finite-dimensional vector space. The following relations follows, rather straightforwardly, from the definition Eq. (4.3)

$$\exp(\mathbf{A})\exp(-\mathbf{A}) = 1 \tag{4.4a}$$

$$\exp(\mathbf{A})^{\dagger} = \exp(\mathbf{A}^{\dagger}) \tag{4.4b}$$

$$\mathbf{B}\exp(\mathbf{A})\mathbf{B}^{-1} = \exp(\mathbf{B}\,\mathbf{A}\mathbf{B}^{-1}) \tag{4.4c}$$

$$\exp(\mathbf{A}+\mathbf{B}) = \exp(\mathbf{A})\exp(\mathbf{B}) \qquad \text{if } [\mathbf{A},\mathbf{B}] = 0 \tag{4.4d}$$

$$\exp(\mathbf{A})\mathbf{B}\exp(-\mathbf{A}) = \mathbf{B} + [\mathbf{A},\mathbf{B}] + \frac{1}{2}[\mathbf{A},[\mathbf{A},\mathbf{B}]] + \cdots \frac{1}{n!}[\mathbf{A},[\mathbf{A},\cdots,[\mathbf{A},\mathbf{B}]\cdots]] \tag{4.4.e}$$

Eq. (4.4) is proven in Exercise 5.

Our use of exponential mappings starts by noting that for an arbitrary hermitian matrix κ, the matrix

$$\exp(i\kappa), \tag{4.5}$$

is unitary since

$$\exp(i\kappa)^{\dagger}\exp(i\kappa) = \exp(-i\kappa)\exp(i\kappa) = 1. \tag{4.6}$$

What remains to be shown is that for any unitary matrix **X**, we can find a hermitian matrix κ so that

$$\mathbf{X} = \exp(i\kappa). \tag{4.7}$$

From the spectral theorem for unitary matrices it is known that an arbitrary unitary matrix can be diagonalized

$$\mathbf{X} = \mathbf{U}\,\varepsilon\,\mathbf{U}^{\dagger} \tag{4.8}$$

where **U** is unitary and ε is a complex diagonal matrix with diagonal elements having modulus 1, i.e.

$$\varepsilon_{ii} = \exp(i\delta_i), \tag{4.9}$$

with δ_i being a real number. Introducing the diagonal matrix δ with δ_i as the diagonal elements gives, using Eq. (4.4c)

$$\mathbf{X} = \mathbf{U}\exp(i\delta)\mathbf{U}^{\dagger} = \exp(i\mathbf{U}\,\delta\mathbf{U}^{\dagger}) \tag{4.10}$$

Since $\mathbf{U}\,\delta\,\mathbf{U}^{\dagger}$ is a hermitian matrix we have thus shown that all unitary matrices can be written as in Eq. (4.7). The attractive feature of writing a unitary matrix as the exponential of i times a hermitian matrix is that the hermitian matrices are very simple to parameterize; just take the elements at or below the diagonal as the independent parameters and use the hermitian condition $\kappa_{ji}^{*} = \kappa_{ij}$ to get the elements above the diagonal.

A general Hermitian matrix, κ, can be written as

$$\kappa = \mathbf{d} + \kappa^{R} + i\kappa^{I}, \tag{4.11}$$

where **d** is a real diagonal matrix, κ^R is a real symmetric matrix with zero diagonal and κ^I is a real antisymmetric matrix. A general unitary matrix can thus be written as

$$\exp(i\,\mathbf{d} + i\,\kappa^{R} - \kappa^{I}) \tag{4.12}$$

If the diagonal matrix **d** is set to zero the corresponding unitary matrix has determinant 1,

$$\det(\exp(i\,\kappa^R - \kappa^I)) = \exp(\mathrm{Tr}(i\,\kappa^R - \kappa^I)$$

$$= 1, \qquad (4.13)$$

since $i\,\kappa^R - \kappa^I$ has vanishing diagonal elements. In obtaining Eq. (4.13) we have used

$$\det(\exp \mathbf{A}) = \exp(\mathrm{Tr}\,\mathbf{A}) \qquad (4.14)$$

that is proven in Exercise 6. The matrices

$$\exp(i\,\kappa^I - \kappa^R) \qquad (4.15)$$

are called special unitary matrices since they have determinant one. It can be shown that a general unitary matrix can be expressed as a unitary diagonal matrix times a special unitary matrix,

$$\exp[i\,\mathbf{d} + i\,\kappa^R - \kappa^I] = \exp[i\,\mathbf{d}']\exp[i\,\kappa'^R - \kappa'^I], \qquad (4.16)$$

where $\mathbf{d}'$, κ'^R, κ'^I in general differ from $\mathbf{d}$, κ^R, κ^I.

The unitary diagonal matrix $\exp[i\,\mathbf{d}']$ induces phase shifts of the orbitals. These phase shifts are redundant in our future derivations where only special unitary matrices (Eq. (4.15)) need to be considered. In time-independent theory it is practise to use real orbitals. All rotations of real orbitals into real orbitals can be written

$$\mathbf{d}\exp[-\kappa^I], \qquad (4.17)$$

where $\mathbf{d}$ is a diagonal matrix with diagonal elements ± 1. Discarding the phase factor, we obtain the set of special orthogonal matrices

$$\exp[-\kappa^I]. \qquad (4.18)$$

Corresponding to a transformation of the orbitals we have a set of transformed creation operators. The one to one mapping between the orbitals and the creation operators implies that the unitary matrix for the orbitals can be used also to transform the creation operators.

$$\tilde{\mathbf{a}}^\dagger = \mathbf{a}^\dagger \exp(i\kappa) \tag{4.19}$$

The transformation of the annihilation operators is determined from Eq. (4.19)

$$\tilde{\mathbf{a}} = \mathbf{a} \exp(-i\kappa^*) \tag{4.20}$$

Using Eqs. (4.19) and (4.20a) it is easily verified that the anticommutation relations hold also for the transformed creation- and annihilation-operators.

In Eq. (4.19) we have determined a unitary matrix that describes the transformation from $\mathbf{a}^\dagger$ to $\tilde{\mathbf{a}}^\dagger$. We now determine an operator that describe the same transformation. To do this we introduce the exponential of an operator analogously to the matrix definition in Eq. (4.3). The relations of Eq. (4.4) also hold for A corresponding to an operator. For a hermitian matrix κ, we introduce the hermitian operator

$$\hat{\kappa} = \sum_{ij} \kappa_{ij}\, a_i^\dagger\, a_j, \tag{4.21}$$

and note that the operator $\exp(i\hat{\kappa})$ is unitary. We consider now the transformation of $a_i^\dagger$

$$\exp(i\hat{\kappa}) a_i^\dagger \exp(-i\hat{\kappa}), \tag{4.22}$$

Expanding Eq. (4.22) using Eq. (4.4) gives

$$\exp(i\hat{\kappa}) a_i^\dagger \exp(-i\hat{\kappa}) = a_i^\dagger + i\left[\hat{\kappa}, a_i^\dagger\right] + \frac{i^2}{2}\left[\hat{\kappa}, \left[\hat{\kappa}, a_i^\dagger\right]\right] + \cdots \tag{4.23}$$

The commutators of Eq. (4.23) are easily shown to be

$$\left[\hat{\kappa}, a_i^\dagger\right] = \sum_k \kappa_{ki}\, a_k^\dagger$$

$$\left[\hat{\kappa}, \left[\hat{\kappa}, a_i^\dagger\right]\right] = \sum_k \left(\kappa^2\right)_{ki} a_k^\dagger, \tag{4.24}$$

and for the n-fold commutator we have

$$\left[\hat{\kappa}, \cdots \left[\hat{\kappa}, a_i^\dagger\right]\right] = \sum_k (\kappa^n)_{ki}\, a_k^\dagger, \tag{4.25}$$

We therefore may write Eq. (4.23) as

$$\exp(i\hat{\kappa}) a_i^\dagger \exp(-i\hat{\kappa}) = \sum_k a_k^\dagger \left(\delta_{ki} + i\kappa_{ki} + \cdots + \frac{i^n}{n!}(\kappa^n)_{ki} + \cdots\right) \tag{4.26}$$

$$= \sum_k a_k^\dagger (\exp(i\kappa))_{ki}.$$

The transformation in Eq. (4.22) is identical to the transformation, Eq. (4.20). Eq. (4.22) thus gives an operator representation of the mapping from $\mathbf{a}^\dagger$ to $\tilde{\mathbf{a}}^\dagger$.

One of the advantages of using the transformation in Eq. (4.22) is that it allows operator manipulations of formulae. Consider for example an occupation number vector $|\mathbf{n}; \kappa\rangle$ with occupation vector n referring to orbitals in the transformed basis. This vector can be written

$$|\mathbf{n}; \kappa\rangle = \prod_i \left(\tilde{a}_i^\dagger\right)^{n_i} |vac\rangle$$

$$= \exp(i\hat{\kappa}) \prod_i \left(a_i^\dagger\right)^{n_i} \exp(-i\hat{\kappa}) |vac\rangle$$

$$= \exp(i\hat{\kappa})\, |\mathbf{n}; 0\rangle, \tag{4.27}$$

where $|n; 0\rangle$ is the occupation number vector with the same occupations, but with the untransformed orbitals. To obtain the last equality in Eq. (4.27) we have used

$$\hat{\kappa} |vac\rangle = 0 \tag{4.28}$$

Writing a general unitary matrix as in (4.11), a general unitary operator becomes

$$\exp\left[i \sum_i d_i\, a_i^\dagger a_i + i \sum_{i>j} \kappa_{ij}^R \left(a_i^\dagger a_j + a_j^\dagger a_i\right) - \sum_{i>j} \kappa_{ij}^I \left(a_i^\dagger a_j - a_j^\dagger a_i\right)\right]$$

Following the discussion after Eq. (4.12), the diagonal elements can be eliminated to yield special unitary operators,

$$\exp\left[i \sum_{i>j} \kappa^R_{ij} \left(a^\dagger_i a_j + a^\dagger_j a_i \right) - \sum_{i>j} \kappa^I_{ij} \left(a^\dagger_i a_j - a^\dagger_j a_i \right) \right] \tag{4.30}$$

If we are interested in only real orbitals, κ can be restricted to i κ^I yielding the special orthogonal operators,

$$\exp\left[- \sum_{i>j} \kappa^I_{ij} \left(a^\dagger_i a_j - a^\dagger_j a_i \right) \right] \tag{4.31}$$

5. From Spin Orbitals to Orbitals

In the previous sections, the occupation number vectors were specified in terms of the occupation of a set of spin orbitals, and the operators were defined by integrals over spin orbitals multiplied with spin orbital excitation operators. The spin orbitals depend on a continuous spatial coordinate, $\mathbf{r}$, and a discrete spin coordinate m_s. The spin coordinate takes two values, $\pm \frac{1}{2}$, so the complete spin basis is spanned by two functions $\sigma(m_s)$, $\sigma = \alpha, \beta$ defined as

$$\alpha\left(\frac{1}{2}\right) = 1 \qquad \beta\left(\frac{1}{2}\right) = 0$$

$$\alpha\left(-\frac{1}{2}\right) = 0 \qquad \beta\left(-\frac{1}{2}\right) = 1 \tag{5.1}$$

The total spin of $\alpha(m_s)$ and $\beta(m_s)$ is $\frac{1}{2}$,

$$\hat{S}^2 \alpha(m_s) = \frac{3}{4} \alpha(m_s)$$

$$\hat{S}^2 \beta(m_s) = \frac{3}{4} \beta(m_s) \tag{5.2}$$

It is sometimes convenient to denote the $\alpha(m_s)$ spin function with $\frac{1}{2}$ and the β (m_s) spin function with $-\frac{1}{2}$.

Integration over the spin coordinates corresponds to

$$\int d\, m_s\, f(m_s) = f\left(\frac{1}{2}\right) + f\left(-\frac{1}{2}\right), \tag{5.2a}$$

and the completeness relation for the spin functions is

$$\sum_{m_s} \sigma_i\,(m_s)\,\sigma_j\,(m_s) = \delta_{\sigma_i \sigma_j} \tag{5.2b}$$

A general spin orbital can be written as

$$\phi_i(\mathbf{r}, m_s) = \phi_{i\alpha}\,(\mathbf{r})\,\alpha(m_s) + \phi_{i\beta}\,(\mathbf{r})\,\beta\,(m_s), \tag{5.2c}$$

In the following we will consider the more restricted form

$$\phi_i(\mathbf{r}, m_s) = \phi_i\,(\mathbf{r})\,\sigma(m_s), \tag{5.2d}$$

so a given spin orbital either has α- or β-spin.

The quantum mechanical operators can be divided according to spatial and spin properties. In first quantization a pure spatial (spin free) operator F^c does not change the spin functions so F^c commutes with the spin function

$$\left[F^c, \sigma(m_s)\right] = 0, \qquad \left(F^c\text{: spin free}\right) \tag{5.2e}$$

A pure spin operator does not change the orbital functions, and F^c commutes with the spatial orbitals

$$\left[F^c, \phi_i(\mathbf{r})\right] = 0. \qquad \left(F^c\text{: pure spin operator}\right) \tag{5.2f}$$

Many operators, like the Fermi-contact operator, changes both the spatial- and spin functions.

In second quantization, the creation-operators corresponding to Eq. (5.2d) are written

$$a_i^\dagger = a_{i\sigma}^\dagger, \; \sigma = \alpha, \beta. \tag{5.2g}$$

A one electron operators in the spin orbital basis can be written

$$\hat{h} = \sum_{ij\sigma_i\sigma_j} h_{i\sigma_i j\sigma_j} a^{\dagger}_{i\sigma_i} a_{j\sigma_j} \tag{5.3}$$

where i and j run over spatial orbitals and σ_i and σ_j run over spin functions. The matrix element of a spin-free one-electron operator is

$$\begin{aligned} h_{i\sigma_i j\sigma_j} &= \int d\underline{r} dm_s \phi^*_i(\underline{r})\sigma_i(m_s) h^c(\mathbf{r})\phi_j(\mathbf{r})\sigma_j(m_s) \\ &= \delta_{\sigma_i\sigma_j} \int d\mathbf{r} \phi^*_i(\mathbf{r}) h^c(\mathbf{r})\phi_j(\mathbf{r}) \\ &= \delta_{\sigma_i\sigma_j} h_{ij} \end{aligned} \tag{5.4}$$

where h_{ij} is a matrix element between spatial orbitals

$$h_{ij} = \int d\mathbf{r} \phi_i(\mathbf{r}) h^c(\mathbf{r}) \phi_j(\mathbf{r}) \tag{5.5}$$

The second quantization representation of a spin-free one-electron operator is

$$\begin{aligned} \hat{h} &= \sum_{i\sigma_i\, j\sigma_j} h_{i\sigma_i\, j\sigma_j} a^{\dagger}_{i\sigma_i} a_{j\sigma_j} \\ &= \sum_{ij} h_{ij} E_{ij}, \end{aligned}$$

where we have introduced the orbital excitation operator

$$E_{ij} = \sum_{\sigma} a^{\dagger}_{i\sigma} a_{j\sigma}. \tag{5.7}$$

One example of a spin free one electron operator is the one electron operator in the exponent of the general unitary operator in Eq. (4.29). Using the above derivation the general unitary operator in Eq. 84.29) becomes

$$\exp\left[i\sum_{i} d_i E_{ii} + i\sum_{i>j} \kappa_{ij}^{R}\left(E_{ij} + E_{ji}\right) - \sum_{i>j} \kappa_{ij}^{I}\left(E_{ij} - E_{ji}\right)\right] \qquad (5.7a)$$

The special unitary operator in Eq. 84.30) becomes

$$\exp\left[i\sum_{i>j} \kappa_{ij}^{R}\left(E_{ij} + E_{ji}\right) - \sum_{i>j} \kappa_{ij}^{I}\left(E_{ij} - E_{ji}\right)\right] \qquad (5.7b)$$

and the special orthogonal operator becomes

$$\exp\left[-\sum_{i>j} \kappa_{ij}^{I}\left(E_{ij} - E_{ji}\right)\right] \qquad (5.7c)$$

A two electron operator is in the spin orbital basis

$$\hat{g} = \frac{1}{2} \sum_{\substack{i\sigma_i \sigma_j \\ k\sigma_k\, l\sigma_l}} g_{i\sigma_i j\sigma_j k\sigma_k l\sigma_l}\, a_{i\sigma_i}^{\dagger} a_{l\sigma_l}^{\dagger} a_{k\sigma_k} a_{j\sigma_j} \qquad (5.7a)$$

The matrix element of a spin-free two-electron operator is

$$g_{i\sigma_i j\sigma_j k\sigma_k l\sigma_l} =$$

$$\int \mathbf{dr}dm_s\mathbf{dr}'dm_s'\phi_i^*(\mathbf{r})\sigma_i(m_s)\phi_j(\mathbf{r})\sigma_j(m_s)g^c(\mathbf{r},\mathbf{r}')\phi_k^*(\mathbf{r}')\sigma_k(m_s)\phi_l(\mathbf{r}')\sigma_l(m_s)$$

$$= \delta_{\sigma_i\sigma_j}\, \delta_{\sigma_k\sigma_l}\, g_{ijkl} \qquad (5.8)$$

where g_{ijkl} is an integral over spatial orbitals,

$$g_{ijkl} = \int \mathbf{dr}\mathbf{dr}'\phi^*(\mathbf{r})_i\phi(\mathbf{r})_j g(\mathbf{r},\mathbf{r}')\phi^*(\mathbf{r}')_k\phi(\mathbf{r}')_l. \qquad (5.9)$$

which have the permutational symmetry

$$g_{ijkl} = g_{klij} \qquad (5.9a)$$

If the orbitals are real the integrals have the additional permutational symmetry

$$g_{ijkl} = g_{jikl} = g_{ijlk} = g_{jill} \tag{5.9b}$$

The second quantization representation of a spin-free two-electron operator can therefore be written as

$$\begin{aligned}\hat{g} &= \sum_{ijkl} g_{ijkl} \sum_{\sigma_1\sigma_2} a^{\dagger}_{i\sigma_1} a^{\dagger}_{k\sigma_2} a_{l\sigma_2} a_{j\sigma_1} \\ &= \sum_{ijkl} g_{ijkl} \left(E_{ij}E_{kl} - \delta_{jk}E_{il}\right) \\ &= \sum_{ijkl} g_{ijkl}\, e_{ijkl}\end{aligned} \tag{5.10}$$

In Eq. (5.10) we have introduced the two-electron operator

$$e_{ijkl} = E_{ij}E_{kl} - \delta_{jk}E_{il} = \sum_{\sigma_1\sigma_2} a^{\dagger}_{i\sigma_1} a^{\dagger}_{k\sigma_2} a_{l\sigma_2} a_{j\sigma_1}$$

The second quantization representation of the Hamiltonian in Eq. (2.11) is

$$\hat{H} = \sum_{ij} h_{ij}E_{ij} + \frac{1}{2}\sum_{ijkl} g_{ijkl}e_{ijkl} \tag{5.10a}$$

The E operators are the elementary operators, the generators, in unitary group theory. Below we give some commutator relations between E operators, e operators, creation and annihilation operators that are important for developing a good strategy for evaluating many of the quantities that show up in the later derivation.

$$\left[E_{mn}, a^{\dagger}_{i\sigma}\right] = \delta_{ni}\, a^{\dagger}_{m\sigma} \tag{5.10b}$$

$$\left[E_{mn}, a_{i\sigma}\right] = -\,\delta_{im}\, a_{n\sigma} \tag{5.10c}$$

$$\left[E_{mn}, E_{ij}\right] = E_{mj}\,\delta_{in} - E_{in}\,\delta_{jm} \tag{5.10d}$$

$$\left[E_{mn}, e_{ijkl}\right] = \delta_{in}e_{mjkl} - \delta_{mj}e_{inkl} + \delta_{kn}e_{ijml} - \delta_{ml}e_{ijkn} \tag{5.10e}$$

Eqs. (5.10b) – (5.10e) are proven in Exercise 9.

Let us now consider a first quantization operator, h^c, that only works in the spin space, so Eq. (5.2f) holds. The second quantization representation, $\hat{h}$, can be written

$$\hat{h} = \sum_{\substack{ij \\ \sigma_i \sigma_j}} \int d\mathbf{r}\, dm_s\, \phi_i^*(\mathbf{r})\sigma_i(m_s) h^c\, \phi_j(\mathbf{r})\sigma_j(m_s)\, a_{i\sigma_i}^\dagger a_{j\sigma_j}$$

$$= \sum_{\sigma_i \sigma_j} \int dm_s\, \sigma_i(m_s) h^c\, \sigma_j(m_s) \sum_i a_{i\sigma_i}^\dagger a_{i\sigma_j} \qquad (5.11a)$$

Three operators $\hat{S}_+$, $\hat{S}_-$, and $\hat{S}_z$ works only in the spin space. From the actions of these operators,

$$S_+^c \beta = \alpha,\ S_+^c \alpha = 0$$

$$S_-^c \beta = 0,\ S_-^c \alpha = \beta$$

$$S_z^c \beta = -\tfrac{1}{2}\beta,\ S_z^c \alpha = +\tfrac{1}{2}\alpha \qquad (5.11)$$

we obtain the matrix elements

$$(S_+)_{i\sigma, j\sigma'} = \int d\mathbf{r} dm_s \phi_i^*(\mathbf{r})\sigma(m_s) S_+^c \phi_j(\mathbf{r})_j\, \sigma'(m_s)$$

$$= \delta_{ij}\delta_{\sigma\alpha}\delta_{\sigma'\beta}$$

$$(S_-)_{i\sigma, j\sigma'} = \int d\underline{r} ds \phi_i(\mathbf{r})\sigma(m_s) S_-^c \phi_j(\mathbf{r})\sigma'(m_s)$$

$$= \delta_{ij}\delta_{\sigma\beta}\delta_{\sigma'\alpha} \qquad (5.12b)$$

$$(S_z)_{i\sigma j\sigma'} = \int d\mathbf{r} dm_s \phi_i(\mathbf{r})\sigma(m_s) S_Z^c \phi_j(\mathbf{r})\sigma(m_s)$$

$$= \frac{1}{2}\delta_{ij}\delta_{\sigma\sigma'}(\delta_{\sigma\alpha} - \delta_{\sigma\beta}) \qquad (5.12c)$$

From Eqs. (5.12) we obtain

$$\hat{S}_+ = \sum_i a^\dagger_{i\alpha} a_{i\beta}$$

$$\hat{S}_- = \sum_i a^\dagger_{i\beta} a_{i\alpha}$$

$$\hat{S}_Z = \frac{1}{2}\sum_i \left(a^\dagger_{i\alpha} a_{i\alpha} - a^\dagger_{i\beta} a_{i\beta}\right) \tag{5.14}$$

$\hat{S}_-$ is readily shown to be the hermitian conjugated of $\hat{S}_+$

$$(S_+)^\dagger = \left(\sum_i a^\dagger_{i\alpha} a_{i\beta}\right)^\dagger$$

$$= \sum_i a^\dagger_{i\beta} a_{i\alpha} \tag{5.15}$$

From $\hat{S}_+$ and $\hat{S}_-$ we determine $\hat{S}_x$ and $\hat{S}_y$

$$\hat{S}_x = \frac{1}{2}(\hat{S}_+ + \hat{S}_-) \tag{5.16a}$$

$$\hat{S}_y = \frac{1}{2i}(\hat{S}_+ - \hat{S}_-) \tag{5.16b}$$

The commutator [S+, S–] is obtained as

$$\left[\hat{S}_+, \hat{S}_-\right] = \sum_i a^\dagger_{i\alpha} a_{i\alpha} - a^\dagger_{i\beta} a_{i\beta}$$

$$= 2\,\hat{S}_z \tag{5.16c}$$

The operator $\hat{S}^2$,

$$\hat{S}^2 = \hat{S}_x^2 + \hat{S}_y^2 + \hat{S}_z^2$$

is a two-body operator working in the spin space. Since we are using a complete basis in the spin space, we can use Eq. (2.23) to obtain the second quantization representation of $S_x^c S_x^c$ as $\hat{S}_x \hat{S}_x$, and similarly for $S_y^c S_y^c$ and $S_z S_z$. The second quantization representation of $\hat{S}^2$ becomes

$$\hat{S}^2 = \hat{S}_x^2 + \hat{S}_y^2 + \hat{S}_z^2$$

$$= \hat{S}_+ \hat{S}_- + \hat{S}_z (\hat{S}_z - 1)$$

$$= \hat{S}_- \hat{S}_+ + \hat{S}_z (\hat{S}_z + 1) \tag{5.16c}$$

5a. Tensor operators in spin space

An irreducible tensor operator of rank S in spin space is defined as a set of 2S+1 operators, T(S,M), with M running from –S to S fulfilling the relations

$$\left[\hat{S}_+, T(S,M)\right] = \sqrt{S(S+1) - M(M+1)}\ T(S,M+1) \tag{5.17a}$$

$$\left[\hat{S}_-, T(S,M)\right] = \sqrt{S(S+1) - M(M-1)}\ T(S,M-1) \tag{5.17b}$$

$$\left[\hat{S}_z, T(S,M)\right] = MT(S,M) \tag{5.17c}$$

In Eqs. (5.17a) and (5.17b)

$$T(S,S+1) = T(S,-S-1) = 0. \tag{5.17d}$$

From Eq. (5.17) we obtain for T(S,M) | vac>

$$\hat{S}_+ T(S,M)\,|\,vac\rangle = \sqrt{S(S+1) - M(M+1)}\ T(S,M+1)\,|\,vac\rangle \tag{5.19a}$$

$$\hat{S}_- T(S,M)\,|\,vac\rangle = \sqrt{S(S+1) - M(M-1)}\ T(S,M-1)\,|\,vac\rangle \tag{5.19b}$$

$$\hat{S}_z T(S,M)\,|\,vac\rangle = MT(S,M)\,|\,vac\rangle \tag{5.19c}$$

If T(S,M) is a tensor operator of rank S then T(S,M)|vac> is a tensor state of rank S. Using Eq. (5.16c) (5.19a) and (5.19b) we obtain

$$\left[\hat{S}^2, T(S,M)\right]|vac\rangle = \hat{S}^2\ T(S,M)|vac\rangle$$

$$= S(S+1)T(S,M)|vac\rangle \tag{5.18}$$

Eq. (5.18) and (5.19c) express that T(S,M)|vac> is an eigenfunction of $\hat{S}^2$ and $\hat{S}_z$ with eigenvalues S and M respectively. Notice that $[\hat{S}^2, T(S,M)]$ does not satisfy any simple commutator equation.

A tensor operator with S=0 is called a singlet operator, an operator with S $=\frac{1}{2}$ is called a doublet operator and an operator with S=1 is a triplet operator. Below we give simple examples of singlet, doublet and triplet tensor operators. Let us initially consider the creation operator $a^\dagger_{i\sigma}$ where the spin function is either alpha spin $\sigma = \frac{1}{2}$ or beta spin $\sigma = -\frac{1}{2}$. Using this notation we obtain

$$\left[\hat{S}_+, a^\dagger_{i\sigma}\right] = \sqrt{\tfrac{3}{4} - \sigma(\sigma+1)}\ a^\dagger_{i\sigma+1}$$

$$\left[\hat{S}_-, a^\dagger_{i\sigma}\right] = \sqrt{\tfrac{3}{4} - \sigma(\sigma-1)}\ a^\dagger_{i\sigma-1}$$

$$\left[\hat{S}_Z, a^\dagger_{i\sigma}\right] = \sigma\ a^\dagger_{i\sigma} \tag{5.20}$$

A comparison with Eq. (5.17) shows that $a^\dagger_{i\frac{1}{2}}, a^\dagger_{i-\frac{1}{2}}$ is a tensor operator of rank $\frac{1}{2}$.

From hermitian conjugation of Eq. (5.20) we obtain the following relations for the annihilation operators

$$\left[\hat{S}_+, a_{i\sigma}\right] = -\sqrt{\tfrac{3}{4} - \sigma(\sigma-1)}\ a_{i\sigma-1}$$

$$\left[\hat{S}_-, a_{i\sigma}\right] = -\sqrt{\tfrac{3}{4} - \sigma(\sigma+1)}\ a_{i\sigma+1}$$

$$\left[\hat{S}_z, a_{i\sigma}\right] = -\sigma a_{i\sigma} \tag{5.21}$$

so that $a_{i\frac{1}{2}}, a_{i-\frac{1}{2}}$ is not a tensor operator. However, if we define

$$b_{i\frac{1}{2}} = a_{i-\frac{1}{2}}$$

$$b_{i-\frac{1}{2}} = -a_{i\frac{1}{2}} \tag{5.22}$$

we obtain from Eq. (5.21)

$$\left[\hat{S}_+, b_{i\sigma}\right] = \sqrt{\frac{3}{4} - \sigma(\sigma+1)}\; b_{i\sigma+1}$$

$$\left[\hat{S}_-, b_{i\sigma}\right] = \sqrt{\frac{3}{4} - \sigma(\sigma-1)}\; b_{i\sigma-1}$$

$$\left[\hat{S}_z, b_{i\sigma}\right] = \sigma b_{i,\sigma} \tag{5.23}$$

which shows that $a_{i-\frac{1}{2}}, -a_{i\frac{1}{2}}$ is a tensor operator with rank $\frac{1}{2}$. The reason that $a_{i-\frac{1}{2}}$ is the component with spin projection $+\frac{1}{2}$ is that $a_{i-\frac{1}{2}}$ removes an electron with spin projection $-\frac{1}{2}$.

New tensor operators can be obtain as linear combinations of strings of creation- and annihilation operators according to the usual angular momentum coupling rules. The coupling of two **different** sets of creation operators $\left(a^{\dagger}_{i\frac{1}{2}}, a^{\dagger}_{i-\frac{1}{2}}\right)$ and $\left(a^{\dagger}_{j\frac{1}{2}}, a^{\dagger}_{j-\frac{1}{2}}\right)$ gives rise to a triplet operator and a singlet operator. In Exercise 11 it is shown that the three components of the triplet operator are

$$Q_{ij}(1,1) = a^{\dagger}_{i\frac{1}{2}}\, a^{\dagger}_{j\frac{1}{2}}$$

$$Q_{ij}(1,0) = \frac{1}{\sqrt{2}}\left(a^{\dagger}_{i\frac{1}{2}}\, a^{\dagger}_{j-\frac{1}{2}} + a^{\dagger}_{i-\frac{1}{2}}\, a^{\dagger}_{j\frac{1}{2}}\right)$$

$$Q_{ij}(1,-1) = a^{\dagger}_{i-\frac{1}{2}}\, a^{\dagger}_{j-\frac{1}{2}} \tag{5.24}$$

and the singlet operator is

$$Q_{ij}(0,0) = \frac{1}{\sqrt{2}}\left(a^{\dagger}_{i\frac{1}{2}}\, a^{\dagger}_{j-\frac{1}{2}} - a^{\dagger}_{i-\frac{1}{2}}\, a^{\dagger}_{j\frac{1}{2}}\right) \tag{5.25}$$

The relations of Eq. (5.17) allow the scaling of all components of a tensor operator with the same factor. The factor used above is the one commonly used. It is noted that if i and j are identical the triplet operator of Eq. (5.24) vanishes.

The coupling of the creation operators $\left(a^{\dagger}_{i\frac{1}{2}},\ a^{\dagger}_{i-\frac{1}{2}}\right)$ and the annihilation operators $\left(b^{\dagger}_{j\frac{1}{2}},\ b^{\dagger}_{j-\frac{1}{2}}\right) = \left(a^{\dagger}_{i-\frac{1}{2}},\ -a^{\dagger}_{i\frac{1}{2}}\right)$ also gives rise to a triplet and a singlet operator. Straightforward substitution of $\left(a^{\dagger}_{j\frac{1}{2}},\ a^{\dagger}_{j-\frac{1}{2}}\right)$ with $\left(b_{j\frac{1}{2}},\ b_{-\frac{1}{2}}\right)$ in Eqs. (5.24) and (5.25) show that (see Exercise 11) the three components of the triplet operator are

$$T_{ij}(1,1) = -a^{\dagger}_{i\frac{1}{2}}\, a_{j-\frac{1}{2}}$$

$$T_{ij}(1,0) = \frac{1}{\sqrt{2}}\left(a^{\dagger}_{i\frac{1}{2}}\, a_{j\frac{1}{2}} - a^{\dagger}_{i-\frac{1}{2}}\, a_{j-\frac{1}{2}}\right)$$

$$T_{ij}(1,-1) = a^{\dagger}_{i-\frac{1}{2}}\, a_{j\frac{1}{2}} \tag{5.26}$$

while the singlet combination is

$$S_{ij}(0,0) = \frac{1}{\sqrt{2}}\left(a^{\dagger}_{i\frac{1}{2}}\, a_{j\frac{1}{2}} + a^{\dagger}_{i-\frac{1}{2}}\, a_{j-\frac{1}{2}}\right) \tag{5.27}$$

The arbitrary scaling factors on the above tensor operators are again chosen in accordance with the general use. It is seen that the triplet operator is defined also for the case where i and j are identical. The singlet operator, $S_{ij}(0,0)$ is proportional to the operator E_{ij}, obtained from considering spin independent one electron operators. The coupling of more elementary operators to strings or linear combinations of strings that transform as irreducible tensor operators in spin space is discussed in Sec. IIIc.

The occupation number vectors can be written as a product of alpha creation operators, an alpha string, times a product of beta creation operators, a beta string

$$|\mathbf{n}_\alpha \mathbf{n}_\beta> = \prod_{i=1}^{m} \left(a_{i\alpha}^{\dagger}\right)^{n_i^\alpha} \prod_{i=1}^{m} \left(a_{i\beta}^{\dagger}\right)^{n_i^\beta} |Vac>. \tag{5.28}$$

The creation operators corresponding to doubly occupied orbitals can alternatively be moved in front of the other creation operators

$$|\mathbf{n}^\alpha \mathbf{n}^\beta> = \text{sign} \prod_{i=1}^{m} \left(a_{i\alpha}^{\dagger} a_{i\beta}^{\dagger}\right)^{n_i^\alpha n_i^\beta} \prod_{i=1}^{m} \left(a_{i\alpha}^{\dagger}\right)^{n_i^\alpha - n_i^\alpha n_i^\beta}$$

$$\prod_{i=1}^{m} \left(a_{i\beta}^{\dagger}\right)^{n_i^\beta - n_i^\alpha n_i^\beta} |Vac>, \tag{5.29}$$

where sign is ±1, depending on the number of transpositions required to move the core, $\prod_{i=1}^{m} \left(a_{i\alpha}^{\dagger} a_{i\beta}^{\dagger}\right)^{n_i^\alpha n_i^\beta}$, in front.

The evaluation of a spin operator times an occupation number vector is faciliated by noting that the core is a singlet spin tensor, since $a_{i\alpha}^{\dagger} a_{i\beta}^{\dagger}$ is a singlet spin operator (Eq. 5.25). The action of $\hat{S}_z$ on $|\mathbf{n}_\alpha \mathbf{n}_\beta>$ becomes

$$\hat{S}_z |\mathbf{n}_\alpha \mathbf{n}_\beta> = \frac{1}{2}\left(\sum_i (n_i^\alpha - n_i^\alpha n_i^\beta) - ((n_i^\beta - n_i^\alpha n_\beta^i)\right) |\mathbf{n}_\alpha \mathbf{n}_\beta>$$

$$= \frac{1}{2} \sum_i \left(n_i^\alpha - n_i^\beta\right) |\mathbf{n}_\alpha \mathbf{n}_\beta> \tag{5.30}$$

The notation in the first line of the RHS of Eq. (5.30) stresses that only singly occupied orbitals need to be included. The action of $\hat{S}_+$ and $\hat{S}_-$ times $|\mathbf{n}_\alpha \mathbf{n}_\beta>$ can be written

$$\hat{S}_+ |\mathbf{n}_\alpha \mathbf{n}_\beta>$$

$$= \sum_j \left[\prod_{i=1}^{m} \left(a_{i\alpha}^{\dagger} a_{i\beta}^{\dagger}\right)^{n_i^\alpha n_i^\beta} \prod_{i=1}^{m} \left(a_{i\alpha}^{\dagger}\right)^{n_i^\alpha - n_i^\alpha n_i^\beta} \left[\left(a_{i\beta}^{\dagger}\right)^{n_1^\beta - n_1^\alpha n_1^\beta} \cdots \left(a_{j\alpha}^{\dagger}\right)^{n_j^\beta - n_j^\alpha n_j^\beta} \right.\right.$$

$$\cdots\left(a^{\dagger}_{m\beta}\right)^{n^{\beta}_{m}-n^{\alpha}_{m}n^{\beta}_{m}}\Bigg]\,|\mathrm{Vac}\rangle\Bigg] \tag{5.31}$$

$$\hat{S}_{-}|\mathbf{n}_{\alpha}\,\mathbf{n}_{\beta}\rangle$$

$$=\sum_{j}\left[\prod_{i=1}^{m}\left(a^{\dagger}_{i\alpha}\,a^{\dagger}_{i\beta}\right)^{n^{\alpha}_{i}n^{\beta}_{i}}\left[\left(a^{\dagger}_{1\alpha}\right)^{n^{\alpha}_{1}-n^{\alpha}_{1}n^{\beta}_{1}}\cdots\left(a^{\dagger}_{j\beta}\right)^{n^{\alpha}_{j}-n^{\alpha}_{j}n^{\beta}_{j}}\cdots\left(a^{\dagger}_{m\alpha}\right)^{n^{\alpha}_{m}-n^{\alpha}_{m}n^{\beta}_{m}}\right]\right.$$

$$\left.\prod_{i=1}^{m}\left(a^{\dagger}_{i\beta}\right)^{n^{\beta}_{i}-n^{\alpha}_{i}n^{\beta}_{i}}\right]|\mathrm{Vac}\rangle \tag{5.32}$$

The number of occupation number vectors generated by the action of $\hat{S}_{+}\left(\hat{S}_{-}\right)$ on $|\mathbf{n}_{\alpha}\,\mathbf{n}_{\beta}\rangle$ equals the number of singly occupied spin orbitals with beta spin (alpha spin), and each vector in the result is obtained by changing the spin function of a singly occupied orbital from beta to alpha (alpha to beta).

The action of $\hat{S}^{2}$ on $|\mathbf{n}_{\alpha}\,\mathbf{n}_{\beta}\rangle$ is obtained as

$$\hat{S}^{2}|\mathbf{n}_{\alpha}\,\mathbf{n}_{\beta}\rangle$$

$$=\left(\hat{S}_{-}\,\hat{S}_{+}+\hat{S}_{z}\left(\hat{S}_{z}+1\right)\right)|\mathbf{n}_{\alpha}\,\mathbf{n}_{\beta}\rangle$$

$$=\left[\frac{1}{2}\sum_{1}^{m}\left(n^{\alpha}_{i}-n^{\beta}_{i}\right)\left(\frac{1}{2}\sum_{1}^{m}\left(n^{\alpha}_{j}-n^{\beta}_{j}\right)+1\right)+\sum_{1}^{m}n^{\beta}_{i}\right]|\mathbf{n}_{\alpha}\,\mathbf{n}_{\beta}\rangle$$

$$+\sum_{jk}\left\{\prod_{i=1}^{m}\left(a^{\dagger}_{i\alpha}a^{\dagger}_{i\beta}\right)^{n^{\alpha}_{i}n^{\beta}_{i}}\left[\left(a^{\dagger}_{1\alpha}\right)^{n^{\alpha}_{1}-n^{\alpha}_{1}n^{\beta}_{1}}\cdots\left(a^{\dagger}_{j\beta}\right)^{n^{\alpha}_{j}-n^{\alpha}_{j}n^{\beta}_{j}}\cdots\left(a^{\dagger}_{m\alpha}\right)^{n^{\alpha}_{m}-n^{\alpha}_{m}n^{\beta}_{m}}\right]\right.$$

$$\left.\left[\left(a^{\dagger}_{j\beta}\right)^{n^{\beta}_{1}-n^{\alpha}_{1}n^{\beta}_{1}}\cdots\left(a^{\dagger}_{k\alpha}\right)^{n^{\beta}_{k}-n^{\alpha}_{k}n^{\beta}_{k}}\cdots\left(a^{\dagger}_{m\beta}\right)^{n^{\beta}_{m}-n^{\alpha}_{m}n^{\beta}_{m}}\right]\right\}$$

$$|\mathrm{Vac}\rangle \tag{5.33}$$

The action of $\hat{S}^2$ on $|\mathbf{n}_\alpha \mathbf{n}_\beta>$ gives thus a term proportional to $|\mathbf{n}_\alpha \mathbf{n}_\beta>$ plus a sum over occupation number vectors, where the spin functions of an alpha orbital and a beta orbital have been flipped. Only permutations of the singly occupied orbitals are included. $|\mathbf{n}_\alpha \mathbf{n}_\beta>$ is thus in general not an eigenfunction for $\hat{S}^2$.

The Hamiltonian is spinfree and it commutes therefore with $\hat{S}^2$ and $\hat{S}_z$, so eigenfunctions of the Hamiltonian can be chosen also as eigenfunctions of $\hat{S}^2$. It is often convenient to restrict the Fock space to a subspace consisting of eigenfunction of both $\hat{S}^2$ and $\hat{S}_z$. An eigenfunction of $\hat{S}_z$ and $\hat{S}^2$ is called a configuration state function (CSF), and is in general a linear combination of occupation number vectors. Since $\hat{S}^2$ commutes with the orbital number operators,

$$\left[\hat{S}^2, N_i\right] = 0 \tag{5.34}$$

$$N_i = a^\dagger_{i\alpha} a_{i\alpha} + a^\dagger_{i\beta} a_{i\beta}, \tag{5.35}$$

the CSF's can be chosen as linear combinations of occupation number vectors with identical eigenvalues for N_i. The set of occupation number vectors having identical eigenvalues of N_i is called an orbital configuration. We will later return to the problem of generating the CSF's as linear combinations of vectors belonging to a given orbital configuration.

6. Expansions of Wave Functions in Fock Space

a. Configuration amplitude parametrization

An N electron wave function can be written as

$$|0> = \sum_i C_i^{(0)} |i>, \tag{6.2}$$

where $\{|i>\}$ is either a basis of N electron CSF's or N electron occupation number vectors. We assume that

$$\sum_i \left|C_i^{(0)}\right|^2 = 1. \tag{6.3}$$

The expansion (6.2) may include all possible vectors in the N electron Fock space but may also contain a subset of these vectors. All vectors having the coefficient 1 for $|0>$ can be written

$$|\tilde{0}> = |0> + \hat{P}\sum_i C_i |i>, \tag{6.4}$$

with $\hat{P}$ being the projection operator

$$\hat{P} = 1 - |0><0|.$$

Assuming $|0>$ is normalized all other normalized vectors with $|0>$ having a positive coefficient can be written

$$|\tilde{0}> = \frac{|0> + \hat{P}\sum_i C_i |i>}{\sqrt{\left(1+\mathbf{C}^t\mathbf{P}\,\mathbf{C}\right)}} \tag{6.5}$$

where the denominator is a normalization factor and **P** is the matrix representation of $\hat{P}$

$$P_{ij} = <i|\hat{P}|j> \tag{6.6}$$

To obtain the normalization factor we have used

$$\hat{P}^2 = \hat{P}$$

The parametrization in Eq. (6.5) is redundant since

$$C \rightarrow C + \alpha\, C^{(0)} \tag{6.7}$$

(α is a constant) does not change $|\tilde{0}\rangle$ from $|0\rangle$. This means

$$\frac{\partial|\tilde{0}\rangle}{\partial\alpha}=\frac{\partial|\tilde{0}\rangle}{\partial C_i}C_i^0$$

$$=0, \tag{6.7a}$$

holds for all choices of $|\tilde{0}\rangle$. For general derivatives we have

$$\frac{\partial^n|\tilde{0}\rangle}{\partial C_1\cdots\partial C_{n-1}\,\partial\alpha}=\sum_i\frac{\partial^n|\tilde{0}\rangle}{\partial C_1\cdots\partial C_{n-1}\,\partial C_i}C_i^0=0 \tag{6.7b}$$

A general parametrization of the wave function space requires that a phase factor is multiplied on Eq. (6.5). As discussed in subsection 4 we do not need to consider the phase factor in the future development.

b. Exponential parametrization

Another representation of the basis $\{|i\rangle\}$ is $\{|0\rangle;\ |0_k\rangle\}$ where $\{|0_k\rangle\}$ is an orthogonal basis for the orthogonal complement to $|0\rangle$. Neglecting the phase factor an arbitrary normalized state vector becomes

$$|\tilde{0}\rangle=C_0'|0\rangle+\sum_{k\neq 0}C_k'|0_k\rangle \tag{6.24}$$

We now show that the state in Eq. (6.24) can be generated by applying the unitary operator

$$\exp(i\,\hat{R})=\exp\left(i\sum_{k\neq 0}\left(R_k|0_k\rangle\langle 0|+R_k^*|0\rangle\langle 0_k|\right)\right) \tag{6.25}$$

on the state $|0\rangle$. The parameters R_k are one less in number than the dimension of $\{|i\rangle\}$ as $k=0$ is excluded in the sum. The term $k=0$ describes the phase factor

$$\exp\left(i\left(R_0|0\rangle\langle 0|+R_0^*|0\rangle\langle 0|\right)\right)$$

$$=\exp(i\,2\mathrm{Re}(R_0))|0\rangle\langle 0|)$$

that may be multiplied on the state as described in subsection 4. The action of the unitary operator in Eq. (6.25) on $|0>$ is determined in Exercise 14.

$$\exp(i\hat{R})\,|0>$$

$$\exp\left(i\left(\sum_{k\neq 0} R_k|0_k><0| + R_k^*|0><0_k|\right)\right)|0>$$

$$= \cos d|0> + i\,\frac{\sin d}{d}\sum_{k=0} R_k\,|0_k> \tag{6.26}$$

where

$$d = \sqrt{\sum_{k\neq 0} |R_k|^2} \tag{6.27}$$

Comparing Eqs. (6.24) and (6.26) show that the coefficients $\{R_k\}$ can be determined from the coefficients $\{C_0^{'},C_k^{'}\}$ as

$$d = \cos^{-1} C_0^{'} \tag{6.29a}$$

$$R_k = \frac{-id}{\sin d}\,C_k^{'} \tag{6.29b}$$

The state $|\tilde{0}>$ in Eq. (6.24) can thus be expressed as

$$|\tilde{0}> = \exp\left(i\left(\sum_{k\neq 0} R_k|0_k><0| + R_R^*|0><0_k|\right)\right)|0> \tag{6.30}$$

with the R_k determined from Eq. 86.29b). Eq. (6.30) gives a non redundant parametrization of all normalized states that can be determined within the basis $\{|i>\}$. The number of parameters $C_k^{'}$ is one larger in number than R_k due to the fact that the $C_k^{'}$ parameters satisfy a normalization condition.

The orthogonal complement set of states to $|\tilde{0}>$ may be obtained operating with the exponential unitary operator on the orthogonal complement set of states. In Exercise 14 it is shown

$$|\tilde{0}_l> = \exp(i\hat{R})|0_l>$$

$$= |0_l> + \frac{\cos d - 1}{d^2} R_l^* \sum_{k\neq 0} R_k |0_k> + \frac{i \sin d}{d} R_l^* |0> \tag{6.31}$$

If an orthogonal basis $\{|0>, |0_k>\}$ is given we can thus use Eqs. (6.30) and (6.31) to generate a unitary transformed basis $\left\{|\tilde{0}>, |\tilde{0}_k>\right\}$.

The coefficients $\{R_k\}$ that generate a unitary transformation of $\{|0>, |0_k>\}$ are complex parameters

$$R_k = R_k^R + i\, R_k^I \tag{6.32}$$

If we restrict the expansion in Eq. (6.24) to have real coefficients we obtain from Eq. (6.29)

$$R_k^R = 0, \tag{6.33}$$

$$R_k^I = -\frac{d}{\sin d} C_k' \tag{6.34}$$

The exponential unitary operator in Eq. (6.25) therefore becomes

$$\exp(i\hat{R}) = \exp\left(-\ R_k^I\left(|0_k><0| + |0><0_k|\right)\right) \tag{6.34a}$$

and the orthogonal transformed states in Eq. (6.26) and (6.31) reads

$$|\hat{0}> = \cos d\ |0> - \frac{\sin d}{d} \sum_{k=0} R_k^I |0_k> \tag{6.35}$$

$$|\hat{0}_l> = |0_l>\ - \frac{\cos -1}{d^2} R_l^I \sum_{k=0} R_l^I |0_k> + \frac{\sin d}{d} R_l^I\ |0> \tag{6.36}$$

where

$$d = \sqrt{\sum_{k \neq 0} \left(R_k^I\right)^2} \tag{6.37}$$

In the above formalism we need the state $|0>$ and an orthogonal basis for the orthogonal complement space. In actual calculations we usually know one state $|0>$ and the construction of an orthogonal complement basis seems a formidable task. In the following we show how the orthogonal basis for the orthogonal complement can straightforwardly be determined from the expansion coefficients to $|0>$. Let us assume that $|0>$ is real and is written as

$$|0> = C_l |l> + \sum_{k \neq l} C_k \, |k>, \tag{6.38}$$

where we have picked out one term $|l>$. The basis $\{|l>, |k>\}$ is just a simple representation of the basis $\{|0>, |0_k>\}$ and we may therefore express $|0>$ in terms of a set of parameters t_k that enter in the exponential mapping

$$|0> = \exp\left(-\sum_{k \neq l} t_k \, |k> <l| - |l><k|\right)|l> \tag{6.39}$$

where

$$d = \cos^{-1} C_l \tag{6.40}$$

$$t_k = \frac{-d}{\sin d} C_k, \tag{6.41}$$

To obtain Eqs. (6.39), (6.40) and (6.41) we have used Eqs. (6.29a) and (6.34). The orthogonal complement set of states can be obtained applying the exponential operator on the states $|k>$ giving

$$\begin{aligned} |0_k> &= \exp\left(-\sum_{m \neq l} t_m \left(|m> <l| - |l><m|\right)\right)|k> \\ &= |k> + \frac{\sin d}{d} t_k |l> + t_k \left(\frac{\cos d - 1}{d^2}\right) \sum_m t_m |m> \end{aligned} \tag{6.42}$$

where we have used Eq. (6.36). Using Eqs. (6.40) and (6.41) we further obtain

$$|0_k\rangle = |k\rangle + \frac{C_k}{(1+C_l)}[|1\rangle + |0\rangle] \tag{6.43}$$

and have thus determined a parametrization of the orthogonal complement set of states from the expansion coefficients in Eq. (6.38).

7. Second Quantization with Non-orthogonal Orbitals

The discussion in the previous sections dealt with the second quantization formalism for orthonormal spinorbitals. In this section, we will generalize the formalism to treat cases where the set of spin orbitals is non-orthogonal. Consider a set of n spin orbitals with the general metric

$$S_{pq} = \int d\mathbf{r}\, d\sigma\, \phi^*(\mathbf{r},\sigma)_p\, \phi(\mathbf{r},\sigma)_q . \tag{7.1}$$

A Fock space corresponding to this set can again be defined as an abstract vector space with basis vectors defined as occupation number vectors $|\mathbf{n}\rangle$, $n_i = 0,1$. The inner product between two occupation number vectors is defined so that for two occupation number vectors with the same number of electrons, the overlap equals the overlap between the corresponding two Slater determinants. Two occupation number vectors with different numbers of electrons are defined to have overlap 0. For two occupation number vectors $|\mathbf{n}\rangle$ and $|\mathbf{m}\rangle$ with n_e and m_e electrons, respectively, we thus define the overlap as

$$\langle \mathbf{n} | \mathbf{m} \rangle = \delta_{n_e, m_e} \det\left(S^{nm}\right)$$

$$S^{nm}_{pq} = S_{n(p),m(q)} \tag{7.2}$$

A set of creation operators $a_i^\dagger$ is defined as in Eq. (1.5). The anticommutation relations between the creation operators and the properties of the hermitian conjugated operators can be deduced from the definition of the creation operators and the general inner product of Eq. (7.2). However we describe a simpler way to obtain these relations. We start by introducing a new set of orbitals

$$\bar{\phi} = \phi \mathbf{C}, \tag{7.3}$$

so that the orbitals $\bar{\phi}$ constitutes an orthonormal set,

$$\bar{S}_{ij} = \int \mathbf{dr}\, d\sigma\, \bar{\phi}^*(\mathbf{r},\sigma)_i\, \bar{\phi}(\mathbf{r},\sigma)_j$$

$$= \left(C^\dagger SC\right)_{ij}$$

$$= \delta_{ij} \tag{7.4}$$

We may thus choose C to be of the form

$$C = S^{-\frac{1}{2}} X, \tag{7.5}$$

where X is a unitary matrix. Corresponding to the orthonormal orbital $\bar{\phi}$, we may introduce the set of creation operators,

$$\bar{\mathbf{a}}^\dagger = \mathbf{a}^\dagger \mathbf{C} \tag{7.6}$$

The operators in the orthonormal basis $\bar{a}^\dagger$ satisfies the usual anticommutation relations. We therefore have

$$a_m^\dagger a_n^\dagger + a_n^\dagger a_m^\dagger = \sum_{pq} C_{pm}^{-1} C_{qn}^{-1} \left(\bar{a}_m^\dagger \bar{a}_n^\dagger + \bar{a}_n^\dagger \bar{a}_m^\dagger\right)$$

$$= 0 \tag{7.7}$$

$$a_n a_m + a_m a_n = \left(a_m^\dagger a_n^\dagger + a_n^\dagger a_m^\dagger\right)^*$$

$$= 0. \tag{7.8}$$

No signs of nonorthogonality showed up in these formulas. The anticommutation relation between a creation operator and an annihilation operator becomes

$$a_m^\dagger a_n + a_n a_m^\dagger = \sum_{pq} C^{-1}_{pm} C^{-1*}_{qn} \delta_{pq}$$

$$= \left(\mathbf{C}^{-1T} \mathbf{C}^{-1*}\right)_{mn}$$

$$= S_{nm}, \tag{7.9}$$

where we have used Eq. (7.4) and that S is hermitian to show that $\mathbf{C}^{-1T} \mathbf{C}^{-1*}$ is $\mathbf{S}^T$. The anticommutator $a_m^\dagger a_n + a_n a_m^\dagger$ is therefore in general nonvanishing for a nonorthogonal basis.

An annihilation operator times the vacuum state still vanishes so the effect of an annihilation operator times an occupation number vector becomes

$$a_q |\mathbf{n}> = \sum_p \Gamma(n)_p\, n_p \left(a_1^\dagger\right)^{n_1} \cdots \left(\left[a_q, a_p^\dagger\right]_+\right)^{n_p} \cdots a_m^\dagger |vac> \tag{7.10}$$

$$= \sum_p \Gamma(\mathbf{n})_p\, S_{qp}\, n_p |n_1 \cdots 0_p \cdots n_m>. \tag{7.11}$$

The action of an annihilation operator times an occupation number vector is thus complicated by the nonorthogonality of the basis set.

Operators can be obtained in the nonorthogonal base from the corresponding operators in the orthonormal base. For a one-electron operator we obtain

$$\hat{h} = \sum_{pq} \bar{h}_{pq}\, \bar{a}_p^\dagger\, \bar{a}_q$$

$$= \sum_{pq} \left(CC^\dagger\, h\, CC^\dagger\right)_{pq} a_p^\dagger\, a_q$$

$$= \sum_{pq} \left(S^{-1}\, h\, S^{-1}\right)_{pq} a_p^\dagger\, a_q, \tag{7.12}$$

where $\bar{h}_{ij}$ is the integral of $\hat{h}$ between orbitals $\bar{\phi}_i^*$ and $\bar{\phi}_j$ and h_{ij} is the integral of $\hat{h}$ between orbitals ϕ_i^* and ϕ_j. In a similar way, a general two-body operator in the nonorthogonal basis can be shown to have the form

$$\hat{g} = \sum_{mnpqm'n'p'q'} S^{-1}_{m'm} S^{-1}_{n'n} S^{-1}_{p'p} S^{-1}_{q'q} g_{mnpq} a^{\dagger}_{m} a^{\dagger}_{p} a_q a_n \tag{7.13}$$

The effect of a one-electron operator times an occupation number vector becomes

$$\hat{h}|\mathbf{n}> = \sum_{pq} \left(S^{-1} h S^{-1}\right)_{pq} a^{\dagger}_{p} a_q |n_1 \cdots n_n>$$

$$= \sum_{pqr} \left(S^{-1} h S^{-1}\right)_{pq} \left(1-n_p+\delta_{rp}\right) \Gamma(\mathbf{n})_p n_r \Gamma(\mathbf{n})_r \varepsilon_{pr} S_{qr} |n_1 \cdots 1_p \cdots 0_r \cdots n_m>$$

$$= \sum_{pr} \left(S^{-1} h\right)_{pr} \left(1-n_p+\delta_{rp}\right) n_r \Gamma(\mathbf{n})_p \Gamma(\mathbf{n})_r \varepsilon_{pr} |n_1 \cdots 1_p \cdots 0_r \cdots n_m> \tag{7.14}$$

where ε_{pm} again is 1 if p is smaller than or equal to m and -1 if p is greater than m. Although the formal development of the action of $\hat{h}$ on $|\mathbf{n}>$ is more complicated in the nonorthogonal case, the final expression looks very much like the orthonormal case, since it was possible to eliminate one summation by modifying the integrals.

The above simplifications can be obtained in a still simpler way. We note that all the complications due to the nonorthogonality arise from the anticommutator, Eq. (7.9). If we instead could work with a transformed set of annihilation operators, $\hat{a}$, that satisfy

$$a^{\dagger}_{i} \hat{a}_j + \hat{a}_j a^{\dagger}_{i} = \delta_{ij}, \tag{7.15}$$

then the operators, $a^{\dagger}_{i} \hat{a}_j$, behave complete like one-electron excitations in an orthonormal basis,

$$a^{\dagger}_{p} \hat{a}_q |\mathbf{n}> = \left(1-n_p+\delta_{pq}\right) n_q \varepsilon_{pq} \Gamma(\mathbf{n})_p \Gamma(\mathbf{n})_q |n_1 \cdots 1_p \cdots 0_q \cdots n_m>._1 \tag{7.16}$$

If we define the operators $\hat{a}$ as

$$\hat{\mathbf{a}} = \mathbf{aB} \tag{7.17}$$

then Eq. (7.9) gives

$$\mathbf{B} = \mathbf{S}^{-1}. \tag{7.18}$$

A general one-electron operator $\hat{h}$ can be expanded in the set of operators $a_i^\dagger \hat{a}_j$

$$\hat{h} = \sum_{mn} \left(S^{-1} h\, S^{-1}\right)_{mn} a_m^\dagger\; a_n$$

$$= \sum_{mn} \left(S^{-1} h\right)_{mn} a_m^\dagger\; \hat{a}_n, \tag{7.19}$$

and this allows the straightforward calculation of $\hat{h}$ times an occupation number vector to give the formulae Eq. (7.12) directly.

The operators $\hat{a}_j$ are said to be biorthogonal to the operators $a_i^\dagger$, since they fulfil the relation Eq. (7.15). The operator $a_i^\dagger$ is not the hermitian conjugate of $\hat{a}_i$.

The above development shows that the effect of operators times an occupation number vector can be evaluated for the case, where the basis is nonorthogonal. The remaining problem in using nonorthogonal orbitals is how to efficiently calculate inner products between occupation number vectors. We will not describe that in any more detail in this book.

Molecular Symmetry

and

Quantum Chemistry

Peter R. Taylor[1]
ELORET Institute
Palo Alto, CA 94303
USA

Dedicated to A. C. Hurley (1926-1988)

December 21, 1991

[1]Mailing address: NASA Ames Research Center, Moffett Field, CA 94035-1000, USA

Chapter 1

Fundamentals: Abstract Group Theory

1.1 Groups

Consider a set of elements $\mathcal{G}$ together with a *binary operation* represented by $\circ$, such that two elements of the set can be combined to form a new quantity

$$x \circ y \tag{1.1}$$

If $x \circ y$ is an element of $\mathcal{G}$ for any choice of $x, y \in \mathcal{G}$, $\mathcal{G}$ is said to be *closed* under $\circ$. If $\mathcal{G}$ (1) is closed under $\circ$; (2) satisfies the requirement of *associativity*

$$(x \circ y) \circ z = x \circ (y \circ z); \tag{1.2}$$

(3) contains an *identity element* E such that

$$x \circ E = E \circ x = x, \ \forall\, x \in \mathcal{G}; \tag{1.3}$$

and (4) contains for each element x an *inverse* x^{-1} such that

$$x \circ x^{-1} = x^{-1} \circ x = E, \tag{1.4}$$

$\mathcal{G}$ and $\circ$ form a *group*. If a group $\mathcal{G}$ satisfies the requirement of *commutativity*:

$$x \circ y = y \circ x, \ \forall\, x, y \in \mathcal{G}, \tag{1.5}$$

it is termed an *Abelian group*. The number of elements in a group is the *order* of the group.

Examples of groups abound. The set of all permutations of N objects is a group of order $N!$ — the *symmetric group* of N objects. Another example can be obtained by considering powers of a generating element:

$$E, x, x^2, x^3, \ldots \ x^{g-1} \tag{1.6}$$

with

$$x^g = E. \tag{1.7}$$

This group of order g is a *cyclic group*. Any cyclic group is Abelian. Further, if g is prime any group of order g is cyclic.

1.2 Multiplication Tables, Subgroups, Generators

One way of characterizing a group is by its *multiplication table*. Consider, for example, the set E, A, B, C, D, F and a binary operation whose results are represented by Table 1.1.

Table 1.1: Group Multiplication Table

	E	A	B	C	D	F
E	E	A	B	C	D	F
A	A	B	E	F	C	D
B	B	E	A	D	F	C
C	C	D	F	E	A	B
D	D	F	C	B	E	A
F	F	C	D	A	B	E

Clearly, this set forms a group $\mathcal{G}_6$ of order six; the group is neither Abelian nor cyclic. A group with the same multiplication table as $\mathcal{G}_6$ (perhaps to within a permutation of rows and columns) is said to be *isomorphic* to $\mathcal{G}_6$.

We may note from the multiplication table that the elements E, A, B themselves obey the requirements for a group: they form a *subgroup* of the group, of order three. This subgroup is cyclic and hence Abelian. It is conventional to use $\mathcal{H} \subseteq \mathcal{G}$, borrowing the symbol $\subseteq$ from set theory, to denote that $\mathcal{H}$ is a subgroup of $\mathcal{G}$. Similarly, there are three subgroups of order two: E, C; E, D; and E, F; each of which is Abelian. All these subgroups are termed *proper*; they obey the stricter relation $\mathcal{H} \subset \mathcal{G}$ but are of order larger than one. The order of any subgroup of $\mathcal{G}$ is a divisor of g (Lagrange's Theorem). Thus it is not possible to find a subgroup of order four of a group of order six, for example.

An alternative way of characterizing a group is by the use of *generators*. These are selected elements $a, b, \ldots$ of the group together with relations like

$$a^m = E \tag{1.8}$$

giving the order m of each generator, and relations like

$$b^i a^j = a^k b^l \tag{1.9}$$

that allow any product of generators to be rewritten in some lexical order (alphabetical here, for example). In this way all distinct lexically ordered products of the generators yield all elements of the point group. For example, if we set $a = A$ and $b = C$, we have $a^3 = E$, $b^2 = E$, and $ba = a^2b$ as the relations that define our group $\mathcal{G}_6$ above. Generators are indispensable in the manipulations of groups of very large order in formal group theory, but are not much used in quantum chemical applications.

1.3 Cosets, Classes

Consider a proper subgroup $\mathcal{H}$ of a group $\mathcal{G}$. If $G \in \mathcal{G}$, but $G \notin \mathcal{H}$, the set $G\mathcal{H}$ is a *left coset* of $\mathcal{H}$ in $\mathcal{G}$. Similarly, the set $\mathcal{H}G$ is a *right coset* of H in G. Note that if $\mathcal{H} = \mathcal{G}$ these sets become simply a reordering of the group $\mathcal{G}$ (Rearrangement Theorem). If $\mathcal{H}$ is a proper subgroup of $\mathcal{G}$, and $G, F \in \mathcal{G}$, the two left (or right) cosets $G\mathcal{H}$ and $F\mathcal{H}$ are either identical or contain no elements in common. Further, any element can occur only once in a given coset. Hence cosets provide a disjoint partitioning of the group elements. Extending this beyond cosets, we can consider two proper subgroups $\mathcal{F}$ and $\mathcal{H}$ and *double cosets* $\mathcal{F}G\mathcal{H}$. Just as for cosets, double cosets are either distinct or identical to one another. However, an element can occur multiple times within a given double coset.

If G and H are elements of $\mathcal{G}$ such that

$$H = XGX^{-1} \text{ for at least one } X \in \mathcal{G}, \tag{1.10}$$

G and H are said to be *conjugate* to one another. A subset of $\mathcal{G}$ whose members are all conjugate to one another is termed a *class*. Note that the identity is in a class by itself, so that no class can be a proper subgroup of $\mathcal{G}$. It is also obvious from Eq. 1.10 that if $\mathcal{G}$ is Abelian each element is in a different class. The group of order six whose multiplication table was given previously has three classes formed by the elements E; A, B; and C, D, F; as can be seen by inspection.

1.4 Transformations of Vectors, Functions and Operators

Our discussions so far have tacitly implied that the order of the group g is a finite number, but this is not a necessary requirement, and we shall in fact deal with a number of *infinite groups*, as well as finite groups. Of most immediate importance to us, however, are groups of transformations that leave certain objects invariant, such as spatial rotations and reflections of a solid or an array of points, or transformations of functions. Before concluding this recapitulation of abstract group theory, therefore, we shall discuss some important aspects of groups of transformations.

Consider first a Cartesian coordinate system with unit *basis vectors* $\vec{e}_1$, $\vec{e}_2$, $\vec{e}_3$. We define the effect of a positive rotation about the z axis through an angle θ, corresponding to a symmetry operation G, say, as a counterclockwise rotation; since this operation leaves z unaffected we can consider only two basis vectors

$$G(\vec{e}_1 \quad \vec{e}_2) = (\vec{e}_1{}' \quad \vec{e}_2{}') \tag{1.11}$$

and the operation G that carries the basis vectors $\vec{e}$ into new vectors $\vec{e}'$. As foreshadowed in Eq. 1.11 we shall express the transformation in matrix form by writing the basis vectors *rowwise* as a matrix:

$$(\vec{e}_1{}' \quad \vec{e}_2{}') = (\vec{e}_1 \cos\theta + \vec{e}_2 \sin\theta \quad -\vec{e}_1 \sin\theta + \vec{e}_2 \cos\theta) \tag{1.12}$$

as illustrated in Fig. 1.1.

Figure 1.1: Transformation of basis vectors

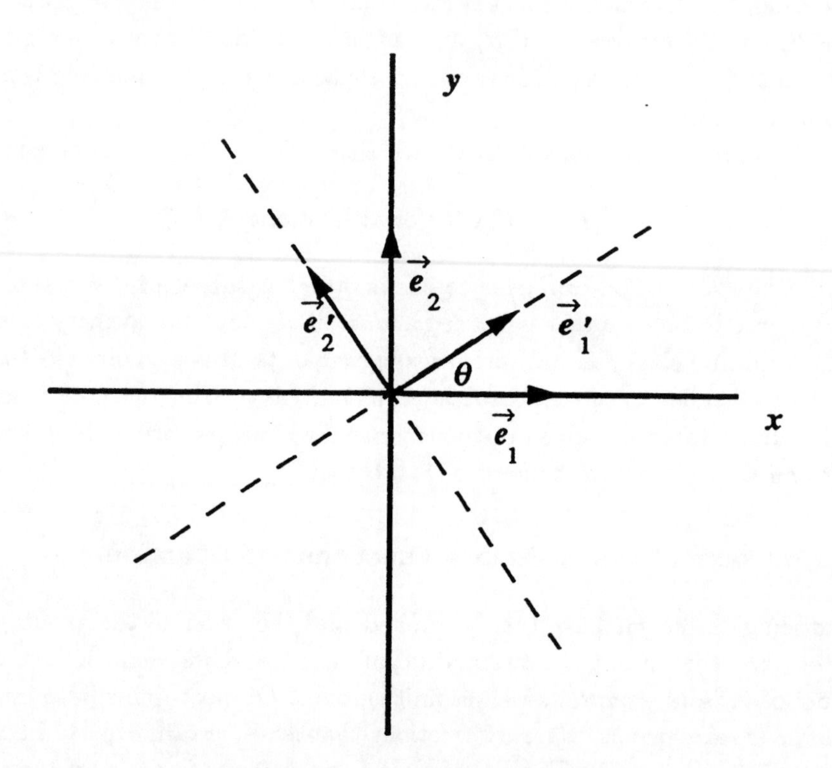

Note that we are representing the symmetry operation as

$$G(\vec{e}_1 \quad \vec{e}_2) = (\vec{e}_1 \quad \vec{e}_2)\mathrm{G}, \tag{1.13}$$

where

$$\mathrm{G} = \begin{pmatrix} \cos\theta & -\sin\theta \\ \sin\theta & \cos\theta \end{pmatrix}. \tag{1.14}$$

An arbitrary point $\vec{r}$ with coordinates x, y, and z can be written in terms of the basis vectors as

$$(\vec{e}_1 \quad \vec{e}_2 \quad \vec{e}_3) \begin{pmatrix} x \\ y \\ z \end{pmatrix}. \tag{1.15}$$

In terms of the transformed basis $\mathbf{e}'$, our arbitrary point can be written $\mathbf{e}'\vec{r}'$, and we must have

$$\begin{aligned} \mathbf{e}\vec{r} &= \mathbf{e}'\vec{r}' \\ &= \mathbf{e}\mathbf{G}\vec{r}' \\ &= \mathbf{e}\mathbf{G}\mathbf{G}^{-1}\vec{r}. \end{aligned} \tag{1.16}$$

Hence

$$\vec{r}' = \mathbf{G}^{-1}\vec{r}, \tag{1.17}$$

and the basis vectors and vector components are said to transform *contragrediently.*

We now consider transformations of a function of position, $f(r)$. Following Wigner [1] and most later authors, we define the effect of a symmetry operation G on the function by introducing an operator O_G satisfying

$$O_G f(Gr) = f(r) \tag{1.18}$$

or, equivalently,

$$O_G f(r) = f(G^{-1}r) \tag{1.19}$$

The reason for this definition can be seen in Fig. 1.2, where we see that the effect of Eq. 1.18 is to rotate the contours of f into those of $O_G f$: the value of the function is preserved by the symmetry operation.

Now, let $f_i(r); i = 1, n$ be a set of n functions that transform among themselves under symmetry operations as

$$G\left(f_1(r), \ldots, f_n(r)\right) = \left(f_1(r), \ldots, f_n(r)\right) \mathbf{D}(G) \tag{1.20}$$

where $\mathbf{D}(G)$ is the $n \times n$ matrix of transformation coefficients. The set of matrices $\mathbf{D}(G)$, $\forall\, G \in \mathcal{G}$, obeys the same multiplication rules as the elements G themselves, and is said to form a *matrix representation* of the group $\mathcal{G}$; $f_i(r); i = 1, n$ is said to form a set of *basis functions* for this representation, or to *carry* this representation. The representation is of dimension, or *degeneracy*, n. When $n = 1$ the representation is termed nondegenerate.

Note that by writing the basis functions as a row vector, they transform among themselves as do the basis vectors, *not* like vector components. Basis functions transform *cogrediently* to basis vectors. With this definition, the $\mathbf{D}(G)$, the functional

Figure 1.2: Transformation of functions

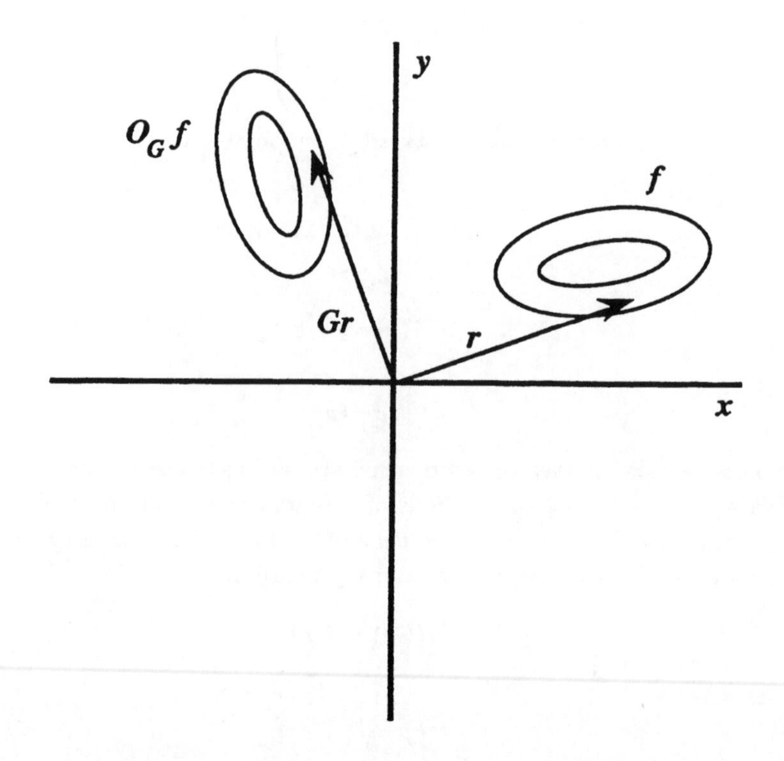

operators O_G, and the symmetry operations G all multiply in the same order; if $GH = F$, for example, $O_G O_H = O_F$ and $\mathbf{D}(G)\mathbf{D}(H) = \mathbf{D}(F)$. Beware of group theory texts that do not explicitly define and follow these conventions: sooner or later they must omit or disguise some form of mathematical inconsistency! In the following we will commonly not distinguish between O_G and G, since there will seldom be any ambiguity, but the distinctions we have drawn in this section should always be borne in mind.

Matrix representations are perhaps the most important objects for practical applications of group theory in quantum chemistry. We have seen how they can be defined in terms of a set of basis functions in Eq. 1.20. Evidently, by finding suitable sets of basis functions that transform among themselves under the operations of the group, we can find matrix representations of arbitrarily large dimension. Furthermore, we can apply an arbitrary similarity transformation $\mathbf{X}$ to our representations, since if $\mathbf{D}(G)\mathbf{D}(H) = \mathbf{D}(F)$, $\mathbf{X}\mathbf{D}(G)\mathbf{D}(H)\mathbf{X}^{-1} = \mathbf{X}\mathbf{D}(F)\mathbf{X}^{-1}$. We first restrict ourselves, therefore, to considering only representations that are *not* equivalent within a similarity transformation. Second, we restrict ourselves to the consideration only of representations by *unitary* matrices, since any representation by (nonsingular) matrices is equivalent to a unitary representation through a similarity transformation. Finally, representation matrices that can be expressed in the block diagonal form

$$\begin{pmatrix} \mathbf{D}^\alpha & 0 & \cdots & \\ 0 & \mathbf{D}^\beta & 0 & \cdots \\ \vdots & & \ddots & \end{pmatrix} \tag{1.21}$$

are said to be *reducible*: any set of matrices $\mathbf{D}(G)$ which cannot be written in this form for at least one operation G is said to be *irreducible*. We can restrict our considerations here to group representations that are unitary, inequivalent, and irreducible, conventionally abbreviated to *irreducible representations* or *irreps*. Abelian groups have only one-dimensional irreps, since in general only matrices of order one (that is, scalars) commute.

The most powerful theorem in group theory, for our purposes, is the great orthogonality theorem (GOT) which states that for irreps $\mathbf{D}^\alpha$ and $\mathbf{D}^\beta$, of respective dimensions n_α and n_β,

$$\sum_G D^\alpha_{ij}(G)^* D^\beta_{kl}(G) = g n_\alpha^{-1} \delta_{\alpha\beta}\delta_{ik}\delta_{jl}. \tag{1.22}$$

This can be used to establish, for example, that

$$\sum_\alpha n_\alpha^2 = g. \tag{1.23}$$

Suppose now we have a reducible representation $\mathbf{D}$ that is composed of several irreps:

$$\mathbf{D} = c_\alpha \mathbf{D}^\alpha \oplus c_\beta \mathbf{D}^\beta \oplus \ldots, \tag{1.24}$$

where we use the symbol $\oplus$ to denote a "direct sum" like the matrix in Eq. 1.21. From the GOT we have

$$g^{-1} n_\alpha \sum_G \mathbf{D}^\alpha(G)^* \mathbf{D}(G) = c_\alpha, \text{ etc.} \tag{1.25}$$

In this way we can readily express any reducible representation as a direct sum of irreps; we can *reduce* such a representation. Similarly, if f is an arbitrary function, the result of the action

$$\begin{aligned} \mathcal{P}^\alpha_{ij} f &= g^{-1} n_\alpha \sum_G D^\alpha_{ij}(G)^* G f \\ &= F^\alpha_{ij} \end{aligned} \tag{1.26}$$

is either zero, or a function that transforms as a basis function for the ith row of irrep α. This is easily verified by operating with H on F^{α}_{ij}:

$$\begin{aligned}
HF^{\alpha}_{ij} &= g^{-1}n_{\alpha}\sum_{G} D^{\alpha}_{ij}(G)^{*}HGf \\
&= g^{-1}n_{\alpha}\sum_{G} D^{\alpha}_{ij}(H^{-1}G)^{*}Gf \\
&= g^{-1}n_{\alpha}\sum_{G}\sum_{k} D^{\alpha}_{ik}(H^{-1})^{*}D^{\alpha}_{kj}(G)^{*}Gf \\
&= \sum_{k} D^{\alpha}_{ik}(H^{-1})^{*}F^{\alpha}_{kj} \\
&= \sum_{k} D^{\alpha}_{ki}(H)F^{\alpha}_{kj},
\end{aligned} \tag{1.27}$$

which is exactly the behaviour exhibited by a basis function for row i of irrep α. A function like F^{α}_{ij} is said to be *symmetry adapted*, belonging to the symmetry species (α, i). An operator

$$\mathcal{P}^{\alpha}_{ii} = g^{-1}n_{\alpha}\sum_{G} D^{\alpha}_{ii}(G)^{*}G \tag{1.28}$$

is called a *projection operator* or *projector*. It satisfies the idempotency condition

$$\mathcal{P}^{\alpha}_{ii}\mathcal{P}^{\alpha}_{ii} = \mathcal{P}^{\alpha}_{ii}. \tag{1.29}$$

If f^{α}_{i} is a basis function for row i of irrep α, then a basis function for row j, a *partner function*, can be obtained from

$$f^{\alpha}_{j} = g^{-1}n_{\alpha}\sum_{G} D^{\alpha}_{ji}(G)^{*}Gf^{\alpha}_{i}. \tag{1.30}$$

Operators such as

$$\mathcal{P}^{\alpha}_{ji} = g^{-1}n_{\alpha}\sum_{G} D^{\alpha}_{ji}(G)^{*}G \tag{1.31}$$

are called *shift operators* or *ladder operators*, or sometimes "step-up" and "step-down" operators, because they can be used to increase or decrease the row indices of the basis functions. Since the operators of Eq. 1.31 are not idempotent they are not projection operators, but the term is often applied indiscriminately to both the genuine projectors and the shift operators.

We have noted that there is a considerable arbitrariness in the choice of representation matrices. For example, we have already recognized that representations related through a similarity transformation are equivalent. It is convenient to have a way of characterizing irreps that is invariant to such transformations, and one such way is to use not the full matrices but the *traces* of the matrices, called *group characters* and conventionally denoted $\chi^\alpha(G)$. First, we can use characters to simplify the GOT as

$$\sum_G \chi^\alpha(G)^* \chi^\beta(G) = g\delta_{\alpha\beta}. \tag{1.32}$$

Second, the character of a given irrep is the same for all operators in the same class of the group $\mathcal{G}$. Hence we can simplify Eq. 1.32 even further to

$$\sum_k^{classes} N_k \chi^\alpha(G_k)^* \chi^\beta(G_k) = g\delta_{\alpha\beta}, \tag{1.33}$$

where the sum is now over the *classes* of $\mathcal{G}$, G_k is an element chosen from class k, and N_k is the number of elements in class k. Third, the number of irreps is equal to the number of classes in the group. Together with Eq. 1.23 this last observation readily allows enumeration of the dimension of all the irreps in the group.

Using characters we can rewrite the equation for reducing a reducible representation χ as

$$c_\alpha = g^{-1} \sum_G \chi(G)^* \chi^\alpha(G) \tag{1.34}$$

(or the equivalent form with a sum over classes only), and express the projection operator from Eq. 1.28 as

$$\mathcal{P}^\alpha = g^{-1} n_\alpha \sum_G \chi^\alpha(G)^* G. \tag{1.35}$$

Note, however, that since we now work with only the trace of the matrix, we have no information about off-diagonal elements of the irrep matrices and hence no way to construct shift operators. The business of establishing symmetry-adapted functions therefore involves somewhat more trial and error than the approach detailed above. Character projection necessarily yields a function that transforms according to the desired irrep (or zero, of course), but application of character projection to different functions will be required to obtain a set of basis functions for a degenerate irrep, and the resulting "basis functions" need not be symmetry adapted for the full symmetry species (irrep and row) obtained above.

Despite some loss of information in using characters rather than full matrix representations, the former are so much simpler that the most common group-theoretical manipulations in quantum chemistry are performed entirely with characters. In this course we shall employ both approaches, as it is useful to acquire some facility with full matrix projectors and shift operators.

Chapter 2

Fundamentals: Molecular Point Groups

2.1 Symmetry Operations and Point Groups

The classification of molecular symmetry operations that we shall follow here is the conventional one (see, e.g., Tinkham [2]), involving *rotations* about a specified axis, denoted C_n for a counterclockwise rotation through $2\pi/n$; *reflections* in a plane, denoted σ; *inversion* through the coordinate origin, denoted i; and *rotation-reflections*, a counterclockwise rotation about a specified axis followed by reflection in a plane perpendicular to that axis, denoted S_n where the rotation is through $2\pi/n$. Note that the rotation and reflection in this compound operation commute, so their order is unimportant. The operation of pure rotation is often termed *proper rotation*, while the others are referred to as *improper rotations.*

For systems with only one axis of rotational symmetry, or systems with only one such axis with $n > 2$, the molecule-fixed z-axis is conventionally aligned with this *principal axis*. One advantage of this strategy is that the (complex or real) spherical harmonics then provide a simple and readily identifiable source of basis functions for the irreps of the group. Groups with a well-defined principal axis are collectively referred to as *axial groups*. If only a single rotation axis is present, such groups are denoted C_n and are of order n. The addition of a reflection plane perpendicular to this axis, denoted σ_h, gives the groups C_{nh}, of order $2n$. Alternatively, addition of a two-fold axis perpendicular to the principal axis gives the groups D_n, of order $2n$, whereas addition of a reflection plane that contains the principal axis, denoted σ_v, gives the groups C_{nv}, of order $2n$. A group defined by a rotation-reflection axis (which requires that 2π is an *even* multiple of the rotation angle) is denoted S_{2n} and is of order $2n$.

Groups of higher order can be obtained by combining these operations further. The groups D_{nh}, of order $4n$, are obtained by adding reflection in a horizontal plane to the group D_n, while the groups D_{nd}, also of order $4n$, are obtained by adding an S_{2n} improper axis to D_n.

Linear systems display the symmetry of one of two groups of infinite order.

The first is $C_{\infty v}$, which has a principal axis about which any rotation is a symmetry operation, plus an infinity of vertical planes containing the principal axis. Systems that in addition have inversion symmetry belong to $D_{\infty h}$.

Lastly, there are several groups with multiple intersecting axes of order greater than two. These are collectively referred to as the *cubic groups*. The simplest is the group T, which comprises the proper rotations of a regular tetrahedron and is of order 12. The proper and improper rotations of the tetrahedron form a group of order 24 denoted T_d. This group is isomorphic to the group of proper rotations of a regular octahedron (or cube), denoted O. Adding a centre of inversion to the group T yields a different group of order 24, denoted T_h. The proper and improper rotations of the octahedron form a group of order 48 denoted O_h. Finally, the proper rotations of a regular icosahedron (or dodecahedron) form a group of order 60, denoted I, while the proper and improper rotations form a group of order 120, denoted I_h. Only the cubic groups can have irreps of dimension greater than two: all the axial groups have at most doubly degenerate irreps. Although systems with cubic symmetry are not very common in chemistry, the cubic groups have a rich structure that has been much explored.

The determination of the point group of a particular molecule or object is a relatively straightforward matter if approached systematically, and an appropriate "recipe" is given in the Supplementary Notes to this course. By far the most common error is the failure to detect two-fold rotation axes perpendicular to the principal axis, so that, for example, a system with D_3 symmetry is incorrectly identified as having only C_3 symmetry.

For convenience in presentation of, for example, character tables, it is useful to regard many point groups as *direct product groups* obtained by combining groups of lower order. For example, the group C_{3v} may be considered as arising from the direct product

$$C_{3v} = C_3 \otimes C_s$$

where C_s comprises the identity and a reflection plane *containing* the principal axis. If we take a different C_s that contains the plane *perpendicular* to the principal axis we obtain a different direct product group, namely C_{3h}. If we take the direct product of C_3 with a group C_2 so that the rotation axes are collinear, we obtain the group C_6, while if the C_2 axis is perpendicular to the C_3 axis we obtain the group D_3. Finally, if we take the direct product of C_3 with the group S_2 (C_i), which contains the inversion operation, we obtain the group S_6. S_6, C_{3h} and C_6 are isomorphic to one another, and C_{3v} and D_3 are isormorphic to each other. We could therefore represent all five groups with only two distinct character tables.

2.2 Irreducible Representations of the Point Groups

Character tables are what most quantum chemists remember best of their group theory, but it is convenient to complete this review of the fundamentals by examining first the character table for a group and then the full matrix irreps for that group. As an iluustration, we shall examine the group C_{3v} (or the isomorphic D_3). Here is the character table in the same format as used in the Tables provided for this course. The

Table 2.1: C_{3v} Character Table

	E	$2C_3$	$3\sigma_v$		
A_1	1	1	1	z	
A_2	1	1	-1		R_z
E	2	-1	0	(x, y)	(R_x, R_y)

group is of order six, it has three classes and therefore three irreps: these must be of dimension one, one and two since the sum of the dimensions squared is the order of the group. (The labels for the irreps follow the notation introduced by Mulliken; full details of this scheme are given in the Supplementary Notes.)

We can see immediately that the direct product representation $A_2 \otimes E$ is the irrep E (direct products with A_1 are of course trivial). The direct product $E \otimes E$ gives the characters $(4, 1, 0)$, a reducible representation that can be reduced to $A_1 \oplus A_2 \oplus E$. Basis functions for the irreps are also given in the table, as well as the behaviour of (R_x, R_y, R_z), corresponding to rotations about the three coordinate axes.

Somewhat more information is provided by the full matrix representations of Table 2.2. It can be verified, for example, that the functions x and y respectively

Table 2.2: C_{3v} Matrix Irreps

	E	C_3	C_3^2	σ_v^a	σ_v^b	σ_v^c
A_1	1	1	1	1	1	1
A_2	1	1	1	-1	-1	-1
$(E)_{11}$	1	$-\frac{1}{2}$	$-\frac{1}{2}$	1	$-\frac{1}{2}$	$-\frac{1}{2}$
$(E)_{21}$	0	$\frac{\sqrt{3}}{2}$	$-\frac{\sqrt{3}}{2}$	0	$-\frac{\sqrt{3}}{2}$	$\frac{\sqrt{3}}{2}$
$(E)_{12}$	0	$-\frac{\sqrt{3}}{2}$	$\frac{\sqrt{3}}{2}$	0	$-\frac{\sqrt{3}}{2}$	$\frac{\sqrt{3}}{2}$
$(E)_{22}$	1	$-\frac{1}{2}$	$-\frac{1}{2}$	-1	$\frac{1}{2}$	$\frac{1}{2}$

transform as basis functions for the first and second rows of the E representation. We can also see immediately what will happen if we consider only the *subgroup* C_s of C_{3v} comprising the identity and the first σ_v operation. This subgroup has two one-dimensional irreps labelled A' and A'', respectively symmetric and antisymmetric with respect to σ_v. Evidently, the C_{3v} A_1 and A_2 irreps correspond to A' and A'' in C_s, while the first row of the E irrep corresponds to A' and the second to A''. Formally, the reduction of an irrep of a group $\mathcal{G}$ to one or more irreps of a subgroup of $\mathcal{G}$ is called *subduction*: the E irrep of C_{3v} subduces to $A' \oplus A''$ in C_s. The distinction between reduction and subduction of a representation is often casually ignored and the term reduction is used for both operations outside the group-theoretical literature, but it is a distinction worth preserving.

2.3 Basis Functions for Irreducible Representations

Character tables usually list the transformation properties of the variables x, y, and z, as well as the dyadic products xy, etc. Hence for groups of low order the character tables will usually provide basis functions for all irreps using these simple functions. However, for groups of order, say, 12 or more, it is not uncommon to find that basis functions are not listed for some irreps. Perhaps the most straightforward source of basis functions is the spherical harmonics, although for groups of high order it may be necessary to go to very high l values ($l = 9$ for the octahedral group, for instance). For the cubic groups, linear combinations of the spherical harmonics like the so-called cubic harmonics used by Slater [3], for example, may be more useful.

2.4 Separable Degeneracy

For some point groups, such as the pure rotation groups C_n, the irreps are necessarily complex, and basis functions for these irreps may also be complex. For example, the group C_3 has three one-dimensional representations (recall that the cyclic groups are Abelian), and the simplest basis functions for the non-totally symmetric irreps are $x - iy$ and $x + iy$. This is inconvenient for calculations involving only real variables and functions. In fact, in the absence of time-reversal operations (or the equivalent effect of magnetic interactions), energy levels belonging to these two irreps are always degenerate, so it is reasonable to consider them as components of a doubly degenerate representation. Such a representation is termed *separably degenerate*. It is important to understand that such a representation is *not* irreducible. Some of the selection rules and symmetry behaviour we shall consider in these lectures do not apply to reducible representations, and hence do not necessarily apply to such separably degenerate representations.

2.5 Linear Groups: Group Integration

Finally, we consider briefly the linear groups. Since these are of infinite order, it should come as no surprise that the sums that appeared in the GOT (Eq. 1.22) or the projection/shift operators (Eqs. 1.28 and 1.31) are replaced by integration over the *group manifold*, the domain of the group operators. For example, in the case of $C_{\infty v}$ we have

$$\mathcal{P}^{\gamma} f = \frac{1}{4\pi} \left\{ \int_{-\pi}^{\pi} \chi^{\gamma}[C(\phi)]^{*} C(\phi) f \, d\phi + \int_{-\pi}^{\pi} \chi^{\gamma}[\sigma_v(\alpha)]^{*} \sigma_v(\alpha) f \, d\alpha \right\} \tag{2.1}$$

for the character projector $\mathcal{P}^{\gamma}$ operating on the function f. There is no particular difficulty in manipulating Eq. 2.1 to obtain symmetry-adapted functions, although it is usually simpler to classify functions by their "angular momentum" about the principal axis (this is very easy for spherical harmonics, for instance). We should note that the replacement of a sum over group elements with group integration is not possible for all groups of infinite order, but only those satisfying certain requirements. This will not be an issue in the present lectures.

Chapter 3

Symmetry Properties of Orbitals

3.1 Symmetry of the Molecular Hamiltonian

From the perspective of non-relativistic quantum mechanics, a collection of electrons and atomic nuclei interact through the Coulomb Hamiltonian

$$H = -\frac{1}{2}\sum_i \nabla_i^2 + \frac{1}{2M_A}\sum_A \nabla_A^2 + \sum_{i>j} r_{ij}^{-1} - \sum_{i,A} Z_A r_{iA}^{-1} + \sum_{A>B} Z_A Z_B r_{AB}^{-1}. \tag{3.1}$$

Note that all the particles are moving here and that we have used atomic units in which the mass and charge of the electron are unity. Since the only dependence on particle variables here involves the interparticle distances, which are scalars, and the differential operators in the kinetic energy terms, this Hamiltonian is unaffected by any arbitrary rotation in space, by coordinate inversion (and thus by any improper rotations), by permutation among identical particles, and by uniform translation in any direction. The symmetry properties of this Hamiltonian are therefore of very high order, as are those of its energy levels, which correspond to individual rotational/vibrational/electronic wave functions. In practice, it is of course much more convenient to adopt a clamped-nucleus Born-Oppenheimer approach to molecular quantum mechanics, thereby separating off nuclear motion and producing the Hamiltonian

$$H = -1/2\sum_i \nabla_i^2 + \sum_{i>j} r_{ij}^{-1} - \sum_{i,A} Z_A r_{iA}^{-1} + \sum_{A>B} Z_A Z_B R_{AB}^{-1}. \tag{3.2}$$

Since the nuclear coordinates appearing in Eq. 3.2 are fixed parameters, as indicated by using the upper-case symbol R for the internuclear distances, the spatial symmetry of the Hamiltonian is reduced to those operations that leave the nuclear framework invariant. (Permutational symmetry among the electrons is retained and will be considered in Chapter 6.)

3.2 Stabilizers of Centres

Consider an arbitrary molecule whose nuclear framework is invariant under the operations of the group $\mathcal{G}$. In general, application of an operator $G \in \mathcal{G}$ will transform a given centre A into a new centre, its *image* $G(A)$. It is possible, however, that some operations will leave the centre A fixed. Such operations form a subgroup $\mathcal{U}$ of $\mathcal{G}$, with the property

$$U(A) = A \, \forall \, U \in \mathcal{U}. \tag{3.3}$$

A subgroup with the property given in Eq. 3.3 is referred to as the *stabilizer* of A in $\mathcal{G}$ [4]. The transformed centres that can be generated from A may be found by applying operators chosen one from each distinct left coset of $\mathcal{U}$ in $\mathcal{G}$. The number of such cosets is the number of such transformed centres and is given by g/u, where g is the order of $\mathcal{G}$ and u is the order of $\mathcal{U}$. The quantity g/u is termed the *index of the stabilizer*, although the term "constituency number" has been used in some quantum chemical applications.

Let us consider an example: a planar symmetric BF_3 molecule (D_{3h} symmetry).

Figure 3.1: Planar Symmetric BF_3 Molecule

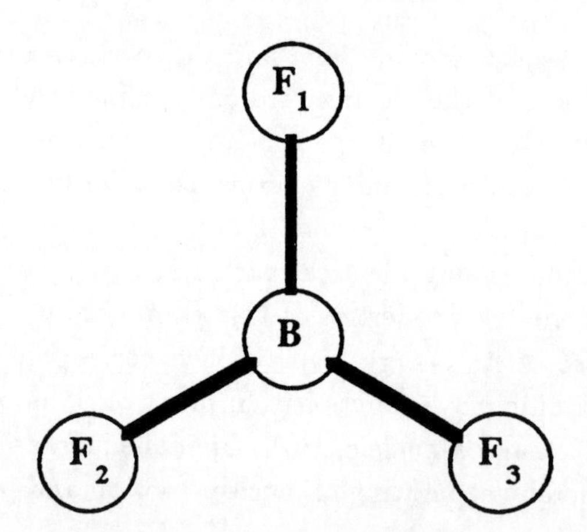

The stabilizer of the origin is obviously the entire group, so the stabilizer of B is D_{3h} itself. Consider now F_1. Reflection in the σ_v plane that contains the BF_1 bond and reflection in the σ_h plane both leave F_1 unchanged; the product of these operations is a C_2 rotation around the BF_1 bond. Hence the stabilizer of F_1 is a C_{2v} subgroup of D_{3h} — we can denote it C_{2v}^1, identifying the C_2 axis direction by a superscript. Since the order of C_{2v} is four and the order of D_{3h} is 12, the index of the

Chapter 3

Symmetry Properties of Orbitals

3.1 Symmetry of the Molecular Hamiltonian

From the perspective of non-relativistic quantum mechanics, a collection of electrons and atomic nuclei interact through the Coulomb Hamiltonian

$$H = -\frac{1}{2}\sum_i \nabla_i^2 + \frac{1}{2M_A}\sum_A \nabla_A^2 + \sum_{i>j} r_{ij}^{-1} - \sum_{i,A} Z_A r_{iA}^{-1} + \sum_{A>B} Z_A Z_B r_{AB}^{-1}. \quad (3.1)$$

Note that all the particles are moving here and that we have used atomic units in which the mass and charge of the electron are unity. Since the only dependence on particle variables here involves the interparticle distances, which are scalars, and the differential operators in the kinetic energy terms, this Hamiltonian is unaffected by any arbitrary rotation in space, by coordinate inversion (and thus by any improper rotations), by permutation among identical particles, and by uniform translation in any direction. The symmetry properties of this Hamiltonian are therefore of very high order, as are those of its energy levels, which correspond to individual rotational/vibrational/electronic wave functions. In practice, it is of course much more convenient to adopt a clamped-nucleus Born-Oppenheimer approach to molecular quantum mechanics, thereby separating off nuclear motion and producing the Hamiltonian

$$H = -1/2\sum_i \nabla_i^2 + \sum_{i>j} r_{ij}^{-1} - \sum_{i,A} Z_A r_{iA}^{-1} + \sum_{A>B} Z_A Z_B R_{AB}^{-1}. \quad (3.2)$$

Since the nuclear coordinates appearing in Eq. 3.2 are fixed parameters, as indicated by using the upper-case symbol R for the internuclear distances, the spatial symmetry of the Hamiltonian is reduced to those operations that leave the nuclear framework invariant. (Permutational symmetry among the electrons is retained and will be considered in Chapter 6.)

Finally, if the effect of operation $G \in \mathcal{G}$ on a set of functions $\phi_1, \ldots, \phi_n$ is given by

$$G\phi = \phi \mathbf{G}, \tag{3.6}$$

we can note immediately that if the centre of the functions ϕ is shifted by G, the trace of $\mathbf{G}$ must be zero, since there can be no non-zero coefficients on the diagonal in $\mathbf{G}$. Hence if a class element G is *not* in the stabilizer of the centre of ϕ, the character of the representation will be zero for that operation. Let us return to BF_3 and illustrate these points explicitly.

Atomic basis functions on B are straightforward to classify. Evidently, an s type function on B will be totally symmetric — an a_1' orbital. A quick inspection of the D_{3h} character table will show that a p set on B, which transforms like the three Cartesian directions, spans the reducible representation $a_2'' \oplus e'$. Functions centred on the F atoms require more effort. Since the operations in the classes containing C_3 and S_3 move all three F atoms, their character is necessarily zero for any functions centred on the F atoms. Consider first a set of s functions on each F. These span a reducible representation with character

E	σ_h	$2C_3$	$2S_3$	$3C_2'$	$3\sigma_v$
3	3	0	0	1	1

Use of Eq. 1.34 indicates that this representation reduces to $a_1' \oplus e'$. For a set of p functions on each F, the representation is

E	σ_h	$2C_3$	$2S_3$	$3C_2'$	$3\sigma_v$
9	3	0	0	-1	1

which reduces to $a_1' \oplus a_2' \oplus a_2'' \oplus 2e' \oplus e''$.

While we have chosen to proceed here by reducing representations for the full group D_{3h}, it would have been simpler to take advantage of the fact that D_{3h} is the direct product of C_{3v} and C_s, where the plane in the latter is perpendicular to the principal axis of the former. The behaviour of any atomic basis functions with respect to the C_s subgroup is trivial to determine, and there are only two classes of non-trivial operations in C_{3v}. In more general cases, it is often worthwhile to look for such simplifications. It is seldom useful, for instance, to employ the full character table for a group that contains the inversion, or a unique horizontal plane, since the symmetry with respect to these operations can be determined by inspection. With these observations and the transformation properties of spherical harmonics given in the Supplementary Notes, it should be possible to determine the symmetries spanned by sets of atomic basis functions for any molecular system. Finally, with access to the appropriate literature the labour can be eliminated entirely for some cases, since a very useful table of how atomic functions on multiple symmetry-related centres transform under common point groups is given as Appendix V of Herzberg [5].

For a simple MO picture of molecular electronic structure, the same procedures can be followed to classify the symmetries spanned by bond and lone-pair orbitals. For instance, we can envisage the electronic structure of BF_3 as involving sp^2 hybridization of the atoms. We would then have B and three F $1s$ core orbitals, two sp^2 and one "π" symmetry lone pair on each F atom, and three sp^2 bond orbitals for the three BF bonds. Our analysis above shows that the four core orbitals comprise 2 a_1' and a doubly degenerate e' orbital; the in-plane lone-pairs transform as a_1', a_2', and 2 e' orbitals; the out-of-plane lone pairs transform as a_2'' and e''; and the three BF bonds as a_1' and e'.

3.4 Projection of Atomic Orbitals

We can go beyond the classification of atomic basis functions of the last subsection to obtain explicit symmetry-adapted basis functions by projection. We can use either the full matrix projection and shift operators (Eqs. 1.28 and 1.31, which are repeated here for convenience):

$$\mathcal{P}_{ii}^{\alpha} = g^{-1} n_{\alpha} \sum_{G} D_{ii}^{\alpha}(G)^{*} G \tag{3.7}$$

and

$$\mathcal{P}_{ji}^{\alpha} = g^{-1} n_{\alpha} \sum_{G} D_{ji}^{\alpha}(G)^{*} G, \tag{3.8}$$

or the character form of Eq. 1.35

$$\mathcal{P}^{\alpha} = n_{\alpha} g^{-1} \sum_{G} \chi^{\alpha}(G)^{*} G. \tag{3.9}$$

In either case it is helpful to have a table of transformation properties of spherical harmonics like that given in the Supplementary Notes. We shall illustrate the procedure by finding symmetry-adapted basis functions arising from an s function on each F atom in BF_3, using full matrix projection operators. We already know that the three atomic basis functions transform as $a_1' \oplus e'$, and since the behaviour with respect to the horizontal plane is already known, we can, without loss of generality, work with the subgroup C_{3v} only. We denote the s functions on F_1, F_2, and F_3 as s_1, s_2, and s_3, respectively, and apply our C_{3v} projection operators

$$\mathcal{P}^{A_1} = \frac{1}{6}\left[E + C_3 + C_3^2 + \sigma_1 + \sigma_2 + \sigma_3\right] \tag{3.10}$$

and

$$\mathcal{P}_{11}^{E} = \frac{2}{6}\left[E - \frac{1}{2}C_3 - \frac{1}{2}C_3^2 + \sigma_1 - \frac{1}{2}\sigma_2 - \frac{1}{2}\sigma_3\right] \tag{3.11}$$

to s_1.

$$\mathcal{P}^{A_1} s_1 = \frac{1}{6}[s_1 + s_2 + s_3 + s_1 + s_3 + s_2] = \frac{1}{3}[s_1 + s_2 + s_3]. \tag{3.12}$$

$$\mathcal{P}^{E}_{11} s_1 = \frac{2}{6}\left[s_1 - \frac{1}{2}s_2 - \frac{1}{2}s_3 + s_1 - \frac{1}{2}s_3 - \frac{1}{2}s_2\right] = \frac{1}{3}[2s_1 - s_2 - s_3]. \tag{3.13}$$

Since we have obtained a basis function for row 1 of the e irrep, we can obtain a basis function for row 2 by applying the shift operator

$$\mathcal{P}^{E}_{21} = \frac{2}{6}\left[\frac{\sqrt{3}}{2}C_3 - \frac{\sqrt{3}}{2}C_3^2 - \frac{\sqrt{3}}{2}\sigma_2 + \frac{\sqrt{3}}{2}\sigma_3\right] \tag{3.14}$$

to our e basis function above.

$$\begin{aligned} &\mathcal{P}^{E}_{21}\frac{1}{3}[2s_1 - s_2 - s_3] \\ &= \frac{\sqrt{3}}{18}[2s_2 - s_3 - s_1 - 2s_3 + s_1 + s_2 - 2s_3 + s_2 + s_1 + 2s_2 - s_1 - s_3] \\ &= \frac{\sqrt{3}}{3}[s_2 - s_3]. \end{aligned} \tag{3.15}$$

As the result of Eq. 3.15 suggests, application of the projection operator

$$\mathcal{P}^{E}_{22} = \frac{2}{6}\left[E - \frac{1}{2}C_3 - \frac{1}{2}C_3^2 - \sigma_1 + \frac{1}{2}\sigma_2 + \frac{1}{2}\sigma_3\right] \tag{3.16}$$

to s_1 would have yielded a zero result, so it would not have been possible to generate a partner function this way. In general, we can apply the projection and shift operators for any values of the column index i and choose as our symmetry-adapted basis any linearly independent set of functions we generate. In the present case, if we operate on s_1 it is clear that the use of $\mathcal{P}^{E}_{22}$ followed by $\mathcal{P}^{E}_{12}$ would generate no information, so there is only one choice of column index i.

We can also consider using the character projector of Eq. 3.9 to obtain our symmetry-adapted basis functions. There is no need to consider the A_1 case again since the irrep is one-dimensional. For $\mathcal{P}^{E}$ we obtain

$$\mathcal{P}^{\alpha} s_1 = \frac{1}{3}[2s_1 - s_2 - s_3] \tag{3.17}$$

by operating on s_1. Clearly, we have exhausted the information we can obtain about the e case from s_1 at this point, so we can attempt to find a partner function by applying the character projector to s_2, say:

$$\mathcal{P}^{\alpha} s_2 = \frac{1}{3}[2s_2 - s_3 - s_1]. \tag{3.18}$$

Inspection of this result shows that it is not equivalent to the partner function Eq. 3.15, and that although it and the result of Eq. 3.17 are linearly independent they are not orthogonal. Explicit Schmidt orthogonalization of the result of Eq. 3.18 to that of Eq. 3.17 yields the same functions as those obtained from the full matrix projection and shift operators. However, without knowledge of the full matrix representations we cannot identify these character projection results with specific rows of the e irrep. In fact, in general the results of character projection will *not* yield basis functions that can be identified with symmetry species.

As was discussed in Chapter 2, the need to have full matrix representations available to obtain basis functions adapted to symmetry species is something of a handicap. Although character projection itself is not adequate for this task, Hurley has shown how the use of a sequence of character projectors for a chain of subgroups of the full point group can generate fully symmetry-adapted functions. Further discussion of this approach is beyond the scope of the present course, but interested readers may care to refer to the original literature [6].

Chapter 4

Integrals and Matrix Elements over Symmetry Orbitals

4.1 Selection Rules

We begin by reviewing perhaps the most fundamental selection rule in quantum chemistry. Let the functions $\{\phi_i^\alpha \; \forall\, i\}$ form a basis of partner functions for irrep α, and similarly $\{\psi_j^\beta\}$ for irrep β. Let $\mathcal{O}$ denote an operator that commutes with all elements of the group $\mathcal{G}$: $\mathcal{O}$ is a totally symmetric operator in the terminology of Sec. 1.4. At this stage, it should be noted, our basis functions can be one- or many-electron functions. Consider now the matrix element

$$\int \phi_i^{\alpha *} \mathcal{O} \psi_j^\beta \, d\tau. \tag{4.1}$$

Since $\mathcal{O}$ is totally symmetric, we can replace it with $G^{-1}\mathcal{O}G$ for any $G \in \mathcal{G}$ to give

$$\int \phi_i^{\alpha *} G^{-1} \mathcal{O} G \psi_j^\beta \, d\tau. \tag{4.2}$$

But from the transformation properties of basis functions Eq. 1.27 and the unitarity of the irreps we have

$$\int \phi_i^{\alpha *} G^{-1} \mathcal{O} G \psi_j^\beta \, d\tau = \sum_k \sum_l D_{ki}^\alpha(G)^* D_{lj}^\beta(G) \int \phi_k^{\alpha *} \mathcal{O} \psi_l^\beta \, d\tau. \tag{4.3}$$

Summing over G and dividing by g gives

$$\begin{aligned} g^{-1} \sum_G \int \phi_i^{\alpha *} G^{-1} \mathcal{O} G \psi_j^\beta \, d\tau &= g^{-1} \sum_G \sum_k \sum_l D_{ki}^\alpha(G)^* D_{lj}^\beta(G) \int \phi_k^{\alpha *} \mathcal{O} \psi_l^\beta \, d\tau \\ &= n_\alpha^{-1} \delta_{\alpha\beta} \delta_{ij} \sum_k \int \phi_k^{\alpha *} \mathcal{O} \psi_k^\alpha \, d\tau, \end{aligned} \tag{4.4}$$

from the GOT. Hence

$$\int \phi_i^{\alpha *} \mathcal{O} \psi_j^\beta \, d\tau = n_\alpha^{-1} \delta_{\alpha\beta} \delta_{ij} \sum_k \int \phi_k^{\alpha *} \mathcal{O} \psi_k^\alpha \, d\tau. \tag{4.5}$$

Note that the "average" over k on the RHS of Eq. 4.5 means that the integral itself is independent of the value of the row index i. Thus the matrix of $\mathcal{O}$ in a basis of symmetry-adapted functions is diagonal on symmetry species, and the diagonal blocks corresponding to rows of the same irrep are the same.

Since the many-electron Hamiltonian, the effective one-electron Hamiltonian obtained in Hartree-Fock theory (the "Fock operator"), and the one- and two-electron operators that comprise the Hamiltonian are all totally symmetric, this selection rule is extremely powerful and useful. It can be generalized by noting that *any* operator can be written in terms of symmetry-adapted operators:

$$\mathcal{O} = \sum_{\gamma} \sum_{k}^{n_\gamma} C_{kt}^{\gamma} \mathcal{O}_{kt}^{\gamma}, \tag{4.6}$$

where the sums run over irreps and rows of irreps, and t is a column index chosen at will. The coefficients C are easily determined by projection. The operators $\mathcal{O}_{kt}^{\gamma}$ are *irreducible tensor operators* that transform as basis functions for the kth row of irrep γ

$$G\mathcal{O}_{kt}^{\gamma}G^{-1} = \sum_{l} \mathcal{O}_{lt}^{\gamma} D_{lk}^{\gamma}(G). \tag{4.7}$$

A general development of matrix elements of an irreducible tensor operator leads to the Wigner-Eckart theorem (see, for example, Tinkham [2] or Chisholm [7]), which relates matrix elements between specific symmetry species to a single *reduced matrix element* that depends only on the irrep labels, but this is beyond the scope of the present course.

4.2 Integrals over Atomic Orbitals and Symmetry Orbitals

The selection rules of Sec. 4.1 are exactly what is required when the functions involved in the matrix elements are symmetry adapted. The operator matrices are then completely blocked by symmetry, and any manipulations of these matrices that can also employ symmetry will benefit from a reduction in effort by exploiting the symmetry. Activities such as integral transformations or computing correlation energies, which have fifth and higherpower dependencies on the size of the problem, benefit especially from the use of symmetry. However, in most quantum chemical calculations the starting point is a set of orbitals on individual atoms (or other centres in the molecule). Such functions are not symmetry adapted. There are two possible approaches to exploiting symmetry in calculating and then manipulating the necessary integrals over these atomic orbitals — the use of symmetry-adapted orbitals, and the use of atomic orbitals but with explicit elimination of redundant atomic orbital integrals. We shall consider the use of symmetry-adapted orbitals, or, shorter, symmetry orbitals in the rest of this chapter, and discuss techniques that work directly in the atomic basis in Chapter 5. Before examining the formulas for symmetry orbital integrals, we need to review some properties of *double cosets*.

4.3 Double Coset Decompositions

We have already referred to double cosets in Sec. 1.3: they are obtained as sets $\mathcal{U}G\mathcal{V}$, where $G \in \mathcal{G}$ and $\mathcal{U}, \mathcal{V} \subseteq \mathcal{G}$. Each double coset is either distinct or equal to another: they thus provide a disjoint partitioning of the elements of $\mathcal{G}$, termed a *double coset decomposition* [8]. Each element within a given double coset occurs λ_G times, where

$$\lambda_G = \left|\mathcal{U} \cap G\mathcal{V}G^{-1}\right|. \tag{4.8}$$

(This formula is given incorrectly by Lomont [8].) A set of *double coset representatives* (DCRs), denoted $\mathbf{R}$, consists of a set of operators R chosen one from each *distinct* double coset; each $R \in \mathbf{R}$ has associated with it a degeneracy factor λ_R, given by Eq. 4.8 (with R replacing G). Using these double coset representatives, we can rewrite expressions involving sums over G in terms of sums over elements of the subgroups $\mathcal{U}$ and $\mathcal{V}$ and elements of $\mathbf{R}$. For example, for a general projection or shift operator we have

$$P_{ir}^{\alpha} = g^{-1}n_\alpha \sum_G D_{ir}^{\alpha}(G)^* G = g^{-1}n_\alpha \sum_U \sum_V \sum_R \lambda_R^{-1} D_{ir}^{\alpha}(URV)^* URV. \tag{4.9}$$

Note that the summation over R implies summation over the elements of the DCR $\mathbf{R}$. In application to symmetry-adapted integrals it is often convenient to substitute U^{-1} for U in 4.9. Since $\mathcal{U}$ is a group this is perfectly legitimate.

We now consider one-electron integrals over symmetry orbitals.

4.4 One-electron Integrals: General Case

We can write the most general form of one-electron integral as

$$\left\langle F_{aAir}^{\alpha} \,|\, O_{kt}^{\gamma} \,|\, F_{bBjs}^{\beta} \right\rangle. \tag{4.10}$$

Here F_{aAir}^{α} is a symmetry orbital obtained by applying projection/shift operators to the atomic orbital f_{aA}:

$$F_{aAir}^{\alpha} = P_{ir}^{\alpha} f_{aA}. \tag{4.11}$$

The orbital f_{aA} is centred on A; a is simply a counting index. F_{bBjs}^{β} is obtained similarly. The operator O_{kt}^{γ} transforms according to the kth row of irrep γ; it can be considered as arising from some prototype operator O by "projection" in

$$O_{kt}^{\gamma} = g^{-1}n_\gamma \sum_G D_{kt}^{\gamma}(G)^* GOG^{-1}. \tag{4.12}$$

We begin by substituting the projection step Eq. 4.11 into Eq. 4.10 and expanding using the explicit form of projection and shift operators given by Eqs. 1.28 and 1.31:

$$\begin{aligned} g^{-1}n_\alpha \sum_G D^\alpha_{ir}(G) \left\langle Gf_{aA}|O^\gamma_{kt}|F^\beta_{bBjs}\right\rangle \\ = g^{-1}n_\alpha \sum_G D^\alpha_{ir}(G) \left\langle Gf_{aA}|O^\gamma_{kt}|GG^{-1}F^\beta_{bBjs}\right\rangle \\ = g^{-1}n_\alpha \sum_G \sum_{\bar{j}} D^\alpha_{ir}(G) D^\beta_{\bar{j}j}(G^{-1}) \left\langle Gf_{aA}|O^\gamma_{kt}|GF^\beta_{bB\bar{j}s}\right\rangle \end{aligned} \tag{4.13}$$

after we have inserted the identity as GG^{-1}. Let us now insert an explicit expression for $F^\beta_{bB\bar{j}s}$ using the DCR form of the projection shift operator, i.e.,

$$F^\beta_{bB\bar{j}s} = g^{-1}n_\beta \sum_U \sum_V \sum_R \lambda_R^{-1} D^\beta_{\bar{j}s}(U^{-1}RV)^* U^{-1}RVf_{bB}. \tag{4.14}$$

We thus obtain

$$\begin{aligned} g^{-2}n_\alpha n_\beta \sum_G \sum_U \sum_V \sum_R \sum_{\bar{j}} D^\alpha_{ir}(G) D^\beta_{j\bar{j}}(G)^* \lambda_R^{-1} D^\beta_{\bar{j}s}(U^{-1}RV)^* \\ \times \left\langle Gf_{aA}|O^\gamma_{kt}|GU^{-1}RVf_{bB}\right\rangle, \end{aligned} \tag{4.15}$$

(using the unitarity of the irrep matrices). We can now use the Rearrangement Theorem to replace G everywhere in Eq. 4.15 with GU, giving

$$\begin{aligned} g^{-2}n_\alpha n_\beta \sum_G \sum_U \sum_V \sum_R \sum_{\bar{j}} D^\alpha_{ir}(GU) D^\beta_{j\bar{j}}(GU)^* \lambda_R^{-1} D^\beta_{\bar{j}s}(U^{-1}RV)^* \\ \times \langle GUf_{aA}|O^\gamma_{kt}|GRVf_{bB}\rangle. \end{aligned} \tag{4.16}$$

But $U \in \mathcal{U}$, the stabilizer of A, hence

$$Uf_{aA} = \sum_{\bar{a}} C_{\bar{a}aA}(U) f_{\bar{a}A}, \tag{4.17}$$

and similarly

$$Vf_{bB} = \sum_{\bar{b}} C_{\bar{b}bB}(V) f_{\bar{b}B}. \tag{4.18}$$

Furthermore,

$$D^\alpha_{ir}(GU) = \sum_{\bar{i}} D^\alpha_{i\bar{i}}(G) D^\alpha_{\bar{i}r}(U), \tag{4.19}$$

and

$$\sum_{\bar{j}} D^\beta_{j\bar{j}}(GU)^* D^\beta_{\bar{j}s}(U^{-1}RV)^* = D^\beta_{js}(GRV)^* = \sum_{\bar{j}} D^\beta_{j\bar{j}}(G)^* D^\beta_{\bar{j}s}(RV)^*. \tag{4.20}$$

Hence Eq. 4.16, the expansion of our original integral (Eq. 4.10), becomes

$$g^{-2}n_\alpha n_\beta \sum_G\sum_U\sum_V\sum_R\sum_{\bar{\imath}}\sum_{\bar{\jmath}} \lambda_R^{-1} D^\alpha_{i\bar{\imath}}(G) D^\alpha_{\bar{\imath}r}(U) D^\beta_{j\bar{\jmath}}(G)^* D^\beta_{\bar{\jmath}s}(RV)^*$$
$$\times \sum_{\bar{a}}\sum_{\bar{b}} C_{\bar{a}aA}(U) C_{\bar{b}bB}(RV) \left\langle Gf_{\bar{a}A}|O^\gamma_{kt}|Gf_{\bar{b}R(B)}\right\rangle. \tag{4.21}$$

We can simplify the notation somewhat by defining

$$\Lambda^{A\alpha}_{\bar{a}a\bar{\imath}r}(G) = g^{-1}n_\alpha \sum_U D^\alpha_{\bar{\imath}r}(GU)^* C_{\bar{a}aA}(GU), \tag{4.22}$$

with the special case

$$\Lambda^{A\alpha}_{\bar{a}a\bar{\imath}r} = g^{-1}n_\alpha \sum_U D^\alpha_{\bar{\imath}r}(U)^* C_{\bar{a}aA}(U), \tag{4.23}$$

and analogous expressions such as

$$\Lambda^{B\beta}_{\bar{b}b\bar{\jmath}s}(G) = g^{-1}n_\beta \sum_V D^\beta_{\bar{\jmath}s}(GV)^* C_{\bar{b}bB}(GV). \tag{4.24}$$

Using this notation Eq. 4.21 becomes

$$\sum_R\sum_{\bar{\imath}}\sum_{\bar{\jmath}}\sum_{\bar{a}}\sum_{\bar{b}} \lambda_R^{-1} (\Lambda^{A\alpha}_{\bar{a}a\bar{\imath}r})^* \Lambda^{B\beta}_{\bar{b}b\bar{\jmath}s}(R) \sum_G D^\alpha_{i\bar{\imath}}(G) D^\beta_{j\bar{\jmath}}(G)^* \left\langle Gf_{\bar{a}A}|O^\gamma_{kt}|Gf_{\bar{b}R(B)}\right\rangle. \tag{4.25}$$

The AO integral

$$\left\langle Gf_{\bar{a}A}|O^\gamma_{kt}|Gf_{\bar{b}R(B)}\right\rangle = \left\langle f_{\bar{a}A}|G^{-1}O^\gamma_{kt}G|f_{\bar{b}R(B)}\right\rangle, \tag{4.26}$$

and since

$$G^{-1}O^\gamma_{kt}G = \sum_{\bar{k}} D^\gamma_{k\bar{k}}(G^{-1})^* O^\gamma_{kt} = \sum_{\bar{k}} D^\gamma_{\bar{k}k}(G) O^\gamma_{\bar{k}t}, \tag{4.27}$$

we can rewrite Eq. 4.25 in a way that exposes the symmetry selection rule more clearly:

$$\sum_R\sum_{\bar{\imath}}\sum_{\bar{\jmath}}\sum_{\bar{k}}\sum_{\bar{a}}\sum_{\bar{b}} \lambda_R^{-1} (\Lambda^{A\alpha}_{\bar{a}a\bar{\imath}r})^* \Lambda^{B\beta}_{\bar{b}b\bar{\jmath}s}(R)$$
$$\times \sum_G D^\alpha_{i\bar{\imath}}(G) D^\beta_{j\bar{\jmath}}(G)^* D^\gamma_{\bar{k}k}(G) \left\langle f_{\bar{a}A}|O^\gamma_{\bar{k}t}|f_{\bar{b}R(B)}\right\rangle. \tag{4.28}$$

The selection rule is embodied in the term

$$\sum_G D^\alpha_{i\bar{\imath}}(G) D^\beta_{j\bar{\jmath}}(G)^* D^\gamma_{\bar{k}k}(G). \tag{4.29}$$

For the case that γ is the totally symmetric irrep, this term reduces by the GOT to the same selection rule as Eq. 4.5.

Eq. 4.28 expresses the desired symmetry integral in terms of distinct AO integrals only — no redundant integrals appear except in certain special cases (see below). The expression is certainly forbidding at first, with its multiple summations, but the individual terms are actually quite simple [9]. The Λ factors are the coefficients of linearly independent transformed AOs, the summation over G is, as we have seen, a selection rule on symmetry species, and the DCR $\mathbf{R}$ can best be viewed as comprising those operators R for which the transformed *charge distributions* $f_{aA}f_{bR(B)}$ are unique. We should note here that when $f_{aA} = f_{bB}$, the definition of the DCRs must be extended from that given in Sec. 4.3 — which would reduce to operators selected from distinct $\mathcal{U}G\mathcal{U}$ — to operators selected from distinct $\mathcal{U}G\mathcal{U} \cup \mathcal{U}G^{-1}\mathcal{U}$. Otherwise some transformed charge distributions $f_{aA}f_{aR(A)}$ will not be unique.

4.5 One-electron Integrals: D_{2h} Case

We could proceed at this point to develop formulas for two-electron integrals over symmetry orbitals using double coset decompositions. However, it will be no surprise now to reveal that the resulting formulas are very complicated, and that they are difficult to implement in a convenient and practical way (to put it mildly). Thus instead of continuing with the general formulation, we will rederive the one-electron integral formula for the special case of D_{2h} and its subgroups. The result could, of course, be obtained by appropriate simplification of Eq. 4.28. But the D_{2h} case is so important in practice, and so useful as an introduction to the two-electron case, that we shall derive the one-electron formula from scratch. Before doing that, a number of special features of D_{2h} and its subgroups should be noted. (We shall hereafter use the term D_{2h} generically, implying that the same situation holds for its subgroups.)

D_{2h} is Abelian, and every element is its own inverse. Under the operation of an element of the stabilizer of A, an atomic orbital transforms as

$$U f_{aA} = p_a(U) f_{aA} \tag{4.30}$$

where $p_a(U) = \pm 1$. (Davidson's notation for this factor [9] is inconsistent with its use — see Taylor [10].) Since every irrep of D_{2h} is one dimensional, an irrep of D_{2h} will be subduced by a single irrep, denoted α_A, of the stabilizer $\mathcal{U}$ of nucleus A. The factor $\Lambda_a^{A\alpha}(G)$ obtained from Eq. 4.22 above, which in this case is

$$\Lambda_a^{A\alpha}(G) = g^{-1} \sum_U \chi^\alpha(GU) p_a(GU), \tag{4.31}$$

will therefore be zero unless the function f_{aA} transforms according to α_A, whereupon

$$\Lambda_a^{A\alpha}(G) = ug^{-1}\chi^\alpha(G)p_a(G). \tag{4.32}$$

Double coset decompositions are simplified for D_{2h}, since

$$\lambda_R = \left|\mathcal{U} \cap R\mathcal{V}R^{-1}\right| = |\mathcal{U} \cap \mathcal{V}|. \tag{4.33}$$

Hence λ_R is independent of R and is the same for all elements of a given set of DCRs $\mathbf{R}$. We shall therefore write it as $\lambda_{\mathbf{R}}$. Similarly, $R\mathcal{V}R^{-1}$, the stabilizer of a transformed centre $R(B)$, is simply $\mathcal{V}$. Finally, we introduce $I_{\alpha\beta\gamma\ldots}$, defined by

$$I_{\alpha\beta\gamma\ldots} = g^{-1}\sum_G \chi^\alpha(G)\chi^\beta(G)\chi^\gamma(G)\cdots . \tag{4.34}$$

This provides a compact notation for factors that arise in selection rules, since $I_{\alpha\beta\gamma\ldots}$ is zero unless $\alpha \otimes \beta \otimes \gamma \cdots$ is the totally symmetric irrep.

We consider the integral

$$\left\langle F_{aA}^\alpha | O^\gamma | F_{bB}^\beta \right\rangle \tag{4.35}$$

and expand the symmetry orbital F_{aA}^α to give

$$g^{-1}\sum_G \chi^\alpha(G)\left\langle Gf_{aA} | O^\gamma | F_{bB}^\beta \right\rangle = g^{-1}\sum_G \chi^\alpha(G)\chi^\beta(G)\left\langle Gf_{aA} | O^\gamma | GF_{bB}^\beta \right\rangle. \tag{4.36}$$

By inserting a double coset decomposition

$$F_{bB}^\beta = g^{-1}\lambda_{\mathbf{R}}^{-1}\sum_U\sum_V\sum_R \chi^\beta(URV)URVf_{bB} \tag{4.37}$$

for the projection operator we obtain

$$g^{-2}\sum_G \chi^\alpha(G)\chi^\beta(G)\lambda_{\mathbf{R}}^{-1}\sum_U\sum_V\sum_R \chi^\beta(URV)\left\langle Gf_{aA} | O^\gamma | GURVf_{bB}\right\rangle. \tag{4.38}$$

Using the Rearrangement Theorem to replace G with GU gives

$$g^{-2}\lambda_{\mathbf{R}}^{-1}\sum_G\sum_U\sum_V\sum_R \chi^\alpha(G)\chi^\alpha(U)\chi^\beta(G)\chi^\beta(U)\chi^\beta(U)\chi^\beta(R)\chi^\beta(V) \times \left\langle GUf_{aA} | O^\gamma | GRVf_{bB}\right\rangle. \tag{4.39}$$

Now,

$$\sum_U \chi^\alpha(U)Uf_{aA} = uf_{aA} \quad \text{(or zero)}, \tag{4.40}$$

$$\sum_V \chi^\beta(V)Vf_{bB} = vf_{bB} \quad \text{(or zero)}, \tag{4.41}$$

and since $\chi^{\beta}(U)\chi^{\beta}(U) = 1$, Eq. 4.39 simplifies to

$$\begin{aligned}
&g^{-2}uv\lambda_{\mathbb{R}}^{-1}\sum_{G}\sum_{R}\chi^{\alpha}(G)\chi^{\beta}(G)\chi^{\beta}(R)\,\langle Gf_{aA}|O^{\gamma}|GRf_{bB}\rangle \\
&= g^{-2}uv\lambda_{\mathbb{R}}^{-1}\sum_{G}\sum_{R}\chi^{\alpha}(G)\chi^{\beta}(G)\chi^{\beta}(R)p_b(R)\left\langle Gf_{aA}|O^{\gamma}|Gf_{bR(B)}\right\rangle \\
&= g^{-2}uv\lambda_{\mathbb{R}}^{-1}\sum_{R}\sum_{G}\chi^{\alpha}(G)\chi^{\beta}(G)\chi^{\gamma}(G)\chi^{\beta}(R)p_b(R)\left\langle f_{aA}|O^{\gamma}|f_{bR(B)}\right\rangle \\
&= uvg^{-1}\lambda_{\mathbb{R}}^{-1}I_{\alpha\beta\gamma}\sum_{R}\chi^{\beta}(R)p_b(R)\left\langle f_{aA}|O^{\gamma}|f_{bR(B)}\right\rangle . \qquad (4.42)
\end{aligned}$$

This is the final form we shall give here for our symmetry orbital integral. It comprises the simple selection rule given by the factor $I_{\alpha\beta\gamma}$, a few numerical factors, and the distinct AO integrals generated by the sum over DCRs R. If the operator O^{γ} depends on the coordinates of a particular centre — more specifically, if the symmetry-adapted form is a linear combination of primitive operators centred at different points in space — a possibility we have ignored up to now, the sum may still contain redundant integrals. These can be eliminated by introducing a further level of double coset decomposition, based on the stabilizers of the operator on the one hand and the transformed charge distributions $f_{aA}f_{bR(B)}$ on the other. Since an analogous step is required in obtaining the two-electron integral formula, we shall proceed to that rather than discuss it here.

4.6 Two-electron Integrals: D_{2h} Case

A two-electron integral over symmetry orbitals for D_{2h} is given by

$$\left(F^{\alpha}_{aA}F^{\beta}_{bB}|F^{\gamma}_{cC}F^{\delta}_{dD}\right). \qquad (4.43)$$

There are several ways to expand this integral. We shall use the most pedestrian here, as it provides the best insight into the use of double cosets, but for the general case a two-step approach originally suggested by Davidson [9] is perhaps more useful. We begin by expanding F^{α}_{aA} and F^{γ}_{cC} explicitly, giving

$$\begin{aligned}
&g^{-2}\sum_{G}\sum_{H}\chi^{\alpha}(G)\chi^{\gamma}(H)\left(Gf_{aA}F^{\beta}_{bB}|Hf_{cC}F^{\delta}_{dD}\right) \\
&= g^{-2}\sum_{G}\sum_{H}\chi^{\alpha}(G)\chi^{\beta}(G)\chi^{\gamma}(H)\chi^{\delta}(H)\left(Gf_{aA}GF^{\beta}_{bB}|Hf_{cC}HF^{\delta}_{dD}\right) \qquad (4.44)
\end{aligned}$$

by inserting GG and HH as the identity. We now expand F^{β}_{bB} and F^{δ}_{dD} using respectively a set of DCRs $\mathbb{R}$ obtained using the stabilizers $\mathcal{U}$ and $\mathcal{V}$ of centres A and B, and $\mathbf{S}$ obtained using the stabilizers $\mathcal{W}$ and $\mathcal{X}$ of centres C and D, giving

$$\begin{aligned}
&g^{-4}\lambda_{\mathbf{R}}^{-1}\lambda_{\mathbf{S}}^{-1}\sum_{G}\sum_{H}\chi^{\alpha}(G)\chi^{\beta}(G)\chi^{\gamma}(H)\chi^{\delta}(H)\\
&\quad\times\sum_{U}\sum_{V}\sum_{R}\sum_{W}\sum_{X}\sum_{S}\chi^{\beta}(U)\chi^{\beta}(R)\chi^{\beta}(V)\chi^{\delta}(W)\chi^{\delta}(S)\chi^{\delta}(X)\\
&\quad\times(Gf_{aA}GURVf_{bB}|Hf_{cC}HWSXf_{dD})\\
&= g^{-4}\lambda_{\mathbf{R}}^{-1}\lambda_{\mathbf{S}}^{-1}\sum_{G}\sum_{H}\sum_{U}\sum_{V}\sum_{R}\sum_{W}\sum_{X}\sum_{S}\chi^{\alpha}(G)\chi^{\alpha}(U)\chi^{\beta}(G)\chi^{\beta}(U)\\
&\quad\times\chi^{\gamma}(H)\chi^{\gamma}(W)\chi^{\delta}(H)\chi^{\delta}(W)\chi^{\beta}(U)\chi^{\beta}(R)\chi^{\beta}(V)\chi^{\delta}(W)\chi^{\delta}(S)\chi^{\delta}(X)\\
&\quad\times(GUf_{aA}GRVf_{bB}|HWf_{cC}HSXf_{dD})
\end{aligned}\tag{4.45}$$

by the Rearrangement Theorem. Eliminating the character products that are unity and applying all operators except G and H to the atomic orbitals gives

$$\begin{aligned}
&g^{-4}\lambda_{\mathbf{R}}^{-1}\lambda_{\mathbf{S}}^{-1}\sum_{G}\sum_{H}\sum_{U}\sum_{V}\sum_{R}\sum_{W}\sum_{X}\sum_{S}\chi^{\alpha}(G)\chi^{\alpha}(U)\chi^{\beta}(G)\chi^{\gamma}(H)\\
&\quad\times\chi^{\gamma}(W)\chi^{\delta}(H)\chi^{\beta}(R)\chi^{\beta}(V)\chi^{\delta}(S)\chi^{\delta}(X)\\
&\quad\times p_a(U)p_b(V)p_c(W)p_d(X)p_b(R)p_d(S)\left(Gf_{aA}Gf_{bR(B)}|Hf_{cC}Hf_{dS(D)}\right).
\end{aligned}\tag{4.46}$$

But

$$\sum_{U}\chi^{\alpha}(U)p_a(U)=u,\ \text{etc.,}\tag{4.47}$$

so Eq. 4.46 becomes

$$\begin{aligned}
&g^{-4}uvwx\lambda_{\mathbf{R}}^{-1}\lambda_{\mathbf{S}}^{-1}\sum_{G}\sum_{H}\sum_{R}\sum_{S}\chi^{\alpha}(G)\chi^{\beta}(G)\chi^{\gamma}(H)\chi^{\delta}(H)\chi^{\beta}(R)\chi^{\delta}(S)\\
&\quad\times p_b(R)p_d(S)\left(Gf_{aA}Gf_{bR(B)}|Hf_{cC}Hf_{dS(D)}\right)\\
&= g^{-4}uvwx\lambda_{\mathbf{R}}^{-1}\lambda_{\mathbf{S}}^{-1}\sum_{G}\sum_{H}\sum_{R}\sum_{S}\chi^{\alpha}(G)\chi^{\beta}(G)\chi^{\gamma}(G)\chi^{\delta}(G)\\
&\quad\times\chi^{\gamma}(H)\chi^{\delta}(H)\chi^{\beta}(R)\chi^{\delta}(S)\\
&\quad\times p_b(R)p_d(S)\left(f_{aA}f_{bR(B)}|Hf_{cC}Hf_{dS(D)}\right),
\end{aligned}\tag{4.48}$$

if we use the Rearrangement Theorem to replace H with GH and then note that the integral $(Gf_{aA}Gf_{bR(B)}|GHf_{cC}GHf_{dS(D)})$ must equal $(f_{aA}f_{bR(B)}|Hf_{cC}Hf_{dS(D)})$, since the integral is a scalar and thus invariant to an overall rotation by G. Using the selection rule factor I of Eq. 4.34, the integral of Eq. 4.43 becomes

$$\begin{aligned}
&g^{-3}uvwxI_{\alpha\beta\gamma\delta}\lambda_{\mathbf{R}}^{-1}\lambda_{\mathbf{S}}^{-1}\sum_{H}\sum_{R}\sum_{S}\chi^{\gamma}(H)\chi^{\delta}(H)\chi^{\beta}(R)\chi^{\delta}(S)\\
&\quad\times p_b(R)p_d(S)\left(f_{aA}f_{bR(B)}|Hf_{cC}Hf_{dS(D)}\right).
\end{aligned}\tag{4.49}$$

Unfortunately, this is not the end of the story, since the summation over H may contain redundant terms. To eliminate these, we use another double coset decomposition, based on the stabilizers $\mathcal{M} = \mathcal{U} \cap \mathcal{V}$, which leaves the *charge distribution* $f_{aA} f_{bR(B)}$ invariant, and $\mathcal{N} = \mathcal{W} \cap \mathcal{X}$, which leaves $f_{cC} f_{dS(D)}$ invariant. (Note that this involves a D_{2h} simplification pointed out above: the stabilizer of $f_{aA} f_{bR(B)}$ is $\mathcal{U} \cap R\mathcal{V}R^{-1}$ in the general case.) Using a set of DCRs $\mathbf{T}$ obtained from $\mathcal{M}$ and $\mathcal{N}$, we can then replace our sum over H to give

$$\begin{aligned}
& g^{-3} uvwx I_{\alpha\beta\gamma\delta} \lambda_{\mathbf{R}}^{-1} \lambda_{\mathbf{S}}^{-1} \lambda_{\mathbf{T}}^{-1} \sum_R \sum_S \sum_M \sum_N \sum_T \chi^\gamma(M) \chi^\gamma(T) \chi^\gamma(N) \\
& \quad \times \chi^\delta(M) \chi^\delta(T) \chi^\delta(N) \chi^\beta(R) \chi^\delta(S) \\
& \quad \times p_b(R) p_d(S) \left(f_{aA} f_{bR(B)} | MTN f_{cC} MTN f_{dS(D)} \right) \\
= \; & g^{-3} uvwx I_{\alpha\beta\gamma\delta} \lambda_{\mathbf{R}}^{-1} \lambda_{\mathbf{S}}^{-1} \lambda_{\mathbf{T}}^{-1} \sum_R \sum_S \sum_M \sum_N \sum_T \chi^\gamma(M) \chi^\gamma(T) \chi^\gamma(N) \\
& \quad \times \chi^\delta(M) \chi^\delta(T) \chi^\delta(N) \chi^\beta(R) \chi^\delta(S) \\
& \quad \times p_b(R) p_d(S) p_c(N) p_d(N) p_c(T) p_d(T) \left(f_{aA} f_{bR(B)} | M f_{cT(C)} M f_{dTS(D)} \right)
\end{aligned} \tag{4.50}$$

after applying the operators T and N. We can now apply M to the entire AO integral giving

$$\begin{aligned}
& g^{-3} uvwx I_{\alpha\beta\gamma\delta} \lambda_{\mathbf{R}}^{-1} \lambda_{\mathbf{S}}^{-1} \lambda_{\mathbf{T}}^{-1} \sum_R \sum_S \sum_M \sum_N \sum_T \chi^\gamma(M) \chi^\gamma(T) \chi^\gamma(N) \\
& \quad \times \chi^\delta(M) \chi^\delta(TS) \chi^\delta(N) \chi^\beta(R) \\
& \quad \times p_b(R) p_d(S) p_c(N) p_d(N) p_c(T) p_d(T) \left(M f_{aA} M f_{bR(B)} | f_{cT(C)} f_{dTS(D)} \right) \\
= \; & g^{-3} uvwx I_{\alpha\beta\gamma\delta} \lambda_{\mathbf{R}}^{-1} \lambda_{\mathbf{S}}^{-1} \lambda_{\mathbf{T}}^{-1} \sum_R \sum_S \sum_M \sum_N \sum_T \chi^\gamma(M) \chi^\gamma(T) \chi^\gamma(N) \\
& \quad \times \chi^\delta(M) \chi^\delta(TS) \chi^\delta(N) \chi^\beta(R) \\
& \quad \times p_b(R) p_d(S) p_c(N) p_d(N) p_c(T) p_d(T) p_a(M) p_b(M) \\
& \quad \times \left(f_{aA} f_{bR(B)} | f_{cT(C)} f_{dTS(D)} \right) .
\end{aligned} \tag{4.51}$$

But

$$\sum_M \chi^\gamma(M) \chi^\delta(M) p_a(M) p_b(M) = m = \lambda_{\mathbf{R}}, \tag{4.52}$$

and

$$\sum_N \chi^\gamma(N) \chi^\delta(N) p_c(N) p_d(N) = n = \lambda_{\mathbf{S}}, \tag{4.53}$$

so we finally obtain

$$\begin{aligned}
& g^{-3} uvwx I_{\alpha\beta\gamma\delta} \lambda_{\mathbf{T}}^{-1} \sum_R \sum_S \sum_T \chi^\beta(R) \chi^\gamma(T) \chi^\delta(TS) \\
& \quad \times p_b(R) p_d(S) p_d(TS) \left(f_{aA} f_{bR(B)} | f_{cT(C)} f_{dTS(D)} \right) .
\end{aligned} \tag{4.54}$$

This formula gives the integral $(F^{\alpha}_{aA}F^{\beta}_{bB}|F^{\gamma}_{cC}F^{\delta}_{dD})$ in terms of non-redundant AO integrals only, at least where all four AOs are distinct. Just as for the one-electron case, we can expand some stabilizers and reduce the dimension of some sets of DCRs when some AOs are the same.

Despite the rather tedious algebra that went into the generation of Eq. 4.54, the formula itself is relatively simple. The operators R, S, and T are just those required to generate all non-redundant shell quadruplets; the rest of the group theory appears in the selection rule I and the parity factors. We may note that the generation of symmetry-adapted integrals requires a loop structure that goes at worst as g^3, not as the g^4 that would result from a naive expansion of the four symmetry orbitals in terms of AOs (to say nothing of the redundancies that would occur). Finally, we can imagine that this approach to symmetry would not be especially difficult to incorporate into an integral program. We would have to extend it to allow the shell structure of integral blocks to be exploited, but otherwise the formula can be programmed more or less as is. It is of some interest to point out that exactly this approach to computing integrals over symmetry orbitals was formulated and implemented almost twenty years ago by Almlöf in the program MOLECULE [11], without any explicit use of double coset decompositions but rather by arriving at the final formula by ingenious arguments involving the generators of D_{2h}.

4.7 Integral Derivatives

The entire apparatus of double coset decompositions can be extended from the case of integrals over symmetry orbitals to derivative integrals, that is, integrals differentiated with respect to the coordinates of the nuclear centres. Such derivative integrals are required for evaluation of the gradient or higher derivatives of the energy with respect to nuclear coordinates. Expressions for both the general case and the D_{2h} case have been given by Taylor [10] for the case of differentiation with respect to symmetry-adapted nuclear coordinates. The (necessarily more complicated) expressions for derivatives of symmetry orbital integrals with respect to individual nuclear coordinates can be obtained from these if desired. However, as we shall discuss in Chapter 8, energy gradients and Hessians with respect to symmetry-adapted nuclear coordinates are usually what are required anyway. Full use is made of D_{2h} symmetry in the program ABACUS which computes MCSCF first and second derivatives [12].

Having raised the issue of using symmetry in computing integral derivatives, we should note that in some situations the symmetry processing can be simplified. For example, in computing the energy gradient, two-electron integral derivatives appear in an expression of the general type

$$\sum_{\alpha}\sum_{\beta}\sum_{\gamma}\sum_{\delta}\sum_{p}\sum_{q}\sum_{r}\sum_{s} P_{pqrs}^{\alpha\beta\gamma\delta}\left(F_p^{\alpha}F_q^{\beta}|F_r^{\gamma}F_s^{\delta}\right)', \tag{4.55}$$

where the prime indicates differentiation of the integral with respect to a nuclear coordinate, and where we have simplified the basis function indices from aA to p, etc. **P** can be regarded here as a two-particle density matrix, so we can view the overall expression as the trace of a matrix product. If we substitute into this trace the formula of Eq. 4.54 (assuming a differentiated integral) we obtain

$$\sum_{\alpha}\sum_{\beta}\sum_{\gamma}\sum_{\delta}\sum_{p}\sum_{q}\sum_{r}\sum_{s} N_d I_{\alpha\beta\gamma\delta} P_{pqrs}^{\alpha\beta\gamma\delta} \times \sum_{R}\sum_{S}\sum_{T} \chi^{\beta}(R)\chi^{\gamma}(T)\chi^{\delta}(TS) N_s \left(f_p f_{R(q)}|f_{T(r)} f_{TS(s)}\right)', \tag{4.56}$$

where we have used N_d and N_s to denote the various numerical factors appearing in Eq. 4.54. If we now define [13]

$$P_{pqrs}^{RST} = \sum_{\alpha}\sum_{\beta}\sum_{\gamma}\sum_{\delta} N_d I_{\alpha\beta\gamma\delta} \chi^{\beta}(R)\chi^{\gamma}(T)\chi^{\delta}(TS) P_{pqrs}^{\alpha\beta\gamma\delta}, \tag{4.57}$$

we can rewrite the trace operation as

$$\sum_{p}\sum_{q}\sum_{r}\sum_{s}\sum_{R}\sum_{S}\sum_{T} P_{pqrs}^{RST} N_s \left(f_p f_{R(q)}|f_{T(r)} f_{TS(s)}\right)'. \tag{4.58}$$

We have thus "back-transformed" the array **P** from a symmetry-adapted basis to a distinct AO basis and taken the trace in the AO basis, rather than explicitly computing symmetry orbital integral derivatives. Since the list of integral first derivatives is about an order of magnitude longer than the list of integrals (or elements of **P**), this results in a considerable reduction in the computational effort. The saving in higher derivative cases is even larger. However, when higher derivatives of the energy are computed the first derivatives of the integrals appear in more complicated expressions than a trace, and then it may be advantageous to employ symmetry-adapted integral derivatives.

It is also possible to generalize the discussion we have given of the two-electron integral case to operators that are not necessarily totally symmetric, such as those arising in various fine-structure integrals (operators such as $x_{12}y_{12}/r_{12}^5$, for example). The simplest generalization is again obtained by considering symmetry-adapted operators.

Chapter 5

Distinct Atomic Orbital Integral Methods

5.1 Transformation of Atomic Orbitals

In the previous chapter we investigated in some detail the construction of integrals over symmetry-adapted atomic orbitals in terms of distinct AO integrals only; as we saw, this procedure becomes rather complicated for degenerate point groups. An alternative approach based on the use of distinct AO integrals only has been popular in quantum chemistry for many years, partly because it lends itself more readily to the use of degenerate point groups. This approach was first derived rigorously for the special case of totally symmetric operators by Dupuis and King [14], and was extended to non-totally symmetric operators by Taylor [15]. Recently, it has been extended again to cover two-electron integral transformations [16], so that it can be employed in most steps of a quantum chemical calculation. We can best begin our description of this approach by considering the transformation properties of a given AO basis as a whole.

Let χ be a row vector of the AOs to be used for a particular molecular calculation. We assume that this AO basis is *closed* under the molecular point group $\mathcal{G}$, so that $G\chi_\mu \in \chi$, $\forall\, G \in \mathcal{G}$. The effect of some $G \in \mathcal{G}$ is then

$$G\chi = \chi\mathbf{G}, \tag{5.1}$$

where $\mathbf{G}$ is in general a *nonunitary* reducible representation matrix. It is nonunitary because the AO basis χ is not orthogonal, and this nonunitarity requires us to pay careful attention to distiguishing between inverses and transposes in what follows.

In practical calculations, the AOs occur in *shells*, where members of a shell have the same radial dependence about a centre but differ in their angular behaviour. As discussed elsewhere, efficiency of integral evaluation usually involves using complete shells of spherical harmonic or Cartesian functions p, d, etc. We shall regard as a shell those functions which transform into themselves or their images on other centres under all operations in the group. Thus

$$GM = M', \tag{5.2}$$

where M' is the shell into which the functions of shell M are mapped under the operation G. We use the term *two-label* to denote a shell pair index such as (MN), and *four-label* for a shell quadruplet such as $(MNPQ)$. We assume that these labels display the same index permutational symmetries as those of one- and two-electron integrals, respectively. We denote the group (of order two) of the two-label permutations as T_2 and the group (order eight) of four-label permutations as T_4. For example, a Hermitian operator has AO integrals whose two-labels transform according to the symmetric irrep of T_2, while for an anti-Hermitian operator the two-labels transform according to the antisymmetric irrep. In group-theoretical terminology, the set of all operations in the product group $T_2 \otimes \mathcal{G}$ that leave the two-label (MN) invariant is the stabilizer of (MN), a subgroup whose order we denote $n(MN)$. The index of this stabilizer is given by

$$q_2(MN) = 2g/n(MN), \tag{5.3}$$

where $2g$ is the order of $T_2 \otimes \mathcal{G}$. In essence, $q_2(MN)$ is the number of *distinct* two-labels that can be generated from the shell pair MN. Dupuis and King [14] introduced the term "constituency number" rather than "index of the stabilizer". We can define an analogous quantity for the four-labels:

$$q_4(MNPQ) = 8g/n(MNPQ). \tag{5.4}$$

Finally, the set of all two-labels in $T_2 \otimes \mathcal{G}$ is referred to as a "grande list", denoted $G2$ (with an analogous $G4$ list), while the list comprising only the distinct elements is termed a "petite list", denoted $P2$ or $P4$.

5.2 Distinct AO Integrals and Skeleton Matrices

Let us now consider integrals over a totally symmetric Hermitian one-electron operator $\mathcal{O}$. Over the AO basis we must have

$$\mathbf{G}^{\dagger}\mathbf{O}\mathbf{G} = \mathbf{O}, \tag{5.5}$$

where $\mathbf{O}$ is the matrix of AO integrals over $\mathcal{O}$ and $\mathbf{G}$ is defined in Eq. 5.1. We define a *skeleton matrix* $\bar{\mathbf{O}}$ corresponding to $\mathbf{O}$ by requiring that the block of $\bar{\mathbf{O}}$ over shells M and N, say, is proportional to the MN shell block of $\mathbf{O}$:

$$\bar{\mathbf{O}}_{MN} = \lambda_{MN}\mathbf{O}_{MN}. \tag{5.6}$$

We require that

$$\sum_{(M'N')} \lambda_{M'N'} = q_2(MN) \tag{5.7}$$

where $(MN) \in P2$ and the sum runs over all elements of $G2$ equivalent to (MN). There are two immediately obvious choices for $\lambda_{M'N'}$:

$$\lambda_{M'N'} = 1 \quad \forall \ (M'N') \in G2, \tag{5.8}$$

or

$$\lambda_{M'N'} = q_2(MN)\delta_{MM'}\delta_{NN'}. \tag{5.9}$$

We now define *symmetrization* [14] of the operator matrix $\mathbf{O}$ as

$$\mathbf{O}_{sym} = (2g)^{-1} \sum_{G} \mathbf{G}^{\dagger}(\mathbf{O} + \mathbf{O}^{\dagger})\mathbf{G}. \tag{5.10}$$

This corresponds to the application of a projection operator corresponding to the totally symmetric irrep to $\mathbf{O}$, and is trivially true by construction for this case. However, we can also establish that

$$\mathbf{O} = \bar{\mathbf{O}}_{sym} = (2g)^{-1} \sum_{G} \mathbf{G}^{\dagger}(\bar{\mathbf{O}} + \bar{\mathbf{O}}^{\dagger})\mathbf{G}, \tag{5.11}$$

as follows. We explicitly expand the matrix multiplications on the right-hand side of Eq. 5.11 and consider just the MNth shell block, giving

$$\begin{aligned} &(2g)^{-1} \sum_{G} \sum_{M'} \sum_{N'} \mathbf{G}^{\dagger}_{MM'}(\bar{\mathbf{O}}_{M'N'} + \bar{\mathbf{O}}^{\dagger}_{M'N'})\mathbf{G}_{N'N} \\ &= (2g)^{-1} \sum_{G} \sum_{M'} \sum_{N'} \mathbf{G}^{\dagger}_{MM'}(\bar{\mathbf{O}}_{M'N'} + \bar{\mathbf{O}}_{N'M'})\mathbf{G}_{N'N} \end{aligned} \tag{5.12}$$

using

$$\bar{\mathbf{O}}^{\dagger}_{M'N'} = \bar{\mathbf{O}}_{N'M'}. \tag{5.13}$$

(Note that here, and in what follows, our notation should be regarded as implying summation over the components of shells M and N.) Recalling that each operator G takes shell M into a single image shell M', the summations over M' and N' are redundant, since only one value of each index can contribute for a given G. We may thus rewrite our formula as

$$(2g)^{-1} \sum_{G} \mathbf{G}^{\dagger}_{MM'}(\bar{\mathbf{O}}_{M'N'} + \bar{\mathbf{O}}_{N'M'})\mathbf{G}_{N'N}. \tag{5.14}$$

But

$$\bar{\mathbf{O}}_{M'N'} = \lambda_{M'N'}\mathbf{O}_{M'N'} \tag{5.15}$$

and

$$\bar{\mathbf{O}}_{N'M'} = \lambda_{N'M'}\mathbf{O}_{N'M'}, \tag{5.16}$$

so we obtain

$$(2g)^{-1}\sum_G(\lambda_{M'N'}+\lambda_{N'M'})\mathbf{G}^\dagger_{MM'}\mathbf{O}_{M'N'}\mathbf{G}_{N'N}=(2g)^{-1}\sum_G(\lambda_{M'N'}+\lambda_{N'M'})\mathbf{O}_{MN}. \quad (5.17)$$

But

$$\sum_G(\lambda_{M'N'}+\lambda_{N'M'})=2g, \quad (5.18)$$

since this is equivalent to a sum over elements of $T_2\otimes\mathcal{G}$. Our symmetrization formula Eq. 5.11 is therefore established.

There are two obvious ways of accomplishing this symmetrization in practice. The first would be to construct a skeleton matrix using the $G2$ list of integrals and Eq. 5.8. The second, and clearly more efficient, is to construct the matrix using the $P2$ list, employing the definition of Eq. 5.9. In this latter case we construct and use only the distinct AO integral blocks. We have not yet specified how to determine the $P2$ list, and one approach, based on the previous chapter, would be to use double coset decompositions. If all we require are the distinct integrals, however, (that is, we are not going on to generate symmetry-adapted integrals from them) this is unnecessarily elaborate. We can use instead what Dupuis and King call the "principle of procrastination" [14]: we loop over all two-labels in $G2$, and operate on each with symmetry operations G. A two-label for which some $G(MN)$ gives a result $M'N'$ that is lexically greater than MN is discarded, so that only the two-label for which all $G(MN)$ are less than MN is retained in the final list $P2$. The name arises from the fact that we keep "putting off" the work of actually handling a two-label until as late as possible.

At this point we have developed a method for handling symmetric one-electron operators: a useful step, but far from adequate to cover all cases we are interested in, since we have said nothing yet about the much more demanding case of two-electron integrals. Let us consider first building the Fock matrix for a closed-shell SCF calculation. This is perhaps most easily expressed in terms of a P supermatrix with elements

$$P(mn|pq)=(mn|pq)-\frac{1}{4}\left[(mp|nq)+(mq|np)\right], \quad (5.19)$$

whereupon we have

$$F_{mn}=h_{mn}+\sum_{pq}d_{pq}P(mn|pq). \quad (5.20)$$

Here $\mathbf{h}$ is the one-electron Hamiltonian and $\mathbf{d}$ is the "density matrix" with elements

$$d_{pq}=2\sum_i^{occ}C_{pi}C_{qi}, \quad (5.21)$$

where we sum over occupied MOs; the MO coefficients (assumed real) are collected into the matrix $\mathbf{C}$. The density matrix has the transformation property

$$\mathbf{d} = \mathbf{G}\mathbf{d}\mathbf{G}^{\dagger}. \tag{5.22}$$

Note that the density matrix transforms contragrediently to operator matrices.

Now, from our manipulations above we can easily see that $\mathbf{F}$ can be constructed from a skeleton matrix obtained using only a $P2$ list, since the Fock operator is totally symmetric. However, this far from ideal, since we would like to avoid the redundancies that arise unless we use the $P4$ list of two-electron integrals, or, here, supermatrix elements. We define first a matrix $\mathbf{V}(IJKL)$ with the property that the IJ block is given by

$$\mathbf{V}(IJKL)_{IJ} = P(IJ|KL)d_{KL} \tag{5.23}$$

and all other blocks are zero. Now the blocks of supermatrix elements transform as

$$P(IJ|KL) = P(I'J'|K'L')G_{I'I}G_{J'J}G_{K'K}G_{L'L} \tag{5.24}$$

and the density matrix transforms as

$$d_{K'L'} = d_{KL}G_{K'K}G_{L'L}. \tag{5.25}$$

Substituting Eqs. 5.24 and 5.25 in Eq. 5.23 gives

$$\mathbf{V}(IJKL)_{IJ} = \mathbf{V}(I'J'K'L')_{I'J'}G_{I'I}G_{J'J}, \tag{5.26}$$

which we could write as

$$\mathbf{V}(IJKL) = \mathbf{G}^{\dagger}\mathbf{V}(I'J'K'L')\mathbf{G}, \tag{5.27}$$

since blocks other than the IJ block are zero anyway. Eq. 5.27 is in exactly the form to substitute into our skeleton symmetrization scheme, except that it does not display the full T_4 index permutational symmetry. However, the symmetrized quantity

$$\mathbf{W}(IJKL) = \mathbf{V}(IJKL) + \mathbf{V}(KLIJ) \tag{5.28}$$

does display this symmetry. Hence we can combine this with Eq. 5.27 to demonstrate that

$$\mathbf{W}(IJKL)_{sym} = \mathbf{W}(I'J'K'L')_{sym} \tag{5.29}$$

if G maps the four-label $IJKL$ into $I'J'K'L'$. Hence we can either process the P-supermatrix by a sum over labels in $G4$:

$$\mathbf{F} = \mathbf{h} + \sum_{\{G4\}} \mathbf{W}(IJKL) \tag{5.30}$$

or by constructing a skeleton matrix

$$\bar{\mathbf{F}} = \mathbf{h} + \sum_{\{P4\}} q_4(IJKL)\mathbf{W}(IJKL) \tag{5.31}$$

and then symmetrizing it. In practice, of course, this is done for the one-electron and two-electron parts together [14].

5.3 Non-totally Symmetric Operators

In the preceding section we treated the case in which integrals over totally symmetric operators were constructed over a "petite list" of shell combinations and then symmetrized to give the full matrix. While this is adequate for the case of SCF calculations, it is far from a complete solution to the problem of making maximum utilization of symmetry equivalent terms. First, there are circumstances in which the operator is not totally symmetric. Second, the alert reader may have perceived that the two-electron integral case that we treated for the Fock matrix was a *partial trace*, that is, we were contracting a four-index tensor with some lower rank tensor (the P-supermatrix with the density matrix, specifically). As we observed in the previous chapter, there are special circumstances to be taken advantage of in such situations — we can effectively move terms arising from symmetry operations on the integral by redefining the tensor with which the integrals are being contracted. In this section we shall address the case of non-totally symmetric operators, by giving the necessary formulas, and extend the range and type of integral/density matrix contractions we can handle. In the next section we will briefly consider the case of the four-index transformation of the integrals.

Since any operator can be written as the sum of Hermitian and anti-Hermitian operators, we can restrict our discussion to these two types only. Further, any operator can be written as a linear combination of irreducible symmetry operators, so we can restrict ourselves to irreducible tensor operators. An operator matrix $\mathbf{O}(\Gamma, K)$ that transforms according to the symmetry (Γ, K) obeys the relationship

$$\mathbf{O}(\Gamma, K) = \sum_{\Lambda} \mathbf{R}\,\mathbf{O}(\Gamma, \Lambda)\mathbf{R}^{\dagger} D^{\Gamma}_{\Lambda K}(R), \tag{5.32}$$

or the equivalent form

$$\mathbf{O}(\Gamma, K) = \sum_{\Lambda} D^{\Gamma}_{K\Lambda}(R)^{*}\mathbf{R}^{\dagger}\mathbf{O}(\Gamma, \Lambda)\mathbf{R}. \tag{5.33}$$

For such an operator our symmetrization formula becomes

$$\mathbf{O}(\Gamma, K)_{sym} = (2g)^{-1} \sum_{G} \sum_{\Lambda} D^{\Gamma}_{K\Lambda}(R)^{*}\mathbf{G}^{\dagger}(\bar{\mathbf{O}}(\Gamma, \Lambda) + p\bar{\mathbf{O}}(\Gamma, \Lambda)^{\dagger})\mathbf{G}. \tag{5.34}$$

A full proof of this can be found in Taylor [15]. The factor p is $+1$ if the operator is Hermitian and -1 if it is anti-Hermitian. We can see an immediate complication relative to our earlier formula Eq. 5.11 in that a full representation matrix for irrep Γ is required. This is considered in more detail below. Additional redundancies in the $P2$ list that arise from the form of particular operators are also treated by Taylor.

In many calculations beyond the Hartree-Fock level a first step is the transformation of at least some integrals. For the simplest such calculation, second-order perturbation theory, integrals with two indices transformed into the occupied MO basis are required. Such integrals appear in many situations, including the MO basis formulation of coupled-perturbed Hartree-Fock theory. We can represent the first phase of this transformation as obtaining "Coulomb" and "exchange" operators:

$$(p|J^{ij}|q) = \sum_r \sum_s (pq|rs) C_{ri} C_{sj} \tag{5.35}$$

and

$$(p|K^{ij\pm}|q) = \sum_r \sum_s \left[(pr|qs) \pm (qr|ps)\right] C_{ri} C_{sj}, \tag{5.36}$$

respectively, where we have written the exchange terms in terms of Hermitian and anti-Hermitian combinations. The sum in these equations is clearly over a $G2$ list and we wish to reduce it so we can process the $P4$ list of integrals only. Let us suppose that the MOs i and j transform as basis functions for the symmetry species (α, a) and (β, b), respectively. (These representations are not required to be irreducible.) We can then fully specify a Coulomb operator matrix as $\mathbf{J}^{ij}(\alpha, \beta; a, b)$; the MNth block of such an operator matrix is given by

$$\mathbf{J}^{ij}_{MN}(\alpha, \beta; a, b) = \sum_P \sum_Q (MN|PQ) \mathbf{C}^{\alpha a}_{iP} \left(\mathbf{C}^{\beta b}_{jQ}\right)^{\dagger}, \tag{5.37}$$

where, for example, $\mathbf{C}^{\alpha a}_{iP}$ is the set of column elements of the MO coefficient matrix connecting MO i with the AOs in shell P and summation over the shell components of P is implied. The contribution of a particular pair of shells P and Q can be written as

$$\mathbf{V}_{MN}(\alpha, \beta; a, b; MNPQ) = (MN|PQ) \mathbf{C}^{\alpha a}_{iP} \left(\mathbf{C}^{\beta b}_{jQ}\right)^{\dagger}. \tag{5.38}$$

This operator matrix transforms as the abth row of the representation $\alpha \otimes \beta$. We can therefore immediately use the analysis for non-totally symmetric operators to devise a skeleton symmetrization scheme based on the $P4$ list. The only problem is that the form of $\mathbf{V}$ shown is not invariant under the index permutations of T_4, but the form

$$\begin{aligned} &\mathbf{W}_{MN}(\alpha, \beta; a, b; MNPQ) \\ &= \frac{1}{2}\left[\mathbf{V}_{MN}(\alpha, \beta; a, b; MNPQ) + \mathbf{V}_{MN}(\alpha, \beta; a, b; PQMN)\right] \end{aligned} \tag{5.39}$$

is, and it is easy then to show that

$$\mathbf{J}^{ij}(\alpha,\beta;a,b) = \sum_{(MNPQ)} \mathbf{W}(\alpha,\beta;a,b;MNPQ)_{sym}, \tag{5.40}$$

where the symmetrization formula is given by Eq. 5.34. We can therefore use either a sum over $(MNPQ)$ in $G4$, or the much more desirable sum over shell quadruplets in $P4$, weighting terms with the appropriate $q_4(MNPQ)$ values. In the latter approach the resulting skeleton matrix is then symmetrized according to Eq. 5.34.

The exchange operators $\mathbf{K}^{ij\pm}$ can be treated in a completely analogous fashion, although different combinations of permutationally inequivalent integrals are required [15].

5.4 Two-electron Integrals

Up to this point we have been concerned with building up operator matrices in the AO basis in terms of distinct AO integrals only. The previous subsection has dealt with the case of a partial AO/MO transformation, but in a more general situation our desire may be to transform from a list of symmetry-distinct AO integrals to a list of symmetry-blocked MO integrals. That is, we exploit the reducible symmetry of the AOs to eliminate redundancies and the irreducible symmetry of the MOs to effect a complete symmetry blocking of the MO integral list. Until recently there were no general computational implementations of such a procedure, although some attempts had been made to "hard-wire" specific cases. One apparent drawback is the need for full matrix representations of the point group. Even with this information the procedure (which can be derived formally by, for example, double coset decompositions) seemed formidably complicated, although the selection rules on the final integrals are given fairly straightforwardly by the Wigner-Eckart theorem.

In 1989, however, a practical resolution of this problem was derived independently by Häser and by Almlöf [16]. They obtain the necessary matrix representation information by utilizing the reducible representation matrices obtained from symmetry transformations on the AO basis as in Eq. 5.1. Full details of their procedure is beyond the scope of this course, but, as would be expected, it has many similarities to the non-totally symmetric operators discussed in the previous section.

Chapter 6

The Many-electron Hamiltonian

6.1 The Born-Oppenheimer Hamiltonian

We have already seen in Sec. 3.1 that before the Born-Oppenheimer separation of nuclear and electronic motion is made, the Coulomb Hamiltonian has very high symmetry, but that the clamped-nucleus Hamiltonian has only the spatial symmetry of the nuclear framework. That is, the Hamiltonian

$$H = -1/2\sum_i \nabla_i^2 + \sum_{i>j} r_{ij}^{-1} - \sum_{i,A} Z_A r_{iA}^{-1} + \sum_{A>B} Z_A Z_B R_{AB}^{-1} \qquad (6.1)$$

commutes with all operations in the molecular point group $\mathcal{G}$. Hence the eigenfunctions of H will be symmetry adapted and will transform as basis functions for symmetry species of $\mathcal{G}$. There are other symmetry operations to which H is invariant. For example, Eq. 6.1 contains no terms that depend on the electron spin variables. As a consequence, many-electron operators such as the total spin $\hat{S}^2$ or its projection along a fixed axis, conventionally $\hat{S}_z$, commute with H, and so eigenfunctions of H will also be eigenfunctions of these operators. Hence our molecular wave functions will be *spin adapted* as well as symmetry adapted, and can be characterized by the expectation values of these spin operators, the spin quantum number S and its projection M_S. A treatment of spin can be developed in terms of the spin operators, leading to projection operators, shift operators, etc., to obtain pure spin eigenfunctions or other M_S components. Alternatively, other approaches that exploit group theory directly can be used. Such methods have gained popularity in molecular quantum chemistry codes over the last fifteen years, and in this treatment we shall consider both group-theoretical methods and those that explicitly consider electron spin. We shall treat the latter first, although both approaches go back to the early days of quantum mechanics. After discussing methods of constructing and using spin eigenfunctions, we will say a little about spatial symmetry adaptation.

6.2 Slater determinants

Probably the best-known approach to the utilization of spin symmetry is that originally developed by Slater and by Fock (see, for example, Hurley [17]). No particular advantage is taken of the spin-independence of the Hamiltonian, at least in the first phase of the construction of the n-particle basis. We take the 1-particle basis to be *spin-orbitals* — products of orthonormal orbitals $\{\phi_k(r)\}$ and the elementary orthornormal functions of the spin coordinate σ

$$\alpha(\sigma) \text{ and } \beta(\sigma), \tag{6.2}$$

which are eigenfunctions of the one-electron spin operators $\hat{s}^2$ and $\hat{s}_z$ with eigenvalues ($s = 1/2, m_s = 1/2$) and ($s = 1/2, m_s = -1/2$), respectively. We thus obtain orthonormal spin-orbitals

$$\phi_i = \phi_i(r)\alpha(\sigma) \text{ and } \bar{\phi}_i = \phi_i(r)\beta(\sigma), \tag{6.3}$$

where the bar signifies beta spin, or "spin down". The n-particle basis is then obtained by taking n-fold products of spin-orbitals, and the antisymmetry required by the Pauli exclusion principle is imposed by antisymmetrizing the product with respect to particle interchanges:

$$\begin{aligned} &|\phi_1(x_1)\phi_2(x_2)\ldots\phi_N(x_N)| \\ &= \frac{1}{\sqrt{N!}}[\phi_1(x_1)\phi_2(x_2)\ldots\phi_N(x_N) - \phi_1(x_2)\phi_2(x_1)\ldots\phi_N(x_N) + \ldots], \end{aligned} \tag{6.4}$$

the familiar Slater determinant, which is commonly represented, as shown, by the diagonal elements. The coordinates x represent both spin and space coordinates, emphasizing again that no attempt has been made to treat the space and spin coordinates of the electrons on other than an identical footing.

One particular advantage of Slater determinants constructed from orthonormal spin-orbitals is that matrix elements between determinants over operators such as H are very simple. Only three distinct cases arise, as is well known and treated elsewhere. It is perhaps not surprising that the simplest matrix element formulas should be obtained from the treatment that exploits symmetry the least, as only the fermion antisymmetry has been accounted for in the determinants. As more symmetry is introduced, the formulas become more complicated. On the other hand, the symmetry reduces the dimension of the problem more and more, because selection rules eliminate more terms. We consider here the spin adaptation of Slater determinants.

Slater determinants constructed from spin-orbitals of the form Eq. 6.3 are eigenfunctions of $\hat{S}_z$, where

$$\hat{S}_z = \sum_i^n \hat{s}_z, \tag{6.5}$$

but not, in general, of the corresponding total spin operator $\hat{S}^2$. Hence while the Hamiltonian matrix over Slater determinants will be blocked by M_S value, we obtain no blocking by S value unless we form spin-adapted linear combinations of Slater determinants that are eigenfunctions of $\hat{S}^2$. One formal approach to obtain such spin eigenfunctions is projection [17], using the operator

$$\mathcal{P}_S = \prod^{S' \neq S} \frac{\hat{S}^2 - S'(S'+1)}{S(S+1) - S'(S'+1)} \tag{6.6}$$

which annihilates all spin components other than the desired one (S) from the function on which it operates. In practice the forms

$$\hat{S}^2 = \hat{S}_+\hat{S}_- + \hat{S}_z^2 - \hat{S}_z \tag{6.7}$$

or

$$\hat{S}^2 = \hat{S}_-\hat{S}_+ + \hat{S}_z^2 + \hat{S}_z \tag{6.8}$$

are the most useful when applying Eq. 6.6. Here $\hat{S}_+$ and $\hat{S}_-$ are the shift operators defined through

$$\hat{S}_+\Psi(S, M_S) = \sqrt{(S - M_S)(S + M_S + 1)}\Psi(S, M_S + 1) \tag{6.9}$$

and

$$\hat{S}_-\Psi(S, M_S) = \sqrt{(S + M_S)(S - M_S + 1)}\Psi(S, M_S - 1), \tag{6.10}$$

where $\Psi(S, M_S)$ is a spin eigenfunction with the given spin quantum numbers.

As an example, we can consider a three-electron determinant $|ab\bar{c}|$ with $M_S = 1/2$ and obtain a doublet spin eigenfunction by applying the operator

$$\frac{\hat{S}^2 - 15/4}{3/4 - 15/4} \tag{6.11}$$

to obtain

$$\frac{1}{3}\left(2|ab\bar{c}| - |a\bar{b}c| - |\bar{a}bc|\right). \tag{6.12}$$

This projection/annihilation approach is probably more useful as an analytical tool, for annihilating the principal spin contaminants from a wave function by hand calculation, for example, than as a computational tool. There is a vast body of literature (see, for example, Paunсz [18]) on generating spin eigenfunctions as linear combinations of Slater determinants, from explicitly precomputed "Sanibel coefficients" to diagonalizing the matrix of $\hat{S}^2$. However, there are other methods that exploit the group theoretical structure of the problem more effectively, and we shall now turn to these.

6.3 Permutational Symmetry and the Symmetric Group

We have already discussed the antisymmetry of the wave function with respect to particle interchange; we can formalize this as requiring that the wave function must transform according to the antisymmetric irrep of the permutation group of N objects, the symmetric group $\mathcal{S}(N)$. The permutations here involve both the space and spin coordinates of the N electrons. We write such permutations as a product $P_\tau P_\sigma$ of separate permutations on the space and spin coordinates, respectively. The key to symmetric group-based approaches to spin adaptation is that the configurations can be written as products of space and spin parts for which the spin parts transform according to *irreducible* representations of the symmetric group:

$$P_\sigma \Theta^N_{S,k} = \sum_l \Theta^N_{S,l} U^S_{lk}, \tag{6.13}$$

where $\mathbf{U}^S$ is an irrep matrix for the group $\mathcal{S}(N)$. This was established by Weyl [19] and by Wigner [1] more than sixty years ago (see also Pauncz [18]). The order of the matrix (and the range of the sum over l) is the degeneracy $f(N,S)$ discussed below; k is a counting index that enumerates the spin functions.

In order to proceed further with this approach we shall briefly describe the representation theory of the symmetric group. Consider the *partitions* of the integer N, that is, sets of non-increasing integers

$$\lambda_1 \geq \lambda_2 \geq \dots \tag{6.14}$$

such that

$$\sum_i \lambda_i = N. \tag{6.15}$$

It can be shown that the number of partitions of N equals the number of classes of the symmetric group $\mathcal{S}(N)$, and hence equals the number of irreducible representations of $\mathcal{S}(N)$. Partitions are conventionally written as $[\lambda_1^{k1} \dots]$, where the "index" $k1$ specifies the number of times the integer λ_1 appears. For example, the possible partitions of three (and the irreps of $\mathcal{S}(3)$) can be written as [3], [2 1] and $[1^3]$. The most common way to represent the partitions is as *Young shapes* [7] as in Fig. 6.1. The rules for constructing Young shapes are very simple — each entry in the partition gives a new row of the shape, so there are as many rows as integers in the partition, with the number of columns in each given by the corresponding integer. If we enter the integers $1 \dots N$ into the boxes of a given Young shape, under the restriction that the numbers must increase down each column and across each row, we obtain a *Young tableau*. The number of ways we can make a Young tableau from a given Young shape gives the degeneracy of the irreducible representation. For example, from Fig. 6.2 we

Figure 6.1: Young Shapes (Irreps) of $\mathcal{S}(3)$

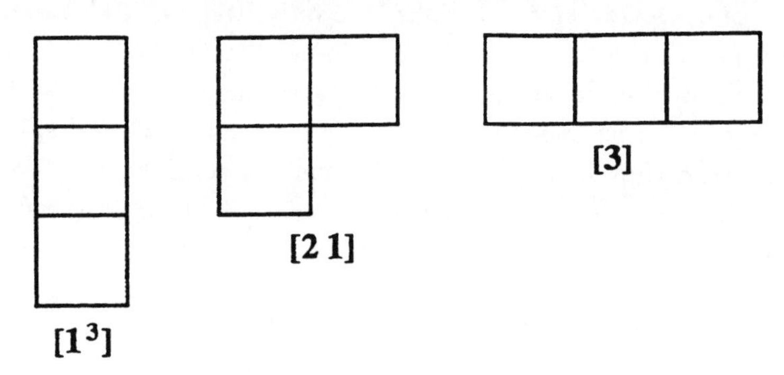

see that the irrep of $\mathcal{S}(5)$ labelled $[3\ 1^2]$ is six-fold degenerate. When enumerating Young tableaux, the sum rule on irrep dimensions is very useful. Finally, we may note that the partition $[N]$ corresponds to the totally symmetric irrep of $\mathcal{S}(N)$, while $[1^N]$ corresponds to the totally antisymmetric irrep. That is, a basis function for $[N]$ is symmetric with respect to permutations of any two objects, whereas a basis function for $[1^N]$ is antisymmetric with respect to such a permutation.

Figure 6.2: Irrep $[3\ 1^2]$ of $\mathcal{S}(5)$

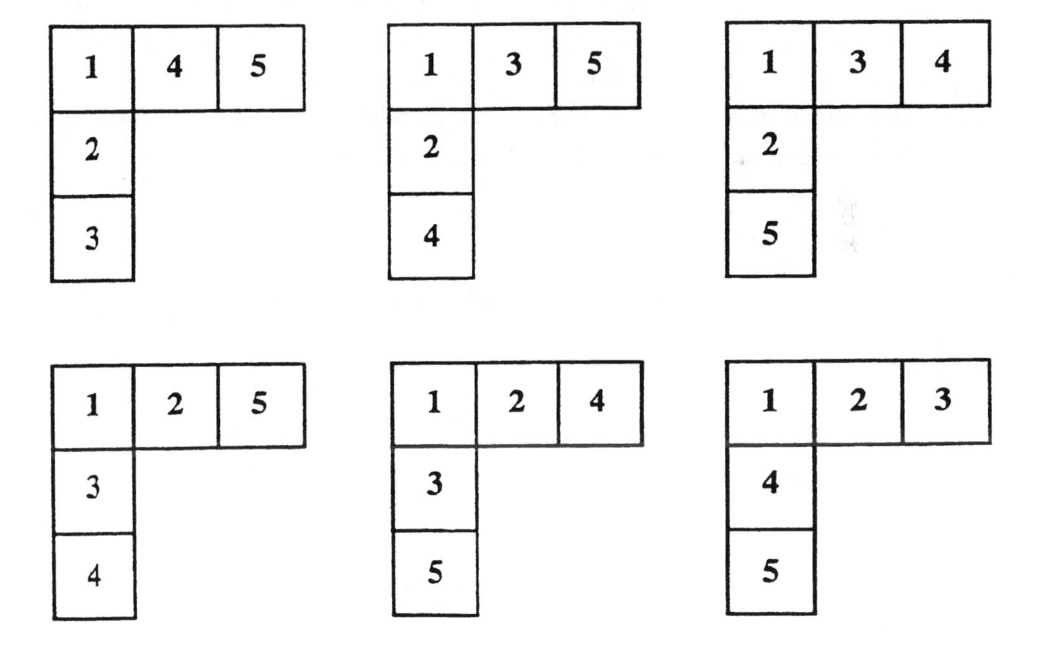

Returning to the case of spin eigenfunctions, we note first that since there are only two one-electron spin functions, the only irreps of $\mathcal{S}(N)$ that are valid as spin eigenfunctions are those whose Young shapes have no more than two rows. For instance, in the three-electron case only the partitions [2 1] and [3] label acceptable spin eigenfunctions, corresponding to doublet and quartet total spin multiplicity, respectively. Note that [2 1] is doubly degenerate, so there are two doublet spin eigenfunctions for three electrons. We also note that an orbital that is maximally occupied (that is, an n-fold degenerate orbital containing $2n$ electrons) does not contribute to the total spin, so that in practical cases we are concerned only with the open-shell electrons. This greatly reduces the number of possible spin eigenfunctions: if we represent the number of spin eigenfunctions with total spin S for N electrons as $f(N,S)$ we find

$$f(N,S) = \frac{(2S+1)N!}{(N/2+S+1)!(N/2-S)!}. \tag{6.16}$$

The classic approach to representing this result is via the *branching diagram* of Fig. 6.3.

The Young shapes form what are sometimes referred to as "standard irreps" of $\mathcal{S}(N)$, since they are adapted to the subgroup chain

$$\mathcal{S}(N) \supset \mathcal{S}(N-1) \supset \ldots \mathcal{S}(2). \tag{6.17}$$

That is, by removing one box from a Young shape of $\mathcal{S}(N)$, say, we obtain a Young shape specifying an irrep of $\mathcal{S}(N-1)$, and so on. The corresponding spin functions are termed *genealogical*, since (unlike spin-projected Slater determinants, for example) the spin functions for $N+1$ electrons are obtained by coupling a single electron to the N-electron spin eigenfunctions. If at the N-electron level we have a spin eigenfunction labelled Θ_S^N, we can obtain two $N+1$-electron spin eigenfunctions from

$$\Theta_{S+\frac{1}{2}}^{N+1} = \Theta_S^N \alpha \tag{6.18}$$

and

$$\Theta_{S-\frac{1}{2}}^{N+1} = -\hat{S}_-\left(\Theta_S^N\right)\alpha + 2S\Theta_S^N\beta. \tag{6.19}$$

The spin functions obtained this way are usually referred to as Yamanouchi-Kotani functions.

In order to obtain CSFs that are antisymmetric to permutations of both space and spin coordinates, we must form combinations of the products

$$\Phi(r)_S^N[\tilde{\lambda}]\Theta_S^N[\lambda] \tag{6.20}$$

of space and spin parts, where the irreps $[\lambda]$ and $[\tilde{\lambda}]$ are said to be *conjugate* or *dual* representations, that is, their product transforms as the antisymmetric representation.

Figure 6.3: Branching Diagram

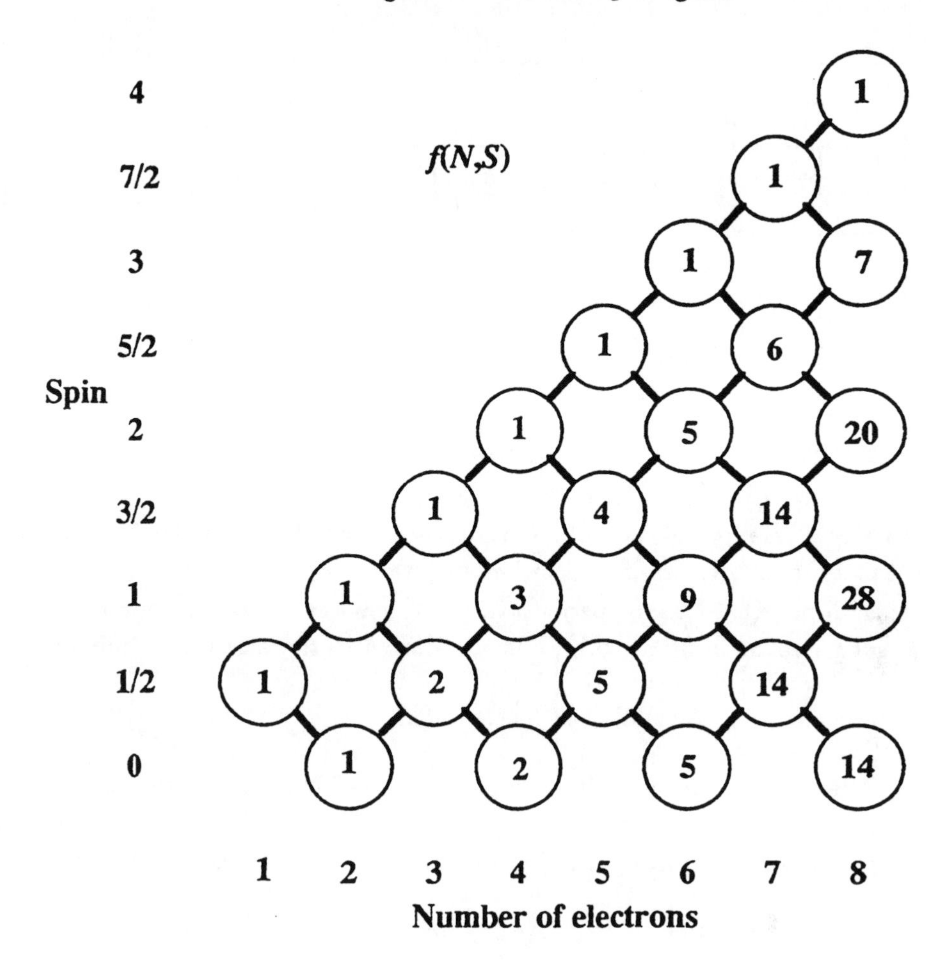

The Young shape of the space part is determined completely as the dual of the spin part, since only one shape can be multiplied by a given spin shape to give the totally symmetric irrep. These shapes are simply related, as shown in Fig. 6.4, so the dual shape is obtained by reflecting the spin function shape across the diagonal of the first square. We can regard this as exactly analogous to obtaining an A_2 basis function in C_{3v} by taking the product $E_xE_y - E_yE_x$ where E_x and E_y are basis functions for the E irrep. In matrix elements of the Hamiltonian, the spin can be integrated out immediately because the spin eigenfunctions are orthonormal, so we can write the matrix elements *entirely* in terms of the space parts of the CSFs:

$$\left\langle \Phi(r)_S^N[\tilde{\lambda}]|H|\Phi(r)_S^N[\tilde{\mu}]\right\rangle . \tag{6.21}$$

Figure 6.4: Dual shapes

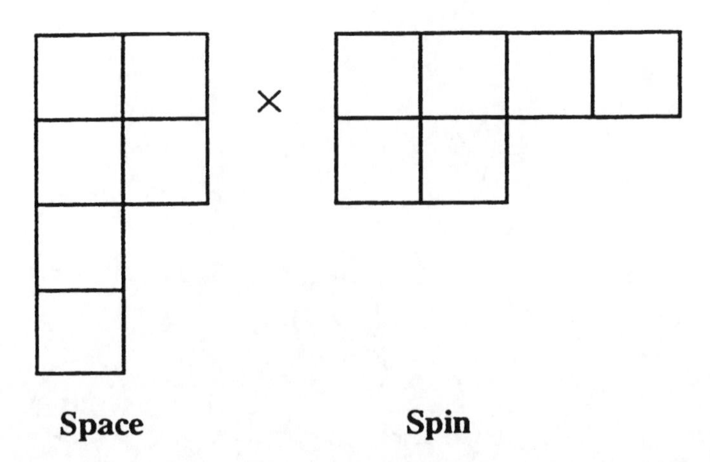

This leads to a *spin-free* formulation of quantum chemistry. The formulas for matrix elements of this type are discussed in detail elsewhere.

Another approach to using permutational symmetry is to consider not the genealogical subgroup chain of Eq. 6.17, but adaptation to the alternative subgroup

$$\mathcal{S}(2) \otimes \mathcal{S}(2) \dots [\otimes \mathcal{S}(1)] \tag{6.22}$$

where the term in square brackets is included for N odd. This is the *Serber* or Jahn-Serber scheme. The spin functions are very simple, corresponding to products of either triplet or singlet pair functions (with one unpaired spin if necessary), like

$$(\alpha\beta - \beta\alpha) \dots (\alpha\beta + \beta\alpha) \dots . \tag{6.23}$$

The important point here is that the ordering of the orbitals in the space part is essentially imposed by the need to couple electrons pairwise. Methods for computing matrix elements of the Hamiltonian over these space functions have been developed by Ruedenberg and co-workers [20], who termed the functions SAAPs: spin-adapted antisymmetrized products.

6.4 The Unitary Group

An approach to constructing CSFs and matrix elements of the Hamiltonian that initially appears quite different from the symmetric group approach can be developed by considering the second-quantized form of the Hamiltonian. If we have an orthonormal basis of n orbitals, the spin-free nature of the Hamiltonian can be explicitly accounted for by defining operators summed over spin:

$$E_{pq} = X^{\dagger}_{p\alpha}X_{q\alpha} + X^{\dagger}_{p\beta}X_{q\beta} \tag{6.24}$$

where $X_{p\alpha}$ ($X^{\dagger}_{p\alpha}$) is an annhilation (creation) operator for spin-orbital $\phi_p(r)\alpha(\sigma)$. As discussed elsewhere the operators E_{pq} obey the commutation relation

$$E_{pq}E_{rs} - E_{rs}E_{pq} = \delta_{qr}E_{ps} - \delta_{ps}E_{rq} \tag{6.25}$$

and are commonly referred to as *generators of the unitary group* $U(n)$, although traditionally these generators would be written in terms of Hermitian and anti-Hermitian combinations of E. We can regard the appearance of $U(n)$ as reflecting the invariance of the Hamiltonian to a unitary transformation on the orbital basis (a full CI calculation is invariant to the choice of orbitals).

This approach, unsurprisingly, leads to consideration of the irreps of $U(n)$ for generating configurations. Note that again we have a spin-free formulation — the spin has been accounted for in the operators E. The structure of groups such as $U(n)$ is a rich and rather complicated area of group theory [7, 21]. However, the fact that a nondegenerate orbital can accommodate at most two electrons immediately comes to our rescue, and we find that we can label the irreps of $U(n)$ of interest once again by partitions of N, the number of electrons. In fact, one way to represent these functions is by the *Weyl shape* of Fig. 6.5, which correspond exactly to our dual shapes for the

Figure 6.5: Weyl shape

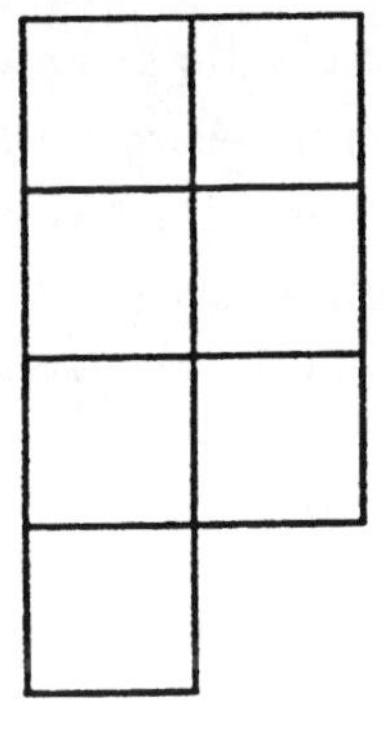

symmetric group case [19]. The configurations we obtain this way, referred to as the Gel'fand-Tsetlin basis, therefore correspond one-to-one with our Yamanouchi-Kotani functions above. The two bases differ by phase factors, as discussed extensively by Paldus and Boyle [22], who, like Shavitt [23] and many other authors, have developed methods for computing Hamiltonian matrix elements between configurations. Shavitt also gives a set of rules by which a given Gel'fand state can be expanded in Slater determinants.

The number of configurations that can be formed from allocating N electrons to n orbitals with spin S is given by a formula due to Weyl and Robinson, but before giving the formula it is useful to establish some similarities and differences to the symmetric group approach. Again, the valid partitions of N (for fermions) must be of the form $[2^a\ 1^b]$, as we saw above, so we immediately have

$$2a + b = N. \tag{6.26}$$

Further, the number of "singly-occupied" orbitals must correspond to the spin, so that

$$b = 2S. \tag{6.27}$$

Finally, of course, and unlike the symmetric group case, we are implicitly keeping track of unoccupied orbitals since we have

$$a + b + c = n, \tag{6.28}$$

where c is the number of unoccupied orbitals. In terms of the parameters a, b, and c, we then have

$$N_{\text{conf}} = \frac{b+1}{n+1} \binom{n+1}{a} \binom{n+1}{c} \tag{6.29}$$

for the number of configurations that can be generated.

We should note another difference between the symmetric group approach and the unitary group approach: the numbers inserted in Young shapes to enumerate the various symmetric group spin eigenfunctions (numbers which correspond in essence to steps across the branching diagram — electrons, in some sense) must increase across the rows and down the columns. However, in the unitary group case, when we enumerate the configurations by inserting numbers into a Weyl shape, we are dealing with all N electrons, not just the open shells. Consequently, we must modify our numbering to be able to include doubly occupied orbitals. The rule for Weyl shapes is that numbers must increase down the columns and must *not decrease* across a row. A typical *Weyl tableau* corresponding to a five-electron doublet configuration might be that shown in Fig 6.6 where orbitals 1 and 3 are doubly occupied and orbital 2 is singly occupied. To within a phase factor, therefore, this configuration is equivalent to the Slater determinant $|1\bar{1}3\bar{3}2|$.

Figure 6.6: Five-electron Weyl tableau

1	1
2	3
3	

6.5 Spatial Symmetry and Configuration State Functions

Whatever method is used in practice to generate spin eigenfunctions, the construction of symmetry-adapted linear combinations, *configuration state functions*, or CSFs, is relatively straightforward. First, we note that all the methods we have considered involve N-particle functions that are products of one-particle functions, or, more strictly, linear combinations of such products. The application of a point-group operator G to such a product is

$$G(\psi_1 \dots \psi_N) = (G\psi_1 \dots G\psi_N), \tag{6.30}$$

that is, the operator is applied to each term of the product. The linear combinations formed to account for overall antisymmetry or spin symmetry all behave the same way under this operation, so it is only necessary to consider one of the products. Using this rule we can easily apply projection or shift operators to our N-particle functions (see, for example, Hurley [24]). Note that if the one-particle functions are symmetry adapted, as is usually the case, the effect of operating on them with G is very simply determined. A final simplification is that any orbital that is maximally occupied transforms according to the totally symmetric irrep and therefore contributes nothing to the final symmetry species. Hence a closed-shell configuration comprising only maximally occupied orbitals transforms as a spin singlet and is totally symmetric.

A simple example illustrates all of these points. Consider a single Slater determinant of the form

$$|e_x\bar{e}_x|, \tag{6.31}$$

where the molecular framework has C_{3v} symmetry. This is a singlet spin eigenfunction, but not a CSF. Application of symmetry projection operators for the symmetry species of C_{3v} shows that this determinant contributes to two (unnormalized) CSFs:

$$\frac{1}{2}\left(|e_x\bar{e}_x| + |e_y\bar{e}_y|\right) \tag{6.32}$$

and

$$\frac{1}{2}\left(|e_x\bar{e}_x| - |e_y\bar{e}_y|\right), \tag{6.33}$$

which transform respectively as ${}^1A_1$ and as the first row of ${}^1E$. The reader may care to apply a shift operator to Eq. 6.33 to generate its partner function, which in this case is also simple to obtain by inspection.

This survey of methods for obtaining configuration state functions has necessarily been brief, since the topic could easily occupy a course of its own. However, we have treated the methods in common use, and much additional material on second quantization techniques and matrix element evaluation will be covered elsewhere.

Chapter 7

Many-electron Configurations

7.1 Equivalent Electrons

We have seen in the previous chapter that exploiting the symmetries of the Hamiltonian allows us to substantially reduce the size of our computational problems. We have already seen how spin- and symmetry-adapted functions can be obtained as antisymmetrized products of spin and space functions, or as linear combinations of Slater determinants. However, a common desire is simply to identify which spin and spatial symmetry configurations can arise from a particular ***orbital occupancy***. The key to simplifying the ***classification*** of these configurations is the recognition that a set of completely occupied orbitals, a closed shell, behaves as a totally symmetric spin singlet, and thus contributes "nothing" to the overall symmetry label. A singly-occupied orbital that transforms as symmetry Γ has the configurational symmetry ${}^2\Gamma$, irrespective of whether Γ is degenerate or not. Consequently, multiple singly-occupied orbitals, say, $r^1 s^1 t^1 \ldots$, give rise to spatial configurations with symmetries obtained by reducing the direct product $\Gamma_r \otimes \Gamma_s \otimes \Gamma_t \otimes \ldots$. The spin values are not determinable from the occupancy alone — each irreducible symmetry appears with all S values from $N/2$, for N singly-occupied orbitals, down to zero (or 1/2).

As an example, consider a tetrahedral molecule in T_d symmetry, with two singly-occupied t_2 symmetry orbitals, say $t_2^1 {t'_2}^1$. The direct product $T_2 \otimes T_2$ reduces to $A_1 \oplus E \oplus T_1 \oplus T_2$, so we obtain singlet states ${}^1A_1$, ${}^1E$, ${}^1T_1$, and ${}^1T_2$, and triplet states ${}^3A_1$, ${}^3E$, ${}^3T_1$, and ${}^3T_2$. A handy check on the correctness of this sort of analysis is to add up the total spin and spatial degeneracies of all the states and verify that it equals the spin and spatial degeneracy of the original orbital product (36 in this case).

Since a spatially degenerate orbital with one "hole" has the same symmetry, for our purposes, as that with one electron, the above analysis can also be used for products of spatially degenerate orbitals that are singly occupied or have one hole. Further, if the orbitals appearing in the product are not degenerate, only a single electronic state spatial symmetry is possible. However, this does not cover all cases. The case of a degenerate orbital occupied by two (or, for triple and higher degeneracies, more than two) electrons, so-called "equivalent electrons", requires some additional analysis. The state symmetries are obviously not given simply by reducing a direct product, because some orbital occupancies will be forbidden by the Pauli exclusion principle. One approach would be to construct all determinantal wave functions compatible with the Pauli principle, and to determine their spin and spatial symmetries. This is simple, but can be time-consuming, and is a waste of effort unless the wave functions are required for some other purpose. A much better approach is to develop a general solution by considering the transformational properties of the determinantal wave functions, as discussed in several texts; Hurley's presentation is perhaps the most pedagogical [17]. We shall quote only the necessary results here.

Suppose that two electrons occupy a degenerate orbital that has character $\chi(G)$ for a given $G \in \mathcal{G}$. The character

$$\chi^0(G) = \frac{1}{2}\left([\chi(G)]^2 + \chi(G^2)\right) \tag{7.1}$$

is the character of the *singlet* states that can be formed, while

$$\chi^1(G) = \frac{1}{2}\left([\chi(G)]^2 - \chi(G^2)\right) \tag{7.2}$$

gives the character of the triplet states. As an example, consider the occupation t_2^2 in the point group T_d. We obtain the following results:

	E	$8C_3$	$6\sigma_d$	$6S_4$	$3C_2$
T_2	3	0	1	−1	−1
$[\chi(G)]^2$	9	0	1	1	1
$\chi(G^2)$	3	0	3	−1	3
$\chi^0(G)$	6	0	2	0	2
$\chi^1(G)$	3	0	−1	1	−1

Using the character tables we can reduce χ^0 to $A_1 \oplus E \oplus T_2$, while χ^1 is the character of the irrep T_1. Hence the occupation t_2^2 gives ${}^1A_1$, ${}^1E$, ${}^1T_2$, and ${}^3T_1$ electronic states. (For those who choose to explore this matter in greater detail, we note that the case we have just considered is tabulated incorrectly by Hurley [17].)

For multiple occupations with equivalent electrons, that is, for products of occupations, one or more of which contains equivalent electrons, we proceed for each occupation as described here, and then couple the various resulting states together in all possible ways to obtain the final states arising from the product of occupations.

If the degeneracy of the open-shell orbital is higher than two, we obtain the same results for the occupation with two holes as for two electrons. However, for such higher degeneracies we can also form open-shell occupations with more than two electrons. For three electrons in a higher than doubly degenerate orbital, we have

$$\chi^{1/2}(G) = \frac{1}{3}\left([\chi(G)]^3 - \chi(G^3)\right) \quad (7.3)$$

for the character of the doublet states, and

$$\chi^{3/2}(G) = \frac{1}{6}\left([\chi(G)]^3 - 3\chi(G)\chi(G^2) + 2\chi(G^3)\right) \quad (7.4)$$

for quartet states. The only molecular cases not covered by Eqs. 7.1–7.4 are systems with icosahedral symmetry and four or five electrons in four-fold or five-fold degenerate orbitals. Specific formulas can be developed for these cases, although it is probably preferable to work with general formulas, devised by Kotani using symmetric group considerations and given also, for example, by Lomont [8].

7.2 Symmetry and Equivalence Restrictions: SCF Calculations

Consider an occupation in which three electrons occupy a doubly degenerate orbital outside a closed shell. We can denote this as

$$\{core\}e^3 \ ({}^2E), \quad (7.5)$$

where we have explicitly indicated the spin and spatial symmetry as ${}^2E$. If we denote the basis functions for the irrep E by row labels x and y, we can consider two Slater determinant wave functions:

$$\left|\{core\}e_x^2 e_y\right| \ ({}^2E_y), \quad (7.6)$$

and

$$\left|\{core\}e_y^2 e_x\right| \ ({}^2E_x), \quad (7.7)$$

where we have indicated the symmetry species on the state designation. Now, if we attempt to obtain an SCF wave function by entering the determinant of Eq. 7.6 into a doublet (i.e., spin-restricted) Hartree-Fock program, our result will display a number of undesirable properties. Not only will the open-shell "e_x" and "e_y" orbitals not be equivalent to one another within a rotation about the principal axis, but, since the open-shell and closed-shell orbitals mix during the SCF procedure, orbitals in the closed shells will not display the expected symmetry degeneracies. In more complicated open-shell cases, the orbitals may not even be of pure symmetry type. This loss of degeneracies is referred to as a loss of *symmetry and equivalence* [25]. A solution that displays symmetry and equivalence has orbitals that transform as symmetry species of the molecular point group (*symmetry*) and comprise full sets of partner functions for all irreps (*equivalence*).

For our SCF calculation the way to obtain such a solution is to optimize the orbitals not for the energy of determinant 7.6, but for the *average* energy of both determinants. This is termed the imposition of ***symmetry and equivalence restrictions***. It involves imposing a constraint on a variational calculation, and consequently the symmetry and equivalence restricted solution will have an energy no lower than the broken symmetry solution; it will usually have a higher energy. We may note that in a UHF calculation we impose neither spin nor spatial symmetry and equivalence restrictions — the terms "restricted" and "unrestricted" were first used in exactly this context of whether to impose symmetry constraints on the wave function.

The averaging of SCF energy expressions to impose symmetry and equivalence restrictions is a straightforward, if sometimes tedious, application of the Slater-Condon rules for matrix elements between determinants of orthonormal orbitals. This matter is discussed in detail elsewhere. The most general SCF programs can handle energy expressions of the form

$$E = \sum_i n_i^{occ} h_{ii} + \sum_i \sum_j (2a_{ij}J_{ij} - b_{ij}K_{ij}) \tag{7.8}$$

where h_{ii} is a matrix element of the one-electron part of the Hamiltonian and J and K are the usual Coulomb and exchange integrals. However, there are many approaches to defining the occupation numbers n_i^{occ} and ***vector coupling coefficients*** a_{ij} and b_{ij}. Usually the occupation number is factored out of the vector coupling coefficients, but there are still many different conventions in use (almost as many as there are programs). Sometimes, converting the average energy expression into vector coupling coefficients is a greater complication in setting up the SCF calculation than deriving the expression in the first place, at least until the user acquires some familiarity with a given program.

7.3 Symmetry and Equivalence Restrictions: Natural Orbitals

It can be necessary and/or desirable to impose symmetry and equivalence restrictions on quantum chemical calculations or results beyond the single-configuration SCF level. For instance, most CI programs generate natural orbitals (NOs) after computing the CI wave function, by forming and diagonalizing the ***first-order reduced density matrix*** or 1-matrix ρ in

$$\sum_{pq} \varphi_p(r_1)\rho_{pq}\varphi_q(r_1')^* = \int \Psi^*(r_1', r_2, \ldots r_N)\Psi^*(r_1, r_2, \ldots r_N)\, dr_2 \ldots dr_N. \tag{7.9}$$

The expression on the left-hand side of Eq. 7.9 is often referred to as the first-order density kernel. (We are assuming here that we have integrated — strictly, summed — over the spin coordinates of *all* N electrons, not just the last $N-1$.)

The NOs $\{\psi_k\}$ are eigenvectors of ρ, and in terms of the NOs we can write the first-order density kernel as

$$\sum_{pq} \varphi_p \rho_{pq} \varphi_q^* = \sum_k \psi_k d_k \psi_k^*, \tag{7.10}$$

where d_k is the *occupation number* of the kth NO. Now the eigenvectors of the matrix of a totally symmetric operator are fully symmetry adapted, and clearly when Ψ in Eq. 7.9 transforms according to a *nondegenerate* irrep Γ of the molecular point group, the 1-matrix must be totally symmetric, since the direct product $\Gamma \otimes \Gamma$ gives the totally symmetric irrep as long as Γ is nondegenerate. However, when Ψ corresponds to, for instance, a component of a degenerate state, like our ^2E_x wave function above, the 1-matrix will *not* transform according to the totally symmetric irrep. Hence the NOs will not be symmetry adapted: they will not display symmetry and equivalence.

The solution to this problem is straightforward [26] — we must obtain a totally symmetric 1-matrix before diagonalizing it, and the appropriate technique is of course projection. For operators the projection becomes

$$\rho^0 = g^{-1} \sum_G G \rho G^{-1}, \tag{7.11}$$

where we need not include the representation matrix/character factor since we are dealing with the totally symmetric irrep and all characters are unity. The correspondence between Eq. 7.11 to give ρ^0, the totally symmetric projection of the 1-matrix, and Eq. 5.11, to symmetrize a skeleton Fock matrix, is obvious — both seek to eliminate any contribution that is not totally symmetric.

In practical implementations the original MO basis $\{\varphi_p\}$ will usually be fully symmetry adapted (say, by performing a symmetry and equivalence restricted SCF calculation). The symmetrization Eq. 7.11 is then very simple to implement. Consider the block of ρ^0 from orbitals that transform according to the symmetry species (α, i) (recall that ρ^0 must be block diagonal on symmetry species)

$$(\rho^0)^{\alpha i,\alpha i} = g^{-1} \sum_G G \rho^{\alpha i,\alpha i} G^{-1} = g^{-1} \sum_G \sum_{jk} \rho^{\alpha j,\alpha k} D_{ij}^{\alpha}(G) D_{ik}^{\alpha}(G)^*. \tag{7.12}$$

But

$$g^{-1} \sum_G D_{ij}^{\alpha}(G) D_{ik}^{\alpha}(G)^* = n_i^{-1} \delta_{jk} \tag{7.13}$$

by the GOT. Hence

$$(\rho^0)^{\alpha i,\alpha i} = n_i^{-1} \sum_j \rho^{\alpha j,\alpha j}. \tag{7.14}$$

Computationally, this can be implemented by discarding all off-diagonal blocks of the original 1-matrix ρ, and averaging the diagonal blocks that belong to different rows of the same irrep. It should be noted that this implementation assumes that the phase relationships between partner functions of a given irrep are the same for all MOs in the irrep, and care should be taken to ensure this is the case in practice.

It is worth noting that expectation values of one-electron operators such as

$$\langle \Psi | \mathcal{O} | \Psi \rangle \tag{7.15}$$

are commonly computed in practice via a density matrix trace:

$$Tr(\mathbf{O}\rho) \tag{7.16}$$

where $\mathbf{O}$ is the matrix of operator $\mathcal{O}$ in the basis in which ρ is computed. Since a non-zero value for Eq. 7.15 can occur only if the operator $\mathcal{O}$ is totally symmetric, we can replace Eq. 7.16 with the form

$$Tr(\mathbf{O}\rho^0), \tag{7.17}$$

using the totally symmetric projection instead. Hence if we transform the matrix of $\mathcal{O}$ to the symmetry-adapted NO basis, we can perform the trace with the corresponding occupation numbers: our symmetry projection on the density matrix has no effect on computed expectation values. This obviates any need to retain the original density matrix or its non-symmetry-adapted NOs for this purpose.

We should point out here that we have, to some extent, played fast and loose with electron spin in this section. We have asserted that spin has been taken care of in some way and hence that we dealt only with spatial symmetry. In fact, if we do not sum over the spin of particle 1 in Eq. 7.9, we obtain a reduced density kernel with spin. The eigenvectors of the corresponding matrix are ***natural spin orbitals*** (NSOs). Since our Hamiltonian contains no spin-dependent terms, the NSOs are of pure spin type (either α or β, corresponding to $m_s = \pm\frac{1}{2}$, respectively). However, not suprisingly, unless the matrix transforms according to the totally symmetric representation for spin, the NSOs do *not* occur in pairs of functions that have the same spatial part but the two different spin functions. We can regard this as a loss of equivalence under the spin group. Only for the case that the wave function Ψ has $M_S = 0$ do we obtain equivalence. Similar to the spatial symmetry case, the easiest way to recover equivalence is to project out the totally symmetric component: this corresponds to averaging the α and β spin blocks of the original matrix, which is exactly how many programs compute the spin-free density matrix. In this case the NSOs are just the NOs multiplied by the two possible spin functions.

More detailed discussions of the spin and symmetry properties of density matrices are given by McWeeny and Kutzelnigg [27] and by Davidson [28].

7.4 Symmetry and Equivalence Restrictions: MCSCF Calculations

In an MCSCF calculation we are interested in optimizing the MOs appearing in some CI wave function. Since the CI energy expression, expressed in terms of reduced density matrices, will in general contain contributions from both the 1-matrix and the 2-matrix (the second-order reduced density matrix), it would appear that to apply symmetry and equivalence restrictions so that our MCSCF orbitals are fully symmetry adapted we must symmetrize both the 1- and 2-matrices. A formula for symmetrizing the latter has been devised [26], but it would lead to a rather complicated computational implementation. We can simplify our work considerably by noting that our desire to impose symmetry and equivalence restrictions on our MCSCF orbitals can be viewed as a desire to allow mixing only between MOs of the same symmetry species. That is, we must ensure that MOs of different species do not mix, and that we obtain full sets of partner functions for each irrep. We can expand the MCSCF energy to second order in orbital rotations as

$$E = E_0 + \sum_{p>q} g_{pq} X_{pq} + \frac{1}{2} \sum_{p>q} \sum_{r>s} H_{pq,rs} X_{pq} X_{rs}, \tag{7.18}$$

where E_0 is the energy with the current MOs, $\mathbf{g}$ is the gradient vector, $\mathbf{H}$ is the Hessian matrix and $\mathbf{X}$ is an antisymmetric matrix used to parametrize the orbital rotations. Making the energy stationary corresponds to solving the Newton-Raphson equations

$$\sum_{r>s} H_{pq,rs} X_{rs} = -g_{pq}. \tag{7.19}$$

Restricting the mixing to symmetry species implies solving the problem

$$\sum_{r>s} H_{pq,rs}^{\alpha i\alpha i,\beta j\beta j} X_{rs}^{\beta j\beta j} = -g_{pq}^{\alpha i\alpha i}, \tag{7.20}$$

with symmetry blocks of the various arrays that are not of these types set to zero. Equivalence restrictions require averaging the blocks of $\mathbf{H}$ over row indices of the various irreps. Full details are given by Bauschlicher and Taylor [26].

While this discussion has concentrated only on the orbital mixing problem, it is perfectly straightforward to include coupling between orbital mixing and the CI coefficients while imposing symmetry and equivalence restrictions. We can best regard the latter as resulting from application of a "symmetrizer" $\mathcal{P}$ to the various components of the calculation, so that the (full) Hessian $\mathbf{H}$ becomes $\mathcal{P}\mathbf{H}\mathcal{P}$, etc. It is simple to cast the averaging procedure discussed in the previous paragraph in this form.

7.5 State Averaging

There is another approach to imposing symmetry and equivalence restrictions on calculations. Instead of averaging density matrix blocks after computing a non-totally symmetric density matrix from one component of a degenerate wave function, we can obtain the same result by averaging density matrices obtained from separate calculations on each component of the degenerate state: *state averaging.* The obvious advantage of this approach is that it requires essentially no additional computer code, compared to the code required for zeroing of some blocks and averaging others and, particularly bookkeeping tasks associated with keeping track of partner functions, etc. However, a major disadvantage is that it requires as many calculations as there are possible components to the degenerate wave function. If a large CI calculation with millions of configurations is involved, this can be a very substantial drawback. Similarly, if we apply this technique in MCSCF calculations, we must average over the appropriate number of roots. While most MCSCF programs have this capability, it can greatly increase the computation time.

There are more subtle disadvantages to using state averaging to impose symmetry and equivalence restrictions than the increase in computation time [26]. The explicit density matrix projection we discussed in Sec. 7.3 can be viewed as producing the same result as averaging the density matrix computed for one component of a degenerate wave function with other components that are related to the first by shift operators. If we employ state averaging, we must *ensure* that the wave functions we are generating are related in this way. This is not necessarily trivial — it is quite possible that the effects of some constraint, such as configuration selection, will vitiate this requirement. Since density matrix projection effectively applies the shift operator for us, we need have no concerns about any constraints of this type we impose.

In one situation, however, state averaging is indispensable as a strategy for obtaining full symmetry and equivalence. If we are specifically considering distortions of a symmetric system, looking at a Jahn-Teller distorted species, say, state averaging provides an excellent means of ensuring that fully symmetry-adapted results are obtained at symmetric geometries. Consider, for example, an X_3 molecule for which the ground state at the symmetric (D_{3h}) geometry is $^2E'$. This state will be Jahn-Teller unstable: if we assume that the nuclear mode that lowers the energy is a variation of one angle away from 60° to give an isosceles triangle, the symmetry distorts to C_{2v}. Thus the $^2E'$ state splits into a $^2A_1$ and a $^2B_2$ state. Now, if we perform separate calculations on these two states, they will *not*, in general, be degenerate with one another at D_{3h} geometries, so features that should be present on our potential surfaces are not there. However, if we perform a single state-averaged calculation on the $^2A_1$ and $^2B_2$ states, we will restore the degeneracy at symmetric geometries. In addition, our orbitals will have the full symmetry and equivalence properties at symmetric geometries, which they would not have if the states were optimized separately. We shall take up this question of state averaging and high symmetry geometries at greater length elsewhere. We should keep one caveat in mind, however. In choosing the states to be averaged, attention needs to be given as to what range of geometries is to be studied. If other points of high symmetry will be considered (if, for example in our X_3 case we were to include some linear geometries), it is possible that additional states will have to be included in the averaging to ensure that *all* higher symmetry and equivalence possibilities are covered.

7.6 Restricted, Unrestricted, Projected, and Extended Wave Functions

Some time ago Löwdin [29] pointed out what he termed a "dilemma" in the use of symmetry in practical calculations (Hurley preferred the term "paradox" [17], which is perhaps even less appropriate!). The issue arises from the fact that the imposition of conditions like symmetry and equivalence restrictions constitutes a reduction in flexibility in the wave function: a variational constraint. Consequently, even though we know that the exact wave function may possess certain symmetry properties, in practical calculations we may achieve a variationally better result, i.e., a lower energy, by not constraining the calculation to display these properties. As we mentioned above, the terms "restricted" and "unrestricted" were first used in relation to Hartree-Fock calculations with and without (spin) symmetry constraints. Of course, we can take an unrestricted result and apply projection operators to restore our desired symmetry properties. Since we have thereby eliminated some contamination in the wave function, our energy will once again decrease. However, this "projected" result is no longer variationally optimum, so we could optimize the projected wave function, yielding the "extended" result. Projection of a single-determinant unrestricted solution will generate a multideterminant wave function, so the "extended" optimization will be a multiconfigurational calculation.

Interest in applying these steps to Hartree-Fock wave functions waxes and wanes with the years: there are periodic resurgences of interest in "extended Hartree-Fock" calculations as a way of improving on single-configuration results (see, for example, Handy and Rice [30]), partly encouraged by the fact that unlike conventional MCSCF calculations there is no need to choose configurations or active orbitals — the configuration space is effectively defined by projection. However, the work (and the configuration space) increases rapidly with the number of electrons involved, just as is the case with more conventional MCSCF methods. In this sense there seems to be little advantage compared to performing a CASSCF calculation, say, from the start.

The other area in which projection of an unrestricted result has received attention is projected Møller-Plesset perturbation theory, the PUMPn methods [31], where n is the order of the perturbation theory. In cases in which the UHF approximation is a poor starting point (considerable spin contamination, for example), the convergence of the MP perturbation expansion can be slow and/or erratic. The PUMP methods apply projection operators to the perturbation expansion, although usually not full projection but simply annihilation of the leading contaminants. This approach has met with mixed success; again, it represents a rather expensive modification to a technique that was originally chosen partly for its economy — seldom a recipe for success.

Chapter 8

Symmetry and Potential Energy Surfaces

8.1 Expansion of the Energy in Displacement Coordinates

Consider expanding the energy of a molecule in displacements from some reference geometry. We denote these displacements by $\{q_a\}$, where a runs over nonredundant displacements ($3N-6$ for a nonlinear reference geometry and $3N-5$ for the linear case). The energy expansion is thus

$$E = E_0 + \sum_a \frac{\partial E}{\partial q_a} q_a + \frac{1}{2} \sum_a \sum_b \frac{\partial^2 E}{\partial q_a \partial q_b} q_a q_b \cdots, \tag{8.1}$$

where E_0 is the energy at the reference geometry and the partial derivatives are evaluated at this geometry. It is convenient to write the expansion in the form

$$E = E_0 + \sum_a g_a q_a + \frac{1}{2} \sum_a \sum_b H_{ab} q_a q_b + \frac{1}{3!} \sum_a \sum_b \sum_c K_{abc} q_a q_b q_c \cdots, \tag{8.2}$$

which defines the *gradient vector* $\vec{g}$ and the *Hessian matrix* $\mathbf{H}$. For future use we have also introduced a cubic term so that the numerical factors appearing in each order are defined.

Now, the individual partial derivatives appearing in Eq. 8.1, or the equivalent matrix elements in Eq. 8.2, are simply scalars, hence they must transform as the totally symmetric irreducible representation of the molecular point group; they are totally symmetric irreducible tensor operators. This immediately tells us that the gradient, for example, can be non-zero only for totally symmetric displacements — symmetry-preserving displacements — because all other displacements would correspond to non-totally symmetric elements of the gradients, and these must be zero. Similarly, the Hessian matrix or force constant matrix will be block diagonal on symmetry species, and within an irrep the diagonal blocks will be the same. We can conveniently rewrite the energy through second order as

$$E = E_0 + \sum_a^{(0)} g_a^0 q_a^0 + \frac{1}{2} \sum_\alpha \sum_i \sum_a^{(\alpha i)} \sum_b^{(\alpha i)} H_{ab}^{\alpha i, \alpha i} q_a^{\alpha i} q_b^{\alpha i}, \tag{8.3}$$

to illustrate these *selection rules* on $\vec{g}$ and $\mathbf{H}$. The summations in Eq. 8.3 reflect the block structure of $\mathbf{H}$, while the superscript 0 (and the corresponding summation limit) indicates restriction to the totally symmetric irrep only. Note that we are assuming in our notation that the displacement coordinates are symmetry adapted. However, the fundamental symmetry properties of the derivative matrices are, of course, independent of the set of displacement coordinates used. We may, for example, use a set of non-symmetry-adapted coordinates in which all elements of the gradient vector are non-zero, but we still have information only about totally symmetric displacements.

8.2 Symmetry of Force Constants

According to the previous section the individual force constants in an expansion of the energy are totally symmetric, and as we have seen this gives rise to a particularly simple selection rule for the quadratic (harmonic) force constants in a basis of symmetry-adapted displacement coordinates. We can develop selection rules for higher force constants in an analogous way. In fact, we could draw on the selection rules for two-electron integrals, for example, to provide the selection rules for quartic force constants, with the simplification that the force constants are equivalent under any permutations of their indices. The two-electron integrals are invariant under only a subgroup of the possible index permutations. Here, we shall give a general treatment of the selection rules for force constants of any order, although for simplicity we shall restrict ourselves to real displacement coordinates and real force constants.

Consider a force constant as a real scalar quantity that can conveniently be denoted $K_{ijk\ldots}^{\alpha\beta\gamma\ldots}$. Such an object would appear multiplying symmetry displacements that are respectively of species (α,i), (β,j), (γ,k),..., for example. First, we observe that unless $\alpha\otimes\beta\otimes\gamma\ldots$ contains the totally symmetric irrep, all such force constants will vanish. Proceeding further, we can operate on K with G, and (using the fact that our irrep matrices are real here) obtain

$$G(K_{ijk\ldots}^{\alpha\beta\gamma\ldots})=\sum_t\sum_u\sum_v\cdots\ D_{ti}^{\alpha}(G)D_{uj}^{\beta}(G)D_{vk}^{\gamma}(G)\ldots K_{tuv\ldots}^{\alpha\beta\gamma\ldots}. \tag{8.4}$$

Summing over G and dividing by g gives

$$K_{ijk\ldots}^{\alpha\beta\gamma\ldots}=g^{-1}\sum_G\sum_t\sum_u\sum_v\cdots\{D_{ti}^{\alpha}(G)D_{uj}^{\beta}(G)D_{vk}^{\gamma}(G)\ldots\}K_{tuv\ldots}^{\alpha\beta\gamma\ldots}. \tag{8.5}$$

This gives all relations between symmetry-related force constants. If the quantity

$$g^{-1}\sum_G D_{ti}^{\alpha}(G)D_{uj}^{\beta}(G)D_{vk}^{\gamma}(G)\ldots \tag{8.6}$$

is zero for all values of $tuv\ldots$, $K^{\alpha\beta\gamma\ldots}_{ijk\ldots}$ is evidently zero. It is easy to see that for the Hessian, for example, this selection rule and the GOT lead immediately to a matrix with the symmetry structure described in Sec. 8.1.

We can consider a more elaborate example as follows. Assume that the molecular point group is C_{3v} and we have a vibrational mode of e symmetry. Which of the possible quartic force constants K^{eeee}_{ijkl} are distinct and non-vanishing? The selection rule is

$$\frac{1}{6}\sum_G \sum_t \sum_u \sum_v \sum_x D^e_{ti}(G) D^e_{uj}(G) D^e_{vk}(G) D^e_{xl}(G) \tag{8.7}$$

and since we are considering just one mode there are five K^{eeee}_{ijkl} that are permutationally distinct. Examining first K^{eeee}_{1111}, use of the representation matrices of C_{3v} and some tedious multiplication leads eventually to the relation

$$K^{eeee}_{1111} = \frac{3}{8}K^{eeee}_{1111} + \frac{3}{4}K^{eeee}_{1122} + \frac{3}{8}K^{eeee}_{2222}. \tag{8.8}$$

Turning to K^{eeee}_{2222} we obtain

$$K^{eeee}_{2222} = \frac{3}{8}K^{eeee}_{1111} + \frac{3}{4}K^{eeee}_{1122} + \frac{3}{8}K^{eeee}_{2222}. \tag{8.9}$$

These two equations together lead to the relations

$$K^{eeee}_{1111} = K^{eeee}_{2222} = 3K^{eeee}_{1122}. \tag{8.10}$$

We note that the same approach could of course be used for force constants coupling different modes of e symmetry. Thus if we have a second mode of e symmetry, denoted by a bar over subscripts, we can obtain expressions for $K^{eeee}_{111\bar{1}}$ or $K^{eeee}_{11\bar{1}\bar{1}}$ the same way. As different modes are introduced, however, the number of terms that are permutationally equivalent decreases, making the algebra ever more tedious.

For point groups of higher order than C_{3v} it may become impractical to carry out these manipulations by hand: for a molecule of tetrahedral symmetry, for example, we might have up to 81 possible quartic force constants involving triply degenerate modes, and for each one a 24-term sum would have to be evaluated. In such cases it would probably be necessary to resort to a computer implementation of the rules. We also note that in some high symmetry cases there are multiple linearly independent force constants arising from the same multiple product of irreps. In the T_d case, for instance, there are two independent quartic constants of the type $t_2t_2t_2t_2$. Finally, we note that care is required in deciding whether some force constants are non-zero in certain situations. Thus in the T_d example the product $e \otimes t_2 \otimes t_2 \otimes t_2$ contains the totally symmetric irrep, so that a non-zero force constant of this form is expected. A full examination of the selection rule, however, shows that such a constant is only non-vanishing when at least two of the three t_2 modes are different from one another. Thus such a constant could be non-zero for an XY_4 tetrahedral molecule, where there are two t_2 modes, but cannot arise for an X_4 tetrahedral molecule, for which there is only one [32].

By exploiting the selection rule Eq. 8.5 the number of force constants that must be determined can be substantially reduced for symmetric molecules. For example, a T_d symmetry X_4 molecule has six vibrational modes: the totally symmetric a_1 mode, a doubly degenerate e mode, and a triply degenerate t_2 mode. However, while the quartic force field for a general tetratomic involves 21 quadratic, 56 cubic and 126 quartic constants, only three quadratic, six cubic, and 11 quartic constants are distinct and non-vanishing by symmetry in our X_4 case. If the force constants were to be computed by finite central differences of energies, this reduction in numbers would be essential to the feasibility of the project. Without symmetry, the energy would have to be evaluated at 1389 geometries, while with symmetry only 46 geometries would be required.

8.3 Calculation of Force Constants

We close this chapter with some remarks about the calculation of force constants in practice, with particular reference to the use of symmetry. First, if analytical derivatives of the energy are available through the order of force constants desired, little further effort on the user's part is required. It is worth noting only that if, say, a Hessian is required to guide a geometry optimization of a walk on a potential energy surface, only the modes that transform according to the totally symmetric irrep are needed. Hence if symmetry-adapted nuclear coordinates are used only a sub-block of the Hessian need be evaluated. However, when all frequencies are needed, to characterize a stationary point perhaps, the full Hessian must be computed. These issues will be treated in detail elsewhere. Commonly, however, analytical derivatives of the desired order are not available, and then some thought must be given to finite-difference methods to obtain the force constants. Since finite differences of energies to obtain property values will be discussed elsewhere, we consider here the use of finite differences of energy *gradients* in the calculation of force constants.

As Pulay, for example, has reviewed in some detail [33], Hessian elements can be conveniently computed by finite differences of gradients:

$$H_{ij} = \frac{g_i(\Delta q_j) - g_i(-\Delta q_j)}{2\Delta q_j}. \tag{8.11}$$

There are several advantages to this approach compared to energy differences (and assuming that analytical second derivatives are not available or are too costly). First, an entire row of the Hessian can be evaluated from one pair of gradient evaluations. Second, some cubic force constants can be evaluated from the same data. Consider now the use of symmetry together with this difference formula. If the nuclear coordinates are symmetry adapted (they could be symmetry-adapted linear combinations of internal coordinates, or symmetry-adapted Cartesian coordinates, for example), we have the relationship

$$g_i(\Delta q_j) = -g_i(-\Delta q_j) \tag{8.12}$$

for any nondegenerate mode that transforms as other than the totally symmetric irrep. (This relationship may also be found for some degenerate modes, but this must be established for each possibility.) Hence for such cases only one, rather than two, gradient evaluations are required. For the molecule H_2O, for example, with two a_1 modes (symmetric stretch and bend) and a b_2 mode (antisymmetric stretch), we would therefore require evaluation of the gradient at two displacements in each a_1 mode and one in the b_2 mode. Since the displacement in the b_2 mode lowers the molecular symmetry from C_{2v} to C_s, we need four gradient evaluations in C_{2v} symmetry and one in C_s symmetry. We can improve on this, however, at least in terms of the number of calculations required. Since the Hessian contains no terms coupling different symmetry species, we can combine calculation of the b_2 displacement with one of the a_1 modes. We would thus displace simultaneously in the first a_1 and the b_2 modes (two calculations in C_s symmetry, as we need positive and negative displacements for the a_1 case but only positive for the b_2 case) and displace separately in the other a_1 mode (two calculations in C_{2v} symmetry). Thus treating each mode separately we need four C_{2v} and one C_s calculation, while by combining the modes we need two C_{2v} and two C_s calculations. Circumstances will dictate which of these approaches is preferable.

The combination of symmetric and non-symmetric mode displacements is a useful one and is applicable quite generally [33]. It is always possible to combine one non-symmetric mode with each symmetric mode in this way to reduce the total number of gradient evaluations (although this will increase the number of lower symmetry calculations that must be performed). In addition, while our discussion has been in terms of internal coordinates, the same sort of analysis can be carried through for Cartesian coordinates. Indeed, a partial motivation for including this material was a remark to the author by Lee that it was apparently not widely known that a complete Hessian for H_2O could be computed by central differences with just four gradient evaluations using Cartesian coordinates. Just as in the analysis above, two C_{2v} and two C_s calculations are required. A more general treatment has recently been given by Stanton [34], although much of the material presented there simply consolidates methods that have been known and used for some time.

Finally, we should note that a number of formulas for anharmonic constants and vibration-rotation interaction constants for symmetric-top molecules given in the spectroscopic literature are incomplete. The problem is that relatively few complete anharmonic analyses have been carried out, and the available examples are not adequate to cover all combinations of degenerate and nondegenerate modes. A detailed discussion is given by Lee and co-workers [35].

Bibliography

[1] E. P. Wigner. *Group Theory and its Application to the Quantum Mechanics of Atomic Spectra.* Academic Press, New York, 1959.

[2] M. Tinkham. *Group Theory and Quantum Mechanics.* McGraw-Hill, New York, 1964.

[3] J. C. Slater. *Quantum Theory of Molecules and Solids, Vol I.* McGraw-Hill, New York, 1963.

[4] J.A. Green. *Sets and Groups.* Routledge and Kegan Paul, London, 1965.

[5] G. Herzberg. *Electronic Spectra of Polyatomic Molecules.* Van Nostrand Reinhold, New York, 1966.

[6] A. C. Hurley. *Chem. Phys. Lett.*, **91**, 163, 1982.

[7] C. D. H. Chisholm. *Group Theoretical Techniques in Quantum Chemistry.* Academic Press, London, 1976.

[8] J.S. Lomont. *Applications of Finite Groups.* Academic Press, New York, 1959.

[9] E. R. Davidson. *J. Chem. Phys.*, **62**, 400, 1975.

[10] P. R. Taylor. *Theoret. Chim. Acta*, **69**, 447, 1986.

[11] J. Almlöf. The MOLECULE integral program. Technical Report 74-29, University of Stockholm, Institute for Theoretical Physics, 1974.

[12] T. Helgaker, H. J. Aa. Jensen, P. Jørgensen, J. Olsen, and P. R. Taylor. ABACUS, a CASSCF and RASSCF energy derivatives program.

[13] P. R. Taylor. *J. Comput. Chem.*, **5**, 589, 1984.

[14] M. Dupuis and H. F. King. *Int. J. Quantum Chem.*, **11**, 613, 1977.

[15] P. R. Taylor. *Int. J. Quantum Chem.*, **27**, 89, 1985.

[16] M. Häser, J. Almlöf, and M. Feyereisen. *Theoret. Chim. Acta*, in press.

[17] A. C. Hurley. *Introduction to the Electron Theory of Small Molecules.* Academic Press, London, 1976.

[18] R. Pauncz. *Spin Eigenfunctions.* Plenum, New York, 1979.

[19] H. Weyl. *The Theory of Groups and Quantum Mechanics.* Dover, New York, 1950.

[20] K. Ruedenberg and R. D. Poshusta. *Adv. Quantum Chem.*, **6**, 267, 1972.

[21] H. Weyl. *The Classical Groups.* Princeton University Press, Princeton, 1946.

[22] J. Paldus and M. J. Boyle. *Phys. Scripta*, **21**, 295, 1980.

[23] I. Shavitt. In J. Hinze, editor, *The Unitary Group.* Springer-Verlag, Berlin, 1981.

[24] A. C. Hurley. *Electron Correlation in Small Molecules.* Academic Press, London, 1976.

[25] R. K. Nesbet. *Proc. Roy. Soc.*, **A230**, 312, 1955.

[26] C. W. Bauschlicher and P. R. Taylor. *Theoret. Chim. Acta*, **74**, 63, 1988.

[27] R. McWeeny and W. Kutzelnigg. *Int. J. Quantum Chem.*, **2**, 187, 1968.

[28] E. R. Davidson. *Reduced Density Matrices in Quantum Chemistry.* Academic Press, New York, 1976.

[29] P. O. Löwdin. *Rev. Mod. Phys.*, **35**, 496, 1963.

[30] N. C. Handy and J. E. Rice. In R. Carbó, editor, *Quantum Chemistry: Basic Aspects, Actual Trends.* Elsevier, Amsterdam, 1989.

[31] H. B. Schlegel. *J. Chem. Phys.*, **84**, 4530, 1986.

[32] A. P. Rendell, T. J. Lee, and P. R. Taylor. *J. Chem. Phys.*, **93**, 7050, 1990.

[33] P. Pulay. In H. F. Schaefer, editor, *Applications of Electronic Structure Theory.* Plenum, New York, 1977.

[34] J. F. Stanton. *Int. J. Quantum Chem.*, **39**, 19, 1991.

[35] T. J. Lee, A. Willetts, J. F. Gaw, and N. C. Handy. *J. Chem. Phys.*, **90**, 4330, 1989.

[36] R. McWeeny. *Symmetry — an Introduction to Group Theory.* Pergamon, Oxford, 1960.

[37] M. Hamermesh. *Group Theory and its Application to Physical Problems.* Addison-Wesley, London, 1962.

[38] B. G. Wybourne. *Classical Groups for Physicists.* Wiley-Interscience, New York, 1974.

[39] P. R. Bunker. *Molecular Symmetry and Molecular Spectroscopy.* Academic Press, New York, 1979.

[40] A. Zee. *Fearful Symmetry.* Collier, New York, 1986.

Molecular Symmetry and Quantum Chemistry: Supplementary Notes

Formulas

Rearrangement Theorem

In any formula involving a sum over group elements G, we can replace G with HG, where H is any group element, since the left coset of $\mathcal{G}$ in $\mathcal{G}$ is the group itself. Thus, for example,

$$g^{-1}n_\alpha \sum_G D^\alpha_{ir}(G)^* G = g^{-1}n_\alpha \sum_G D^\alpha_{ir}(GH)^* GH$$

Great Orthogonality Theorem (GOT)

If α and β are two unitary inequivalent irreducible representations of $\mathcal{G}$, of dimension n_α and n_β respectively,

$$\sum_G D^\alpha_{ij}(G)^* D^\beta_{kl}(G) = g n_\alpha^{-1} \delta_{\alpha\beta}\delta_{ik}\delta_{jl}.$$

In terms of group characters,

$$\sum_G \chi^\alpha(G)^* \chi^\beta(G) = g\delta_{\alpha\beta}.$$

Projection and Shift Operators

Full matrix projector for symmetry species (α, i):

$$\mathcal{P}^\alpha_{ii} = g^{-1}n_\alpha \sum_G D^\alpha_{ii}(G)^* G.$$

Shift operator to generate a partner function of species (α, j) from (α, i):

$$\mathcal{P}^\alpha_{ji} = g^{-1}n_\alpha \sum_G D^\alpha_{ji}(G)^* G.$$

Character projector for irrep α:

$$g^{-1}n_\alpha \sum_G \chi^\alpha(G)^* G.$$

Reduction of a Representation

An arbitrary representation with character $\chi(G)$ can be expanded as

$$\chi(G) = \sum_{\alpha} c_\alpha \chi^\alpha(G),$$

where

$$c_\alpha = g^{-1} \sum_{k}^{classes} N_k \chi(G_k)^* \chi^\alpha(G_k).$$

Here the sum runs over classes: N_k is the number of elements in class k and G_k is any element chosen from that class.

Point groups

Preliminary classification

First identify any proper rotational axes of order higher than two. If there are two or more of these the molecule must belong to one of the cubic groups (tetrahedral, cubic/octahedral or icosahedral system). If there is a single rotation axis of infinite order the molecule is linear. (These steps are given for completeness — it is assumed that most readers are capable of recognizing a linear molecule and that they are unlikely to stumble across an icosahedral system unknowing and unprepared.) If there is a rotation axis of finite order greater than two the molecule is axially symmetric.

Cubic molecules

A molecule with four threefold axes and no proper fourfold axes is tetrahedral. If there is a centre of inversion (unusual but possible, for example, for an $M(H_2O)_6^{n+}$ cluster) the system has point group T_h. If the molecule has three proper twofold axes and six planes of reflection the point group is T_d, otherwise it is T (the latter is very unusual).

A molecule with four threefold axes and three proper fourfold axes is cubic or octahedral. If it has a centre of inversion the point group is O_h, otherwise it is O (very unusual).

A molecule with six fivefold axes is icosahedral. If it has a centre of inversion the point group is I_h, otherwise it is I (very unusual). I (I_h) is sometimes denoted P (P_h).

Linear molecules

The point group of a linear molecule is $C_{\infty v}$ unless the molecule has a centre of inversion, in which case it is $D_{\infty h}$.

Axial molecules

First identify the order n of the principal axis. If there is a plane of reflection perpendicular to this axis, check also for planes containing the axis. If they are present the point group is D_{nh}; if they are not the point group is C_{nh} (unusual). If there is no plane perpendicular to the axis, check for check for a rotation-reflection axis collinear with the principal axis. If this is present and there are reflection planes containing the principal axis the group is D_{nd}, otherwise it is S_{2n}. If there is no rotation-reflection axis but planes that contain the principal axis the point group is C_{nv}. If there are twofold axes perpendicular to the principal axis and no other symmetry operations the point group is D_n. Finally, if the only symmetry is the principal axis the point group is C_n (unusual).

Particular care should be taken to check for two-fold axes perpendicular to the principal axis. Overlooking these is probably the commonest error made in identifying point groups. Note also that the group that results from the product of a centre of inversion with D_n is D_{nh} when n is even but D_{nd} when n is odd.

Other cases

If the principal axis is of order two, check for twofold axes perpendicular to it. If they are present and the molecule has a centre of inversion the point group is D_{2h}, otherwise it is D_2. If they are absent, check for a reflection plane perpendicular to the principal axis. If present the point group is C_{2h}. If absent check for reflection planes that include the principal axis: if they are present the point group is C_{2v}, otherwise it is C_2. We should note that by excluding the case of a principal axis of order two from the axial molecule category it is necessary to check here for an S_4 improper axis and planes containing it. If all these elements are present the point group is D_{2d}, if only the improper axis is present the point group is S_4. Finally, if the molecule has only a plane of reflection, the point group is C_s, while if it has only a centre of inversion the point group is S_2, also written C_i. The point group of a molecule with no elements of symmetry is C_1.

Notes

In the above classification scheme I have indicated point groups that are "unusual", that is, that are rarely encountered. While these are not impossible, special care should be taken to check the identification procedure if a molecule appears to belong to one of these point groups.

If a molecule has no improper symmetry operations it will not be superposable on its mirror image, so it may be optically active. Note that this does *not* require the molecule to have no symmetry.

Transformation properties of spherical harmonics

Definitions

Consider the real spherical harmonics defined by

$$Y_{l,+m}(\theta,\phi) = NP_l^{|m|}(\cos\theta)\cos m\phi$$

and

$$Y_{l,-m}(\theta,\phi) = NP_l^{|m|}(\cos\theta)\sin m\phi$$

where $P_l^{|m|}(\cos\theta)$ is the associated Legendre polynomial, θ and ϕ are the usual polar angular coordinates (the principal axis of the point group is the polar axis), and N is a normalization constant. The range of m is

$$-l, -(l-1), \ldots, 0, \ldots, l-1, l.$$

We shall examine the behaviour of these functions under various symmetry operations. We need only consider the behaviour of functions centred at the origin, as functions centred elsewhere display the same behaviour together with a possible translation of the centre that is easily determined. Cubic groups are excluded as the spherical harmonics are less well suited as basis functions for these cases.

In order to allow explicit transformation formulas to be derived, we give here the transformation matrices for common operations. These are shown operating on basis vectors: we recall that functions transform cogrediently to basis vectors. Note that those vectors not shown are unchanged by the operation under consideration.

Rotation through angle α about the principal axis:

$$(\vec{e}_1 \quad \vec{e}_2)\begin{pmatrix} \cos\alpha & -\sin\alpha \\ \sin\alpha & \cos\alpha \end{pmatrix}.$$

Coordinate inversion:

$$(\vec{e}_1 \quad \vec{e}_2 \quad \vec{e}_3)\begin{pmatrix} -1 & 0 & 0 \\ 0 & -1 & 0 \\ 0 & 0 & -1 \end{pmatrix}.$$

Reflection in a horizontal plane:

$$-\vec{e}_3$$

Reflection in a vertical plane $\sigma_v(\alpha)$ that makes an angle α with the positive x axis. This operation can be written as either $\sigma_v(0)C(\alpha)$, that is, rotation through an angle α followed by reflection in the xz plane, or as $C(2\pi - \alpha)\sigma_v(0)$. In either case the result is

$$(\vec{e}_1 \quad \vec{e}_2) \begin{pmatrix} \cos\alpha & -\sin\alpha \\ -\sin\alpha & -\cos\alpha \end{pmatrix}.$$

Rotation about a twofold axis at an angle α to the positive x axis and perpendicular to the principal axis. This operation is conveniently written as $\sigma_h\sigma_v(\alpha)$, or

$$(\vec{e}_1 \quad \vec{e}_2 \quad \vec{e}_3) \begin{pmatrix} \cos\alpha & -\sin\alpha & 0 \\ -\sin\alpha & -\cos\alpha & 0 \\ 0 & 0 & -1 \end{pmatrix}.$$

Rotations about the principal axis

Under a rotation C_n the functions $Y_{l,+m}$ and $Y_{l,-m}$ (we can drop the arguments) transform into a linear combination in general, as is easily seen from the transformation matrix given above. However, for certain rotation angles (such as π) and m values they may transform into a single function. The function $Y_{l,0}$ is unaffected by C_n, and thus provides a basis function for some nondegenerate irrep of the molecular point group. The behaviour of $Y_{l,+m}$ and $Y_{l,-m}$ for the particular rotation angle will determine whether each provides a basis for a nondegenerate irrep, or both provide a basis for some doubly degenerate irrep. The latter can often be identified by comparison with the linear molecule case, where every $\pm m$ pair provides a basis for a different doubly degenerate irrep: $\Pi, \Delta, \ldots$.

Centre of inversion

The behaviour under inversion is very simple: $Y_{l,\pm m}$ is even (*gerade*) if l is even and odd (*ungerade*) if l is odd.

Horizontal plane

Under a σ_h plane perpendicular to the principal axis we have

$$\sigma_h Y_{l,\pm m} = (-1)^{l+m} Y_{l,\pm m}$$

Vertical plane

Under a σ_v plane $Y_{l,0}$ is unchanged, while $Y_{l,+m}$ and $Y_{l,-m}$ transform into a linear combination: this is easily determined from one of the transformation matrix expressions given above.

Twofold axis perpendicular to principal axis

This case is most easily viewed as reflection in a vertical plane followed by reflection in a horizontal plane, as shown above. The behaviour of $Y_{l,0}$ is given by the horizontal reflection alone, of course, while $Y_{l,+m}$ and $Y_{l,-m}$ transform into a linear combination.

Notation for irreps

The above transformation properties lead directly to Mulliken's notation for irreps of point groups. First, the degeneracy of the irrep is indicated by a letter, A or B for a nondegenerate irrep, E for doubly degenerate, and T for triply degenerate. Note that F is sometimes used (especially by spectroscopists) for a triply degenerate irrep. While it would be logical to use U and V for fourfold and fivefold degenerate irreps, respectively, of the icosahedral groups, usage seems to favour G and H.

A nondegenerate irrep that is symmetric with respect to the principl axis is denoted A, while B indicates antisymmetry with respect to this axis. In point groups with a horizontal plane of reflection, primes $'$ and $''$ respectively indicate symmetry and antisymmetry with respect to the plane, while g and u indicate symmetry and antisymmetry with respect to inversion. For doubly degenerate irreps a subscript m indicates which spherical harmonics $Y_{l,\pm m}$ form basis functions for that irrep. Numerical subscripts are used on nondegenerate irreps to distinguish them where necessary: the numbers indicate the first of the vertical planes or perpendicular twofold axes (in the order specified in the character table) with respect to which the irrep is antisymmetric.

Full matrix representations

We list here full matrix representations for several groups. Abelian groups are omitted, as their irreps are one-dimensional and hence all the necessary information is contained in the character table. We give C_{3v} (isomorphic with D_3) and C_{4v} (isomorphic with D_4 and D_{2d}). By employing higher l value spherical harmonics as basis functions it is straightforward to extend these to C_{nv} for any n, even or odd. We note that the even n C_{nv} case has four nondegenerate irreps while the odd n C_{nv} case has only two.

There is no value in listing matrix irreps for the following direct product groups (note that the group C_s here is $\{E, \sigma_h\}$):

$$D_{nh} = D_n \otimes S_2(n \text{ even})$$
$$D_{nh} = D_n \otimes C_s(n \text{ odd})$$
$$D_{nd} = D_n \otimes S_2(n \text{ odd}).$$

The irrep matrices are duplicated for the additional operations for those direct product group irreps derived from the symmetric irrep of S_2 or C_s, while they are duplicated with a change of sign for irreps derived from the antisymmetric irrep of S_2 or C_s.

Matrix irreps are given also for the groups $C_{\infty v}$ and T_d (isomorphic to O). (It should be noted that the full matrix irreps given by Slater [3] for T_d are not correct.) Irreps for $D_{\infty h}$ can be obtained from the direct product of $C_{\infty v}$ with S_2. The matrix irreps for T_d can be used to find those of T by dropping the improper rotations. (Note that the E representation of T is separably degenerate.) Matrix irreps of T_h can be found from the direct product of T with S_2 — again the E_g and E_u representations are separably degenerate. Matrix irreps for the group O_h can be found from the direct product of O with S_2.

C_{3v} and D_3 Matrix Irreps

D_3		E	C_3	C_3^2	C_2^a	C_2^b	C_2^c
	C_{3v}	E	C_3	C_3^2	σ_v^a	σ_v^b	σ_v^c
A_1	A_1	1	1	1	1	1	1
A_2	A_2	1	1	1	-1	-1	-1
$(E)_{11}$	$(E)_{11}$	1	$-\frac{1}{2}$	$-\frac{1}{2}$	1	$-\frac{1}{2}$	$-\frac{1}{2}$
$(E)_{21}$	$(E)_{21}$	0	$\frac{\sqrt{3}}{2}$	$-\frac{\sqrt{3}}{2}$	0	$-\frac{\sqrt{3}}{2}$	$\frac{\sqrt{3}}{2}$
$(E)_{12}$	$(E)_{12}$	0	$-\frac{\sqrt{3}}{2}$	$\frac{\sqrt{3}}{2}$	0	$-\frac{\sqrt{3}}{2}$	$\frac{\sqrt{3}}{2}$
$(E)_{22}$	$(E)_{22}$	1	$-\frac{1}{2}$	$-\frac{1}{2}$	-1	$\frac{1}{2}$	$\frac{1}{2}$

C_{4v}, D_4 and D_{2d} Matrix Irreps

D_4			E	C_2	C_4	C_4^3	C_2^x	C_2^y	C_2^{135}	C_2^{45}
	D_{2d}		E	C_2	S_4	S_4^3	C_2^x	C_2^y	σ^{135}	σ^{45}
		C_{4v}	E	C_2	C_4	C_4^3	σ^{xz}	σ^{yz}	σ^{135}	σ^{45}
A_1	A_1	A_1	1	1	1	1	1	1	1	1
A_2	A_2	A_2	1	1	1	1	−1	−1	−1	−1
B_1	B_1	B_1	1	1	−1	−1	1	1	−1	−1
B_2	B_2	B_2	1	1	−1	−1	−1	−1	1	1
$(E)_{11}$	$(E)_{11}$	$(E)_{11}$	1	−1	0	0	1	−1	0	0
$(E)_{21}$	$(E)_{21}$	$(E)_{21}$	0	0	1	−1	0	0	−1	1
$(E)_{12}$	$(E)_{12}$	$(E)_{12}$	0	0	−1	1	0	0	−1	1
$(E)_{22}$	$(E)_{22}$	$(E)_{22}$	1	−1	0	0	−1	1	0	0

The numbers 45 and 135 refer to the angles (in degrees) the particular axis makes with the positive x axis, or the plane makes with the xz plane.

$C_{\infty v}$ Matrix Irreps

$C_{\infty v}$	E	$C(\alpha)$	$C(-\alpha)$	$\sigma_v(\alpha)$
A_1	1	1	1	1
A_2	1	1	1	-1
$(E_n)_{11}$	1	c	c	$(-1)^n c$
$(E_n)_{21}$	0	s	$-s$	$(-1)^n s$
$(E_n)_{12}$	0	$-s$	s	$(-1)^n s$
$(E_n)_{22}$	1	c	c	$-(-1)^n c$

$c = \cos\alpha \quad s = \sin\alpha.$
The usual "spectroscopic" notation is $A_1 \equiv \Sigma^+$, $A_2 \equiv \Sigma^-$, and $E_n(n = 1, 2, 3\ldots) \equiv \Pi, \Delta, \Phi\ldots.$

T_d and O Matrix Irreps

		x	y	z	xyz	$\bar{x}y\bar{z}$	$\bar{x}\bar{y}z$	$x\bar{y}\bar{z}$	xyz	$\bar{x}y\bar{z}$	$\bar{x}\bar{y}z$	$x\bar{y}\bar{z}$	$\bar{x}y$	xy	$\bar{x}z$	xz	$\bar{y}z$	yz	z	z	y	y	x	x
T_d	E	C_2	C_2	C_2	C_3	C_3	C_3	C_3	C_3^2	C_3^2	C_3^2	C_3^2	σ	σ	σ	σ	σ	σ	S_4	S_4^3	S_4	S_4^3	S_4	S_4^3
O	E	C_2	C_2	C_2	C_3	C_3	C_3	C_3	C_3^2	C_3^2	C_3^2	C_3^2	C_2	C_2	C_2	C_2	C_2	C_2	C_4^3	C_4	C_4^3	C_4	C_4^3	C_4
A_1	1	1	1	1	1	1	1	1	1	1	1	1	1	1	1	1	1	1	1	1	1	1	1	1
A_2	1	1	1	1	1	1	1	1	1	1	1	1	−1	−1	−1	−1	−1	−1	−1	−1	−1	−1	−1	−1
$(E)_{11}$	1	1	1	1	c	c	c	c	c	c	c	c	1	1	c	c	c	c	1	1	c	c	c	c
$(E)_{21}$	0	0	0	0	s	s	s	s	$-s$	$-s$	$-s$	$-s$	0	0	s	s	$-s$	$-s$	0	0	s	s	$-s$	$-s$
$(E)_{12}$	0	0	0	0	$-s$	$-s$	$-s$	$-s$	s	s	s	s	0	0	s	s	$-s$	$-s$	0	0	s	s	$-s$	$-s$
$(E)_{22}$	1	1	1	1	c	c	c	c	c	c	c	c	−1	−1	$-c$	$-c$	$-c$	$-c$	−1	−1	$-c$	$-c$	$-c$	$-c$
$(T_1)_{11}$	1	1	−1	−1	0	0	0	0	0	0	0	0	0	0	0	0	−1	−1	0	0	0	0	1	1
$(T_1)_{21}$	0	0	0	0	1	−1	1	−1	0	0	0	0	−1	1	0	0	0	0	−1	1	0	0	0	0
$(T_1)_{31}$	0	0	0	0	0	0	0	0	1	−1	−1	1	0	0	−1	1	0	0	0	0	1	−1	0	0
$(T_1)_{12}$	0	0	0	0	0	0	0	0	1	1	−1	−1	−1	1	0	0	0	0	1	−1	0	0	0	0
$(T_1)_{22}$	1	−1	1	−1	0	0	0	0	0	0	0	0	0	0	−1	−1	0	0	0	0	1	1	0	0
$(T_1)_{32}$	0	0	0	0	1	−1	−1	1	0	0	0	0	0	0	0	0	−1	1	0	0	0	0	−1	1
$(T_1)_{13}$	0	0	0	0	1	1	−1	−1	0	0	0	0	0	0	−1	1	0	0	0	0	−1	1	0	0
$(T_1)_{23}$	0	0	0	0	0	0	0	0	1	−1	1	−1	0	0	0	0	−1	1	0	0	0	0	1	−1
$(T_1)_{33}$	1	−1	−1	1	0	0	0	0	0	0	0	0	−1	−1	0	0	0	0	1	1	0	0	0	0
$(T_2)_{11}$	1	1	−1	−1	0	0	0	0	0	0	0	0	0	0	0	0	1	1	0	0	0	0	−1	−1
$(T_2)_{21}$	0	0	0	0	1	−1	1	−1	0	0	0	0	1	−1	0	0	0	0	1	−1	0	0	0	0
$(T_2)_{31}$	0	0	0	0	0	0	0	0	1	−1	−1	1	0	0	1	−1	0	0	0	0	−1	1	0	0
$(T_2)_{12}$	0	0	0	0	0	0	0	0	1	1	−1	−1	1	−1	0	0	0	0	−1	1	0	0	0	0
$(T_2)_{22}$	1	−1	1	−1	0	0	0	0	0	0	0	0	0	0	1	1	0	0	0	0	−1	−1	0	0
$(T_2)_{32}$	0	0	0	0	1	−1	−1	1	0	0	0	0	0	0	0	0	1	−1	0	0	0	0	1	−1
$(T_2)_{13}$	0	0	0	0	1	1	−1	−1	0	0	0	0	0	0	1	−1	0	0	0	0	1	−1	0	0
$(T_2)_{23}$	0	0	0	0	0	0	0	0	1	−1	1	−1	0	0	0	0	1	−1	0	0	0	0	−1	1
$(T_2)_{33}$	1	−1	−1	1	0	0	0	0	0	0	0	0	1	1	0	0	0	0	−1	−1	0	0	0	0

$c = \cos(120°) = -\frac{1}{2}$ $\quad s = \sin(120°) = \frac{\sqrt{3}}{2}$.

The symbols above the symmetry operations indicate their orientation relative to the axes: $\bar{x} = x$, etc.

The Multiconfigurational (MC) Self-Consistent Field (SCF) Theory

by

Björn O. Roos
Department of Theoretical Chemistry
Chemical Centre
P.O.B. 124
S-221 00 Lund
Sweden

1. Introduction

The molecular orbital (MO) is the most fundamental quantity in contemporary quantum chemistry. Almost all of our understanding of "what the electrons are doing in molecules" is based on the molecular orbital concept. Also most of the computational methods used today start by a calculation of the MO´s of the system.

The molecular orbital describes the "motion" of one electron in the electric field generated by the nuclei and some average distribution of the other electron. It is in the simplest model occupied by zero, one, or two electrons. In the case of two electrons occupying the same orbital, the Pauli principle demands that they have opposite spin.

Such an approach leads to a total wave function for the system, which is an anti-symmetrized product of molecular spin orbitals (spin orbital = molecular orbital times a spin function). The Hartree-Fock method is obtained by applying the variation principle to the corresponding energy functional.

The success of the Hartree-Fock method in describing the electronic structure of most closed-shell molecules has made it natural to analyze the wave function in terms of the molecular orbitals. The concept is simple and has a close relation to experiment through Koopmans´ theorem. The two fundamental building blocks of Hartree-Fock (HF) theory are the molecular orbital and its occupation number. In closed-shell systems each occupied molecular orbital carries two electrons, with opposite spin. The occupied orbitals themselves are only defined as an occupied one-electron subspace of the full space spanned by the eigenfunctions of the Fock operator. Transformations between them leave the total HF wave function invariant. Normally the orbitals are obtained in a delocalized form as solutions to the canonical HF equations. This formulation is the most relevant one in studies of spectroscopic properties of the molecule, that is, excitation and ionization. The invariance property, however, makes a transformation to localized orbitals possible. Such localized orbitals can be valuable for an analysis of the chemical bonds in the system.

The Hartree-Fock method can be extended to open-shell systems in two ways. In the restricted Hartree-Fock (RHF) method the open-shell orbitals are added to the closed-shell orbitals and the resulting wave function is projected to have the correct spin symmetry. In some cases this procedure leads to a wave function comprising more than one Slater determinant. An alternative formulation is the unrestricted Hartree-Fock (UHF) method, where the space parts of the orbitals for a closed-shell electron pair are no longer assumed to be equal. As a consequence the method is capable of describing the spin polarization of paired electrons in the presence of unpaired spins. The method has for this reason been used extensively in studies of the spin polarization of radicals. In contrast to the closed-shell HF method the UHF method gives a qualitatively correct description of bond dissociation. The serious drawback of the UHF wave function is, that it cannot easily be projected to give a wave function which is an eigenfunction of the total spin. If such a projection is attempted prior to a variational treatment, very complex equations for the molecular orbitals are obtained. The simplicity of the Hartree-Fock method is lost and the gain in accuracy is not large enough to make such an approach useful in practical applications.

All variants of the Hartree-Fock method lead to a wave function in which all the information about the electron structure is contained in the occupied molecular orbitals (or spin orbitals) and their occupation numbers, the latter being equal to 1 or 2.

The concept of the molecular orbital is, however, not restricted to the Hartree-Fock model. Sets of orbitals can also be constructed for more complex wave functions, which include correlation effects. They can be used to obtain insight into the detailed features of the electron structure. One choice of orbitals are the natural orbitals, which are obtained by diagonalizing the spinless first-order reduced density matrix. The occupation numbers (η) of the natural orbitals are not restricted to 2, 1, or 0. Instead they fulfill the condition:

$$0 < \eta < 2 \tag{1:1}$$

If the Hartree-Fock determinant dominates the wave function, some of the occupation numbers will be close to 2. The corresponding MO´s are closely related to the canonical HF orbitals. The remaining natural orbitals have small occupation numbers. They can be analyzed in terms of different types of correlation effects in the molecules. A relation between the first-order density matrix and correlation effects is not immediately justified, however. Correlation effects are determined from the properties of the second-order reduced density matrix. The most important terms in the the second-order matrix can, however, be approximately determined from the occupation numbers of the natural orbitals. Electron correlation can be qualitatively understood using an independent electron-pair model. In such a model the correlation effects are treated for one pair of electrons at a time, and the problem is reduced to a set of two-electron systems. As was first shown by Löwdin and Shull, the two-electron wave function is determined from the occupation numbers of the natural orbitals. The second-order density matrix can then be specified by means of the natural orbitals and their occupation numbers. Consider as an example the following simple two-configurational wave function for a two-electron system:

$$\Phi = C_1(\varphi_1)^2 - C_2(\varphi_2)^2 \qquad (C_2 > 0)\ , \tag{1:2}$$

where $(\varphi_1)^2$ is the HF wave function. The MO´s φ_1 and φ_2 are the natural orbitals and have the occupation numbers $\eta_1 = 2C_1^2$ and $\eta_2 = 2C_2^2$. The diagonal of the second-order density matrix is with real orbitals obtained as,

$$\rho_2(1,2) = \tfrac{1}{2}\eta_1\varphi_1(1)^2\varphi_1(2)^2 + \tfrac{1}{2}\eta_2\varphi_2(1)^2\varphi_2(2)^2 - \sqrt{\eta_1\eta_2}\,\varphi_1(1)\varphi_2(1)\varphi_1(2)\varphi_2(2) \tag{1:3}$$

The last term is the one which describes the correlation of the electron pair by the orbital φ_2. If φ_1 is a bonding MO and φ_2 is the corresponding anti-bonding orbital, the last term in (1:3) ensures that there is one electron on each atom when the bond dissociates. If, as another example, φ_1 = 2s and φ_2 = 2p as in the Be atom, the last term in (1:3) decreases the probability of finding both electrons on the same side of the nucleus and increases the probability of finding them on opposite sides (angular correlation). This effect is very large in beryllium and accounts for 95% of the correlation energy in the valence shell.

For systems with more than one electron pair, the simple picture illustrated above obviously breaks down. The approximate validity of the independent electron-pair model, however, still makes it possible to estimate different correlation effects also in many-electron systems from an inspection of the natural orbital occupation numbers.

The correlation error is normally defined as the difference between the exact eigenvalue of the non-relativistic Hamiltonian for the molecule and the restricted HF energy. While this works well for closed-shell systems, it becomes less meaningful when degeneracies, or near-degeneracies, occur between different electronic configurations. For example, the RHF energy for the H_2 molecule at large internuclear distances is in error by more than 7 eV. Obviously the electrons of two non-interacting hydrogen atoms are not correlated. The reason for the error is, that the RHF model breaks down in cases where several configurations become degenerate or near-degenerate. This is the case in most bond dissociation processes, and along the reaction path for a symmetry-forbidden chemical reaction, just to mention two examples. Thus while the Hartree-Fock model is a valid approximation for most (but not all) molecules around their equilibrium geometry (in their ground electronic state), it cannot in general be used as a qualitatively correct model for energy surfaces.

We shall in these lectures describe methods for determining the molecular orbitals in cases where the wave function is not well described by the HF approximation. Before starting on the more formal development of these methods we shall illustrate the breakdown of the RHF model by a few examples taken from different bonding situations.

1.2 Exercise

1.1 Suppose that $\varphi_1(-\mathbf{r}) = \varphi_1(\mathbf{r})$, but $\varphi_2(-\mathbf{r}) = -\varphi_2(\mathbf{r})$ in (1:3). Show that in this case $\rho_2(\mathbf{r},-\mathbf{r}) > \rho_2(\mathbf{r},\mathbf{r})$.

2. Examples of multiconfigurational wave functions

2.1 The hydrogen molecule

Let us consider as a first example the dissociation of a single bond; the hydrogen molecule, H_2. The molecular orbitals for this molecule can always be written in the form,

$$\varphi_i = N_i(\chi_{iA} \pm \chi_{iB}), \tag{2:1}$$

where χ_{iA} and χ_{iB} are two functions located at the two nuclei, with the property

$$\hat{\imath}\chi_{iA} = \chi_{iB}, \tag{2:2}$$

where $\hat{\imath}$ is the inversion operator. N_i is a normalization constant. The ground state wave function for H_2 is at the equilibrium geometry dominated by the electronic configuration $(\varphi_1)^2$, where the molecular orbital φ_1 is doubly occupied. φ_1 is a bonding orbital built from two atomic orbitals χ_{1A} and χ_{1B} with their main contribution from the hydrogen 1s orbital: $\chi_{1A} = 1s_A$ + "small corrections".

The "small corrections" are important in a quantitative description of the H_2 chemical bond, but for the present qualitative discussion we can neglect them and write

$$\varphi_1 = N_1(1s_A + 1s_B) \tag{2:3}$$

This molecular orbital is in the RHF model assumed to be doubly occupied leading to a total wave function of the form:

$$\Phi_1 = \varphi_1(\mathbf{r}_1)\varphi_1(\mathbf{r}_2)\Theta_{2,0} \tag{2:4}$$

which we shall simply denote as $(\varphi_1)^2$. $\Theta_{2,0}$ is the singlet (S = 0) spin function for two electrons:

$$\Theta_{2,0} = \sqrt{\tfrac{1}{2}}(\alpha_1\beta_2 - \beta_1\alpha_2) \tag{2:5}$$

The RHF model leads to a reasonably accurate description of H_2 around the equilibrium geometry: computed bond distance is 0.735 Å (exp 0.746 Å) and the bond energy is 84 kcal/mol (exp 109 kcal/mol). It is typical for the RHF model that it is able to describe closed shell systems around their equilibrium geometry rather well. The correlation energy is only a small fraction of the total energy., but it is strongly distance dependent, which explains the error in

the computed bond energy (there is no correlation energy at all for two separated hydrogen atoms).
The bond energy given above has been obtained by subtracting from the RHF energy for H_2 the energy of two separated hydrogen atoms (-627 kcal/mol). Suppose instead that we would use the RHF model to compute the potential curve for the dissociation of H_2. The first thing to notice is that the form of the molecular orbital (2:3) is independent of the internuclear distance. The same form of the wave function (2:4) is thus obtained also for the separated atoms. Let us expand this wave function as products of the atomic orbitals $1s_A$ and $1s_B$:

$$\Phi_1 = N_1^2[1s_A(\mathbf{r}_1)1s_A(\mathbf{r}_2) + 1s_A(\mathbf{r}_1)1s_B(\mathbf{r}_2) + 1s_B(\mathbf{r}_1)1s_A(\mathbf{r}_2) + 1s_B(\mathbf{r}_1)1s_B(\mathbf{r}_2)]\Theta_{2,0}$$

We notice that this wave function contains terms where both electrons are located at the same atom. These terms are clearly unphysical at large separations, since they correspond to the dissociation to $H^+ + H^-$, which has an energy around 320 kcal/mol above H + H. It is only the middle terms in the wave function that describe correctly the dissociated products.

It is a typical feature of the RHF model to include these "ionic structures" in fixed proportions into the wave function. Consequently the model cannot be used to describe dissociation processes resulting in products with open shells. The potential curve corresponding to the wave function Φ_1 will actually end up with an energy around 160 kcal/mol above the true energy at the separated limit.

Is there a remedy to this problem? Well, the most straightforward solution is maybe to introduce coefficients in front of the different terms in Φ_1 and write

$$\Psi_{VB}=C_{Ion}\Phi_{Ion}+C_{Cov}\Phi_{Cov}, \qquad (2:6)$$

where

$$\Phi_{Ion} = N_{Ion}\{1s_A(\mathbf{r}_1)1s_A(\mathbf{r}_2) + 1s_B(\mathbf{r}_1)1s_B(\mathbf{r}_2)\}\Theta_{2,0}$$

and

$$\Phi_{Cov} = N_{Cov}\{1s_A(\mathbf{r}_1)1s_B(\mathbf{r}_2) + 1s_B(\mathbf{r}_1)1s_A(\mathbf{r}_2)\}\Theta_{2,0}$$

The coefficients C_{Ion} and C_{Cov} in (2:6) can be varied to yield the correct wave function at the separated limit ($C_{Ion} = 0$). At equilibrium $C_{Ion} \approx C_{Cov}$, which reflects the fact that the RHF determinant dominates the wave function. The formulation (2:6) forms the basis for the valence bond description of the chemical bond. In its description of the molecular wave function in terms of atomic structures it has an appealing chemical appearance. On the other hand, the formulation in terms of non-orthogonal (atomic) basis functions (orbitals) leads to a complicated mathematical structure, which sofar has prevented large scale applications in chemistry. Instead we shall look for a formulation in terms of orthogonal molecular orbitals. We introduce in addition to φ_1, (2:3), the anti-bonding orbital

$$\varphi_2=N_2(1s_A-1s_B). \qquad (2:7)$$

In terms of φ_1 and φ_2 we can now write the wave function (2:6) as

$$\Psi_{MC}=C_1\Phi_1+C_2\Phi_2, \qquad (2:8)$$

where Φ_2 is the electronic configuration $(\varphi_2)^2$. This is the multiconfigurational molecular orbital formulation of the wave function for the chemical bond in H_2. It will correctly describe the entire potential curve. Close to equilibrium $C_1\sim 1$ and $C_2 \sim 0$, while at large separations $C_1 \sim - C_2$. *The quantum chemical description of a chemical bond thus involves both the bonding and the anti-bonding orbital.*

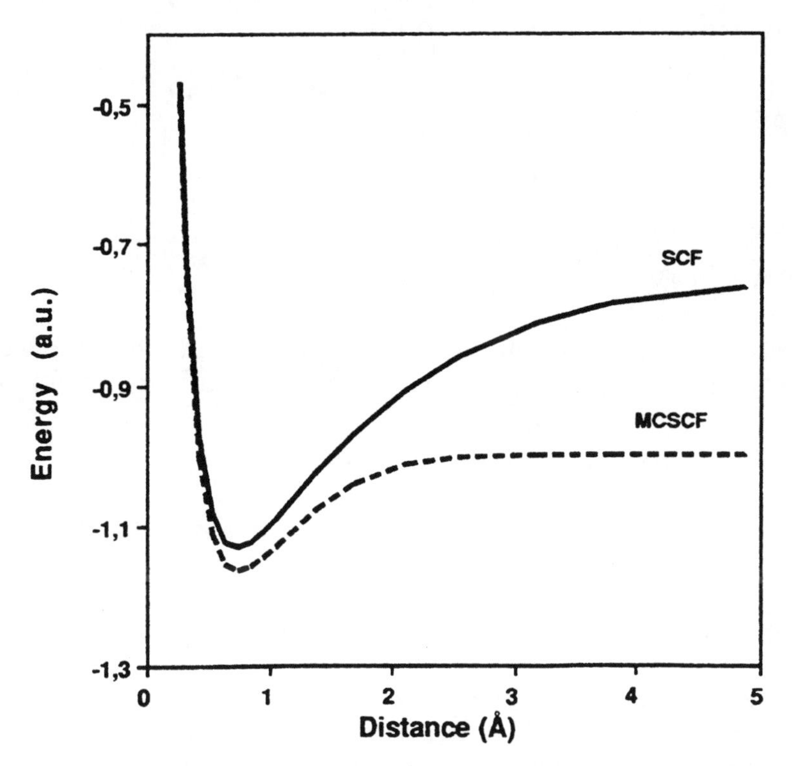

Figure 2.1 Potential curves for H_2 showing the erratic behaviour of the SCF curve as a function of the internuclear distance. For comparison the MCSCF curve which dissociates correctly is also given.

Another way of viewing the multiconfigurational wave function (2:8) is to note that the two configurations Φ_1 and Φ_2 are degenerate at infinite separation. Since the interaction between them is different from zero, strong mixing will occur with $C_1 = \pm C_2$. It is clear that the RHF model will not work in cases where more than one electronic configuration has the same, or nearly the same, energy. Below we will show several examples of such situations, where near degeneracy occur between different electronic configurations. A multiconfigurational treatment is then needed in order to obtain a qualitatively correct description of the electronic structure.

2.2 Multiple bonds

The dissociation of a single bond can be described by a wave function composed of two electronic configurations. A more complex situation occurs in processes involving the simultaneous dissociation of several bonds. As an example consider the triple bond in the nitrogen molecule N_2. The valence bond description at large separation arises from the coupling of two quartet nitrogen atoms into an overall singlet. The corresponding MC wave function is obtained by a transformation from the 2p AO basis to the molecular orbitals

$$
\begin{aligned}
&3\sigma_g \propto (2p_{Az} - 2p_{Bz})\\
&3\sigma_u \propto (2p_{Az} + 2p_{Bz})\\
&1\pi_{ux} \propto (2p_{Ax} + 2p_{Bx})\\
&1\pi_{gx} \propto (2p_{Ax} - 2p_{Bx})\\
&1\pi_{uy} \propto (2p_{Ay} + 2p_{By})\\
&1\pi_{gy} \propto (2p_{Ay} - 2p_{By})
\end{aligned}
\qquad (2:9)
$$

It is clear that the MC function will contain electronic configurations which are up to sextuply excited with respect to the basic RHF configuration for N_2:

$$\Phi_{RHF}=(1\sigma_g)^2(1\sigma_u)^2(2\sigma_g)^2(2\sigma_u)^2(3\sigma_g)^2(1\pi_u)^4 \qquad (2:10)$$

We shall not perform the somewhat elaborous calculation of the MC wave function in detail. A somewhat simpler example is the dissociation of a double bond and it is given as an exercise (exercise 2). Here we only note that the number of configuration state functions (CSF's) will increase very quickly with the number of 'active' orbitals. In most cases we do not have to worry about the exact construction of the MC wave function that leads to correct dissociation. We simply use all CSF's that can be constructed by distributing the electrons among the active orbitals. This is the idea behind the *Complete Active Space SCF (CASSCF) method.* The total number of such CSF's is for N_2 175 for a singlet wave function. A further reduction is obtained by imposing spatial symmetry. All these CSF's are not included in a wave function constructed from the valence bond structure for two ground state nitrogen atoms. A few terms are missing, since they correspond to different spin couplings. Experience shows, however, that spin recouplings easily occur when a bond is formed, and it may therefore be dangerous to exclude these terms from the wave function, even if they are not needed for a correct description of the asymptotic wave function. The general conclusion is then that the MC wave function should comprise all the possible CSF's as is done in the CASSCF method.

2.3 Molecules with competing valence structures

In the examples above we considered the degeneracy effects which arises at the dissociation limit for a covalent chemical bond. Near degeneracy can also occur, for example, when one or more virtual orbitals have low energies. The beryllium atom was mentioned in the introduction as an example. Similar situations occur for some molecules in their ground state close to the equilibrium geometry. The RHF wave function for the ground state of a molecule can often be inferred from the chemical structure formula: each chemical bond and each lone-pair corresponds to a doubly occupied (localized) orbital. The full closed shell electronic configuration is obtained by adding orbitals for the core electrons. When the molecule possesses symmetry it is usually straightforward to transform the occupied MO's to symmetry adapted form.

However, there are systems for which it is not clear how to write down the structure formula. Such situations occur when all valences cannot be filled in a natural way. Several valence structures are then possible. It is not unexpected to find that the RHF model then also runs into similar problems. Consider as an example the ozone molecule O_3. The most important valence structures for this molecule are:

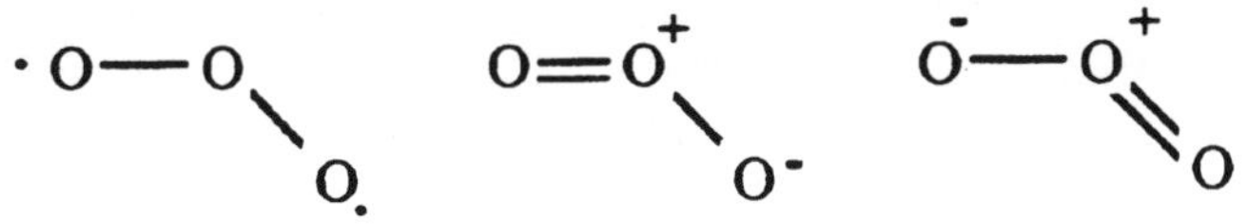

The two last of these structures have an oxygen-oxygen double bond, at the expense of moving one electron to the end atom. The first structure corresponds to a di-radical where two of the π electrons are unbound. The near degeneracy between these valence structures can in MO theory be traced to the near degeneracy between the two upper π molecular orbitals; π_2 and π_3 (the molecule is labelled $O_BO_AO_C$):

$$\pi_1 = c_{11}\pi_A + c_{12}(\pi_B + \pi_C)$$
$$\pi_2 = N_2(\pi_B - \pi_C) \qquad (2:11)$$
$$\pi_3 = c_{31}\pi_A + c_{32}(\pi_B + \pi_C)$$

The RHF electronic configuration is $(\pi_1)^2(\pi_2)^2$, but an MCSCF calculation reveals that the configuration $(\pi_1)^2(\pi_3)^2$ also has a large weight in the wave function:

$$\Phi \approx 0.89(\pi_1)^2(\pi_2)^2 - 0.45(\pi_1)^2(\pi_3)^2.$$

An analysis in terms of VB structures (see exercise 3) shows that this configurational mixing corresponds to approximately 40% diradical character in the wave function for ozone. The RHF wave function, on the other hand, contains only 12% of the diradical VB structure (the result was obtained using Hückel values for the coefficients of the orbitals (2:11)). It is clear from these considerations that a correct treatment of the electronic structure for the ozone molecule must be based on a multiconfigurational wave function.

A similar case obtains for the NO_2 radical. The important feature of the valence structures in this system is a delocalization of the unpaired electron to all three atoms in the system. NO_2 thus contains valence structures where the nitrogen lone-pair is doubly occupied. This has consequences, for example, for the formation of the bond between two NO_2 moieties. Early quantum chemical studies of N_2O_4 consequently gave an NN distance which was 0.1 - 0.2 Å too short. The reason was that the wave functions used did not take into account the delocalization of the odd electron properly. An MCSCF wave function that includes configurations describing the redistribution of the lone-pair electrons, results in a bond distance in good agreement with experiment (around 0.02 Å too long). It is obvious that many systems will exhibit these types of near-degeneracies. For this reason it is important to understand for each system which VB structures are important, or in MO language, which active orbitals are important.

2.4 Transition states on energy surfaces

The calculation of energy surfaces for chemical reactions is an important challenge for quantum chemistry. Such surfaces often exhibit a saddle point somewhere between the minimum energy structure of the reactants and that of the products. The barrier is almost always due to a change of the dominating electronic configuration in the wave function, as shown in the figure below:

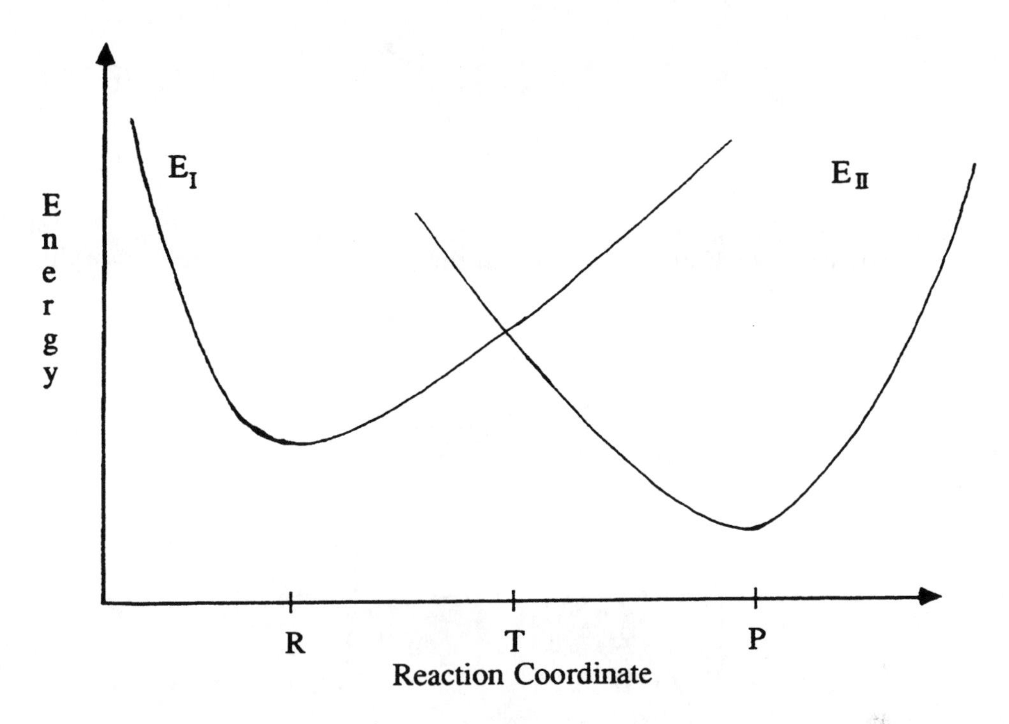

Here E_I is the energy of the electronic configuration for the reactants (point R on the energy surface). This energy has a minimum around R and is repulsive outside this region. E_{II} is the corresponding energy curve for the electronic configuration of the products. A shift of dominating electronic configuration will take place, when the system passes through the saddle point region. At one point the two energies become degenerate and the total wave function is here approximately given as

$$\Psi = \Phi_I - \Phi_{II} \tag{2:12}$$

It is clear that we need a multiconfigurational wave function in order to describe this process properly.

As an explicit illustration let us study the formation of cyclobutane from two ethene molecules:

$$C_2H_4 + C_2H_4 \rightarrow C_4H_8. \tag{2:13}$$

We will assume that the reaction proceeds along a reaction path with D_{2h} symmetry, where the four carbon atoms are rectangularly oriented:

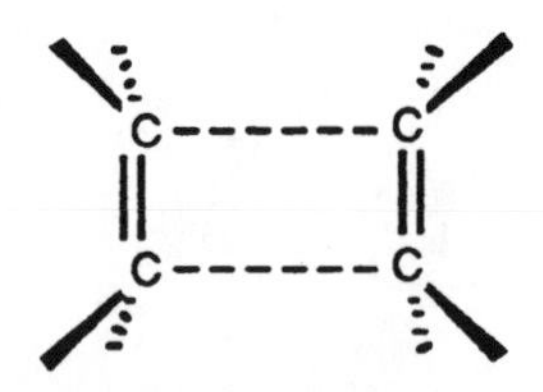

The molecular orbitals directly involved in the reaction are the four π orbitals of the two ethene molecules:

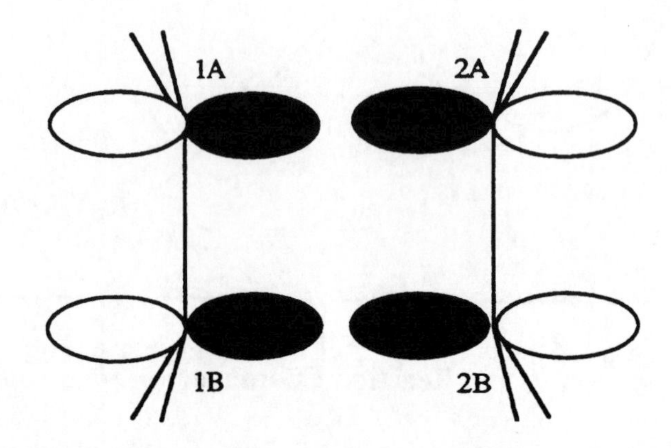

At large distances R they will form two pairs of MO's: π_1, π_1^* and π_2, π_2^*. The electronic configuration is then:

$$\Psi_{R=\infty} = \ldots (\pi_1)^2 (\pi_2)^2 \tag{2:14}$$

or using the full D_{2h} symmetry:

$$\Psi_{R=\infty} = \ldots (\pi_+)^2 (\pi_-)^2 \tag{2:15}$$

where

$$\pi_+ = \frac{1}{\sqrt{2}} (\pi_1 + \pi_2) = \frac{1}{2}(\pi_{1A} + \pi_{1B} + \pi_{2A} + \pi_{2B})$$
$$\pi_- = \frac{1}{\sqrt{2}} (\pi_1 - \pi_2) = \frac{1}{2}(\pi_{1A} + \pi_{1B} - \pi_{2A} - \pi_{2B})$$

For simplicity we have assumed that all orbitals are orthogonal. Note that (2:12) and (2:13) are identical wave functions. At small values of R the π bonds between atoms A and B are broken and replaced with sigma bonds between atoms on the moieties 1 and 2:

$$\sigma_A \propto (\pi_{1A} + \pi_{2A})$$
$$\sigma_B \propto (\pi_{1B} + \pi_{2B}).$$

The electronic configuration is now:

$$\Psi_{R \approx R_e} = ...(\sigma_A)^2(\sigma_B)^2 \tag{2:16}$$

or using symmetry,

$$\Psi_{R \approx R_e} = ...(\sigma_+)^2(\sigma_-)^2 \tag{2:17}$$

where

$$\sigma_+ = \frac{1}{\sqrt{2}}(\sigma_A + \sigma_B) = \frac{1}{2}(\pi_{1A} + \pi_{2A} + \pi_{1B} + \pi_{2B}) = \pi_+$$

$$\sigma_- = \frac{1}{\sqrt{2}}(\sigma_A - \sigma_B) = \frac{1}{2}(\pi_{1A} + \pi_{2A} - \pi_{1B} - \pi_{2B}) = \pi_-^*$$

The electronic configuration is thus different. We can illustrate this by a correlation diagram for the orbitals:

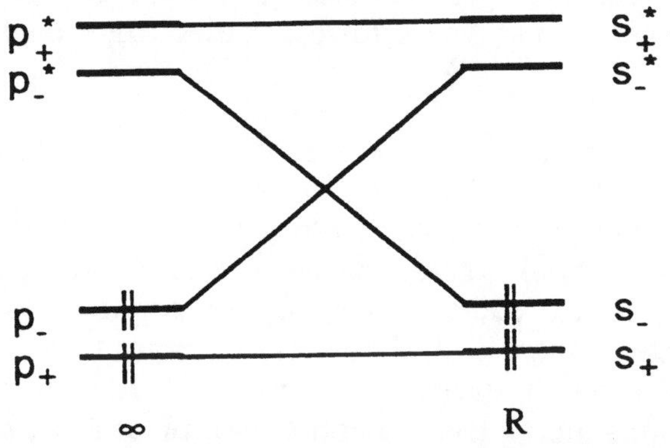

As this correlation diagram illustrates, the configuration (2:12) will become repulsive when the two ethene molecules approach each other, while (2:14) increases in energy in the reverse direction. We are thus led to the situation illustrated in the figure on page 11, and the reaction is expected to proceed via an energy barrier. This is an example of the so called Woodward-Hoffman rules, which are used to predict the existence of energy barriers for concerted reactions in organic chemistry.

Clearly a quantum chemical calculation of the energy surface for this reaction would have to be based on a multiconfigurational wave function, with four active orbitals, the π orbitals of the two ethene molecules, and four active electrons. However, a complication appears: cyclobutane is a quadratic molecule with all four carbon-carbon bonds equal. Our wave function does not have this property. The CC bonds between atoms A and B are treated using

electrons. However, a complication appears: cyclobutane is a quadratic
molecule with all four carbon-carbon bonds equal. Our wave function does no
have this property. The CC bonds between atoms A and B are treated using
closed shells, while those between 1 and 2 include an occupation of the anti-
bonding orbitals. As a consequence this wave function will not yield an energy
minimum at the quadratic geometry. The remedy is of course to include in
addition the σ bonds in the active orbital space, leading to a wave functior
obtained by distributing eight electrons among eight orbitals. Such
complications rather often occur, when choosing the active orbitals for ar
MCSCF calculation.

2.5 Other cases of near-degeneracy effects

The examples of multiconfigurational mixing discussed above refer to
situations where one or more chemical bonds are broken, either completely in
a dissociation process, or partly at a transition state on an energy surface for a
chemical reaction. In the ozone case too the mixing was due to a partia
breaking of the OO π bond in favour of a diradical state. The complete
dissociation process leads to the formation of open shell configurations, while
in concerted chemical reactions one closed shell is broken and transformed
into another closed shell. In both these cases multiconfigurational mixing
occurs as a result of decoupling paired electrons. Near-degeneracy effects can
occur also in other situations, where different electronic configurations have
similar energies.

It is, for example, not unusual to find strong configurational mixing in excited
states of molecular systems. Different configurations, singly excited with
respect to a RHF ground state, are often close in energy. When they are of th
same symmetry, even a small interaction term leads to strong mixing. Typica
examples are the, so called, alternant hydrocarbons. These are plana
unsaturated molecules, where if each second carbon atom is labelled with
star, no starred atom has a starred neighbor, and vice versa. Examples o
alternant hydrocarbons are benzene, naphtalene, butadiene, etc. A simple π
electron treatment, for example the Pariser-Parr-Pople method, yields fo
these molecules singly excited configurations that occur in pairs with equa
energy. A two by two configuration interaction treatment then gives two state
(the plus (+) and minus (-) states), which contain the two degenerat
configurations with equal weight. All of the transition intensity goes into th
plus state, while the transition moment for the minus state is zero in th
simplified model (see exercise 6 for details). These properties of some of th
excited states in alternant hydrocarbons remain approximately valid in mor
accurate models.

Another important case of near-degeneracy occurs in compounds containin
transition metal atoms. There are two reasons for strong configuration
mixing in such molecules. First, the chemical bonds are often weak, leading t
a substantial occupation of the anti-bonding orbital. The second reason

related to near-degeneracies in the transition metal atom. The electronic configurations $d^n s^2$, $d^{n+1}s$, and d^{n+2} often contain multiplets with very similar energies. An example is the nickel atom, where the states d^8s^2, ^3F and d^9s, ^3D only differ in energy by 0.6 kcal/mol. The d^9s, ^1D state is located 9.7 kcal/mol above the ^3F ground state, while the energy difference to d^{10}, ^1S is 42.1 kcal/mol. These closely spaced atomic energy levels will influence the chemical bond in nickel compounds considerably, leading to strong mixing of different atomic configurations. It should be emphasized that a correct calculation of the atomic splittings is very difficult and has to include extensive dynamic correlation (see below) and relativistic effects. These features carries over to some extent to the molecular case, and studies of the chemical bond involving a transition metal atom are complicated. The inclusion of the near-degeneracy effects through an MCSCF calculation is only the first step.

2.6 Closing remarks

We have in this chapter discussed different situations, where the simple single-configurational Hartree-Fock description of the molecular system breaks down. It has been shown that a model can be devised that gives a qualitatively correct wave function in cases where several electronic configurations are close in energy. In the next chapter we shall develop the tools needed to compute such wave functions.

The discussion above has focused on the configurational mixing occurring in near-degenerate systems. One might argue that to account for these effects it is enough to perform a configuration interaction calculation, using a predetermined set of molecular orbitals, obtained for example from an RHF calculation. Thus there should be no need for a multiconfigurational SCF theory, which includes orbital optimization. This is, however, a dangerous approach. It is clear that the molecular orbitals will also be strongly affected. If for example the MO's for the hydrogen molecule were determined from an SCF calculation, one would not obtain atomic hydrogen orbitals at large internuclear separation. The RHF determinant contains, as was shown in the beginning of this chapter, a large ionic component. Thus the orbitals will be intermediate between the orbitals for a hydrogen atom and those of a hydrogen negative ion. Similar errors would occur in more complex systems. For example, multiconfigurational effects often modifies the polarity in a chemical bond, leading to considerable modifications of the MO's involved. In many cases strong correlation effects make the electron density more compact. In general it is necessary to optimize the orbitals using a qualitatively correct wave function. Thus the need for an MC*SCF* model.

Correlation effects in molecules are normally partitioned into near-degeneracy effects (static correlation) and dynamic correlation. Qualitatively they differ in the way they separate the electrons. Static correlation leads to a large separation in space of the two electrons in a pair, for example on two different atoms in a dissociation process. Dynamic correlation on the other hand deals with the interaction between two electrons at short inter-electronic distance, the so called cusp region. It should be emphasized that MCSCF methods deal primarily with the near-degeneracy effects. Other methods are used to treat dynamical correlation. These includes large scale configuration interaction methods, coupled cluster methods, and the use of perturbation theory. Such methods will be treated in other parts of the present course. There is often no need for orbital optimization in calculations of dynamical correlation effects, since the electron density is only weakly affected. This is, however, a rule with several exceptions. One concerns the electron structure of transition metal compounds, where dynamic correlation in the 3d shell can lead to a substantial modification of the electronic structure, for example by changing the polarity in chemical bonds. Another example relates to the calculation of the electron affinity of molecules. It may here happen that the extra electron becomes bound only when large portions of the dynamic correlation energy is included in the calculation. It follows that the MO's have to be determined on that level of approximation. A similar situation obtains for some excited states in molecules, especially those which contain substantial contributions from ionic VB structures in the wave function. A typical example is the V state of the ethene molecule, which in the single determinant approximation is completely ionic. In these cases much too diffuse orbitals are obtained if dynamical correlation effects are not included in the wave function.

The partitioning of the correlation energy into static (near-degenerate) and dynamic is in most cases not obvious. Consider, for example, again the H_2 molecule. At large internuclear separations it is clear that the two configurations Φ_1 and Φ_2 in (2:8) are nearly degenerate. But at shorter distances there are other configurations that have similar weights as Φ_2 in the wave function (for example $(1\pi)^2$), and the effect of Φ_2 is here better treated as dynamical (left-right) correlation. In practice this means that we have to define a configuration as near degenerate if it has a large weight somewhere on the section of the energy surface under study. If the MCSCF calculation is used only as a starting point for more elaborate calculations including dynamical correlation effects, this uncertainty is normally not a problem. However when only MCSCF calculations are performed serious balance problems may occur, where different portions of the dynamical correlation energy is included on different parts of an energy surface.

Normally full valence MCSCF calculations (choosing all valence orbitals as active) represent a balanced treatment of correlation. However this is not always the case, especially not in systems containing lone-pair electrons. For example, a full valence MCSCF calculation for the water molecule yields less accurate values for the bond distance, the bond angle, and the dipole and quadrupole moment than an SCF calculation. The reason is that there are only two orbitals available for correlating the eight valence electrons (the $4a_1$ and the $2b_2$ orbitals). Thus correlation is only introduced into the lone-pair orbital

$3a_1$ and the OH bonding orbital $1b_2$. The orbitals $2a_1$ and $1b_1$ are left uncorrelated, leading to an unbalanced treatment of the correlation of the OH bonds. In fact, if symmetry was relaxed the molecule would in this approximation become unsymmetrical with two unequal OH bond lengths. The remedy to this problem is to introduce to more active orbitals: $5a_1$ and $2b_1$. It turns out that the occupation numbers for these two orbitals (which do not belong to the valence shell!) is not smaller than those of the correlating orbitals within the valence shell. In addition, much improved results for the properties are obtained. We shall return to the problem of choosing the active orbital space for MCSCF calculations later in these lectures.

Finally one notices that the definition of the correlation energy becomes less clear-cut when near-degeneracy effects appear in the wave function. Conventionally the correlation energy has been defined as the difference between the (restricted) Hartree-Fock energy and the exact eigenvalue of the non-relativistic Hamiltonian. When the RHF energy looses its meaning, so does this definition. The correlation energy for H_2 at large internuclear distance is not 160 kcal/mol (see page 6), it is zero. It is not enough to use the energy obtained as the sum of the RHF energies for the two dissociation products instead of the energy of the super-molecule, since we need a definition which is valid for the whole potential curve. To use instead the unrestricted Hartree-Fock method as a basis for the definition of correlation energy is not satisfactory. This method does give the right dissociation products, but fails to give a qualitatively correct description of the potential curve in the region intermediate betwen equilibrium and dissociation. The UHF wave function is, for example, not an eigenfunction of the total spin.
One possibility would be to base the definition on a full valence MCSCF calculation, instead of Hartree-Fock. A full valence MCSCF calculation will always lead to the correct dissociation products. The difference between the corresponding energy and the exact energy could then be used as a definition of the dynamical correlation energy. There are two difficulties with this approach: first it is possible to perform full valence CI calculations only for small systems, containing less than four atoms. Secondly, since there is no clear-cut partitioning of dynamical and static correlation effects, the definition partly looses its meaning. Obviously, we are here trying to fight a battle with windmills. It is better not to. It is important to understand the effects of electron correlation in different bonding situations and to be able to calculate these effects, when needed. It is possible to manage without a formal definition of the correlation energy.

2.7 Exercises

2.1 Show that the wave functions Ψ_{VB}, (2:6), and Ψ_{MC}, (2:8), are identical, for example, by performing explicitly the transformation from AO to MO basis. Find the relation between the coefficients (C_{Ion}, C_{Cov}) and (C_1, C_2). Show also that $C_1 = - C_2$ at infinite separation of the two hydrogen atoms.

2.2 Which are the electronic configurations needed to describe the dissociation of the ethene molecule into two triplet methylene radicals. Start by constructing the VB function for two triplets coupled to an overall singlet function. Then transform to molecular orbitals. The dissociation is assumed to proceed by stretching the CC bond, keeping D_{2h} symmetry. What is the weight of the Hartree-Fock configuration at infinite separation?

2.3 Write down the electronic configurations for the ground states of the molecules: water, carbon dioxide, formaldehyde, ethene, benzene, and the nitrogen dioxide radical. The occupied MO's shall be given using the full point group symmetry of the molecule.

2.4 Consider the three atomic π orbitals in the ozone molecule: π_A, π_B, π_C, where A is the central oxygen atom. Set up the VB wave functions corresponding to the three valence structures:

·O—O—O· (I) O=O⁺—O⁻ (II) ⁻O—O⁺=O (III)

I II III

Then make a transformation to an MO basis using the Hückel orbitals:

$$\pi_1 = \sqrt{\tfrac{1}{2}}\pi_A + \tfrac{1}{2}(\pi_B + \pi_C)$$

$$\pi_2 = \sqrt{\tfrac{1}{2}}(\pi_B - \pi_C)$$

$$\pi_3 = \sqrt{\tfrac{1}{2}}\pi_A - \tfrac{1}{2}(\pi_B + \pi_C)$$

Determine the MC wave function, which corresponds to the pure diradical valence structure I. Compare with the optimized MCSCF wave function:

$$\Phi = 0.89(\pi_1)^2(\pi_2)^2 - 0.45(\pi_1)^2(\pi_3)^2,$$

What are the weights of the three VB structures in this wave function.

2.5 The ring closure of cis-butadiene to form cyclobutene can take place either in a con-rotatory or in a dis-rotatory way. In the first place the CH_2 planes of the end groups rotate in the same direction keeping a C_2 axis of symmetry:

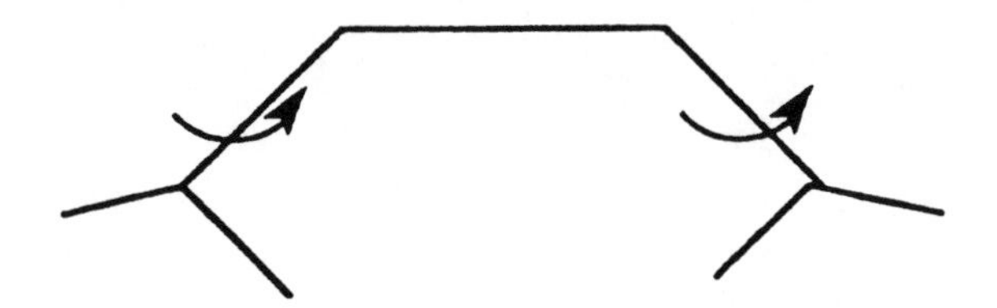

while in the second case the rotations are in opposite directions, now keeping a mirror plane of symmetry. Study the electronic configurations in the ground state of the reactant and the product. Use symmetry arguments to predict which reaction path is most favourable (leading to the lowest barrier). This reaction is another application of the Woodward-Hoffman rules.

2.6 In simple π-electron theory the *alternant hydrocarbons* have some special features. In these planar unsaturated hydrocarbons each second carbon atom is labelled with a star (*), resulting in a division of the atoms into two sets, the *starred* and the *unstarred*, with no two atoms of the same set neighbors. One feature is the so called Coulson-Rushbrooke theorem, or the pairing theorem: the bonding (occupied) π-orbitals are given in the form,

$$\varphi_i = \sum_{p(*)} c_{ip}\pi_p + \sum_{p()} c_{ip}\pi_p$$

and the corresponding anti-bonding (virtual) orbitals are obtained as,

$$\varphi_i' = \sum_{p(*)} c_{ip}\pi_p - \sum_{p()} c_{ip}\pi_p$$

Here p(*) and p( ) means summation over starred and unstarrred atoms respectively. π_p are the atomic π-orbitals. As a consequence of the above relations the orbital energies are paired such that $\varepsilon_i + \varepsilon_{i'} = 0$.
Use the zero differential overlap (ZDO) approximation for the two-electron integrals to show that the singly excited configurations $\Phi_{i \to j'}$ and $\Phi_{j \to i'}$ have the same energies. Show also that the transition dipoles are equal. When they are allowed to interact two states are formed:

$$\Phi^{+}_{i\to j} = \sqrt{\frac{1}{2}}(\Phi_{i\to j'} + \Phi_{j\to i'})$$

$$\Phi^{-}_{i\to j} = \sqrt{\frac{1}{2}}(\Phi_{i\to j'} - \Phi_{j\to i'})$$

Which of these two states has the lowest energy and what is the transition intensity to the two states. These simple properties of excited states of alternant hydrocarbons remain approximately valid in more accurate theories, at least for the lower excited states.

3. The MCSCF wave function and energy expression

3.1 Introduction

In the preceding chapter we showed a number of electronic structure problems, where a proper wave function had to be constructed as a linear combination of several electronic configurations. In this chapter we will discuss the technical aspects in constructing these MCSCF wave functions and in determining the variational parameters - the CI coefficients and the molecular orbitals.

It may be appropriate to start by reminding you about the method normally used to determine the molecular orbitals in Hartree-Fock theory. The HF operator is dependent on the molecular orbitals themselves via the two-electron repulsion term, which contains the one-electron density matrix. Iterative methods therefore have to be used in solving these equations: A guess is made of the density matrix. The Fock matrix is then constructed and diagonalized. The MO's obtained are used to construct a new, and hopefully improved density. The process is iterated until the change in the density matrix is below a certain preset threshold. The solution is then said to be *self - consistent*. The iterative process converges normally very slowly, or not at all. Interpolation and extrapolation methods have therefore been devised to improve convergence, and most SCF programs used today are very efficient in this respect.

Similar iterative schemes were used to determine the MO's for multiconfigurational wave functions, in the early implementations. Fock-like operators were constructed and diagonalized iteratively. The convergence problems with these methods are, however, even more severe in the MCSCF case, and modern methods are not based on this approach. The electronic energy is instead considered to be a function of the variational parameters of the wave function - the CI coefficients and the molecular orbital coefficients. Second order (or approximate second order) iterative methods are then used to find a stationary point on the energy surface.

The MCSCF optimization process is only the last step in the computational procedure that leads to the MCSCF wave function. Normally the calculation starts with the selection of an atomic orbital (AO) basis set, in which the molecular orbitals are expanded. The first computational step is then to calculate and save the one- and two-electron integrals. These integrals are commonly processed in different ways. Most MCSCF programs use a super-matrix (as defined in the closed shell HF operator) in order to simplify the evaluation of the energy and different matrix elements. The second step is then the construction of this super-matrix from the list of two-electron integrals. The MCSCF optimization procedure includes a step, where these AO integrals are transformed to MO basis. This transformation is most effectively performed with a symmetry blocked and ordered list of AO integrals. Step three in the calculation is then the construction of such an ordered list.This step is needed since most integral codes produce integrals in some other order. Finally we need, before the actual MCSCF calculation can start, also a set of trial molecular orbitals. They can be obtained in a number of different ways. The most straightforward procedure is to perform an SCF calculation on the system under study, or a closely related system. In most cases these orbitals are good enough to ensure convergence in the MCSCF optimization process. If a series of calculations are performed on an energy surface, one can use orbitals obtained from a nearby point on the surface as starting orbitals. More advanced methods can, if necessary, be used, as for example perturbation theory, or CI calculations, where the natural orbitals are used as input for the MCSCF calculation. In the discussion of the MCSCF optimization procedures, we shall assume that we have available a list of ordered and symmetry-blocked one- and two-electron integrals, a list of super-matrix elements, and a set of trial molecular orbitals.

3.2 Annihilation and creation operators

Most formulations of MCSCF theory are based on the second quantization formalism. We therefore review briefly in this section the basic definitions of the annihilation and creation operators, and the expansion of quantum mechanical operators in products of them.

Suppose that we have at our disposal an ordered set of orthonormalized spin-orbitals:

$$\{\phi_i;\ i = 1,2,....,2n\}, \tag{3:1}$$

where each spin-orbital is constructed as a product of a space orbital and a spin function. The number of space orbitals is n. When we use the phrase "molecular orbital" in the text below, we normally refer to the space part of the spin-orbitals. Using the spin-orbitals we can construct an N-electron basis consisting of B(2n,N) Slater determinants, where the binomial coefficient is defined in the usual way as B(i,j)=i!/(j!(i-j)!). Let us write these determinants in the occupation number representation as a series of 1 and 0 indicating which spin-orbitals are occupied in the determinant, for example:

$$|1,1,0,0,0,1,0,1,1\rangle,$$

showing that spin-orbitals 1,2,6,8, and 9 are occupied. In general we write the representation of a determinant as

$$|\, m_1,m_2,m_3,........m_{2n}\rangle, \tag{3:2}$$

where m_i is the occupation number (0 or 1) for spin-orbital i, and we have assumed that the spin orbitals are ordered in some predetermined way. The antisymmetry of the Slater determinant implies a change of sign of the ket (3:2), when two occupied spin-orbitals are transposed as a result of a reordering of the spin orbitals. The occupation number representation of the Slater determinants defines an orthonormal vector space, the *Fock space,* comprising determinants with all possible occupation numbers, including the vacuum state, where all occupations are zero:

$$|vac\rangle = |0_1, 0_2, ...0_{2n}\rangle$$

We now define an annihilation operator, $\hat{a}_i$, acting on the elements of the Fock space, such that it removes an electron from spin-orbital i to give a ket with N-1 electrons occupied:

$$\hat{a}_i|m_1,m_2,...m_i,...m_{2n}\rangle = m_i(-1)^{p_i}|m_1,m_2,...0_i,...m_{2n}\rangle \qquad (3:3)$$

where p_i is the number of transpositions needed to move orbital i to the first position in the ket. Note that the effect of the annihilation is zero, when the orbital i is not occupied. We define in addition the corresponding creation operator, $\hat{a}_i^\dagger$, which adds an electron in orbital i:

$$\hat{a}_i^\dagger|m_1,m_2,...m_i,...m_{2n}\rangle = (1-m_i)(-1)^{p_i}|m_1,m_2,...1_i,...m_{2n}\rangle \qquad (3:4)$$

It may be shown that the two operators are the adjoint of each other. The annihilation and creation operators fulfill the following anti-commutator relations:

$$\begin{aligned} &\hat{a}_i\hat{a}_j + \hat{a}_j\hat{a}_i = 0 \\ &\hat{a}_i^\dagger\hat{a}_j^\dagger + \hat{a}_j^\dagger\hat{a}_i^\dagger = 0 \\ &\hat{a}_i^\dagger\hat{a}_j + \hat{a}_j\hat{a}_i^\dagger = \delta_{ij} \end{aligned} \qquad (3:5)$$

These operator relations allow manipulating the operators independently of the function they are operating on. In general we will work with products of the operators. These can then often be simplified by the use of (3:5) or relations derived from them. Important operator products are those that preserve the number of particles. They always contain equally many annihilation and creation operators. A basic operator of this kind is the single excitation operator, which excites an electron from orbital i to orbital j:

$$\hat{a}_i^\dagger\hat{a}_j|m_1,m_2,...0_i,...1_j,...\rangle = (-1)^{p_j-p_i}|m_1,m_2,...1_i,...0_j,...\rangle \qquad (3:6)$$

where we assume j≥i. In the derivation of the MCSCF energy expression we will assume that the Hamiltonian does not contain any spin-dependent terms. It is then possible to formulate the theory in terms of spin summed excitation operators $\hat{E}_{ij}$, defined as:

$$\hat{E}_{ij} = (\hat{a}^{\dagger}_{i\alpha}\hat{a}_{j\alpha} + \hat{a}^{\dagger}_{i\beta}\hat{a}_{j\beta}) \tag{3:7}$$

The indices i and j in (3:7) now refer to the n molecular orbitals without the spin factor.These operators fulfill the same commutator relation as the generators of the unitary group of dimension n, and are often referred to as generators. The commutator relation has the following form:

$$[\hat{E}_{ij},\hat{E}_{kl}] = \hat{E}_{il}\delta_{jk} - \hat{E}_{kj}\delta_{il} \tag{3:8}$$

The definition of the operator (3:7) leads to the following relation for the adjoint operator

$$\hat{E}^{\dagger}_{ij} = \hat{E}_{ji} \tag{3:9}$$

We note that if $\hat{E}_{ij}$ operates on a ket where orbital j is unoccupied the result zero is produced. In the same way ,when orbital i is doubly occupied, zero is obtained if i is different from j. When they are equal the result is two times the ket. In general:

$$\hat{E}_{ii}|m\rangle = n_i|m\rangle \tag{3:10}$$

where n_i is the occupation number for the molecular orbital i (0, 1, or 2).

3.3 Operators and matrix elements

The number conserving products of the annihilation-creation operators form an operator basis in which we can represent the quantum mechanical operators in the space spanned by all Slater determinants generated in the spin-orbital space (3:1). For a one-electron operator, $\hat{F}$, we obtain:

$$\hat{F} = \sum_{i,j} F_{ij}\hat{a}^{\dagger}_i\hat{a}_j \tag{3:11}$$

where F_{ij} is the matrix element of the operator over the spin-orbital basis:

$$F_{ij} = \int \phi_i{}^*(x)\hat{F}(x)\phi_j(x)dx \tag{3:12}$$

x here denotes the space and spin coordinates: $x = \{\mathbf{r},\sigma\}$. As an exercise, the reader is suggested to prove that equation (3:11) is correct by showing that the correct expressions are obtained for matrix elements of this operator in the N-electron basis of Slater determinants. For a spin independent operator we can sum the terms in (3:11) pair-wise over the spin quantum number and obtain the result in terms of the excitation operators $\hat{E}_{ij}$:

$$\hat{F} = \sum_{i,j} F_{ij}\hat{E}_{ij} \tag{3:13}$$

where the sum is now over the molecular orbitals and the integrals F_{ij} are defined in the molecular orbital basis. A matrix element of this operator between two Slater determinants $|m\rangle$ and $|n\rangle$ can be written in the form:

$$\langle m|\hat{F}|n\rangle = \sum_{i,j} F_{ij}\langle m|\hat{E}_{ij}|n\rangle = \sum_{i,j} F_{ij}D_{ij}^{mn} \tag{3:14}$$

where $D_{ij}^{mn} = \langle m|\hat{E}_{ij}|n\rangle$ are called one-electron coupling coefficients. For Slater determinants they have the values -1, 0, 1, or 2. The diagonal elements equals the occupation number of orbital i if m = n. Otherwise they are zero. This shows a relation between the coupling coefficients and the first order reduced density matrix. In fact for a wave function of the form:

$$|\Psi\rangle = \sum_{m} c_m|m\rangle \tag{3:15}$$

the first-order reduced density matrix can be written in the form

$$D_{ij} = \langle\Psi|\hat{E}_{ij}|\Psi\rangle = \sum_{m,n} c_m^{*}c_n D_{ij}^{mn} \tag{3:16}$$

A comparison with the one-particle matrix elements (3:14) immediately shows the validity of (3:16). We can now proceed to the two-electron operators. A general such operator can in the operator basis be represented as

$$\hat{G} = \sum_{i,j,k,l} g_{ijkl}\hat{a}_i^{\dagger}\hat{a}_k^{\dagger}\hat{a}_l\hat{a}_j \qquad \text{where}$$

$$g_{ijkl} = \iint \phi_i^{*}(x_1)\phi_k^{*}(x_2)\hat{G}(x_1,x_2)\phi_j(x_1)\phi_l(x_2)dx_1dx_2 \tag{3:17}$$

and the integration is over the space and spin variables of the two interacting electrons. The only two-electron operator $\hat{G}$ that will occur in the present treatment is the inter-electronic repulsion $1/r_{12}$. This operator is spin-independent and we can therefore sum (3:17) over the spin variables. After some manipulations, where we use the symmetry properties of the integrals g_{ijkl} and the anti-commutator relations for the creation-annihilation operators, we obtain:

$$\hat{G} = \frac{1}{2}\sum_{i,j,k,l} g_{ijkl}(\hat{E}_{ij}\hat{E}_{kl} - \delta_{jk}\hat{E}_{il}) \tag{3:18}$$

where the summation is now over molecular orbitals and the integral is defined as:

$$g_{ijkl} = \int\int \varphi_i^*(r_1)\varphi_j(r_1)(1/r_{12})\varphi_k^*(r_2)\varphi_l(r_2)dV_1dV_2 \tag{3:19}$$

In the same way as for the one-electron operator we can form matrix elements between Slater determinants for the operator (3:18)

$$<m|\hat{G}|n> = \sum_{i,j,k,l} g_{ijkl}P^{mn}_{ijkl} \tag{3:20}$$

where the two-electron coupling coefficients, P^{mn}_{ijkl}, are defined as

$$P^{mn}_{ijkl} = \frac{1}{2}<m|\hat{E}_{ij}\hat{E}_{kl} - \delta_{jk}\hat{E}_{il}|n> \tag{3:21}$$

The commutator relation (3:8) for the excitation operators together with the adjoint relation (3:9) leads to the following symmetry properties for the coupling coefficients (for real-valued wave functions):

$$P^{mn}_{ijkl} = P^{mn}_{klij} = P^{nm}_{lkji} = P^{nm}_{jilk} \tag{3:22}$$

By means of (3:20) and (3:21) we can write down the second order reduced density matrix with elements P_{ijkl} for the wave function (3:15) as

$$P_{ijkl} = \sum_{m,n} c^*_m c_n P^{mn}_{ijkl} \tag{3:23}$$

We can now use the formalism given above to write down the Hamiltonian in the operator basis constructed from the excitation operators $\hat{E}_{ij}$:

$$\hat{H} = \sum_{i,j} h_{ij}\hat{E}_{ij} + \frac{1}{2}\sum_{i,j,k,l} g_{ijkl}(\hat{E}_{ij}\hat{E}_{kl} - \delta_{kj}\hat{E}_{il}) \tag{3:24}$$

Here h_{ij} are the one-electron integrals including the electron kinetic energy and the electron-nuclear attraction terms, and g_{ijkl} are the two-electron repulsion integrals defined by (3:19). The summations in (3:24) are over the molecular orbital basis, and the definition is, of course, only valid as long as we work in this basis. Notice that the number of electrons does not appear in the definition of the Hamiltonian. All such information is found in the Slater determinant basis. This is true for all operators in the second quantization formalism.

For a normalized CI wave function of the type (3:15), expanded in the determinant basis, we obtain the energy as the expectation value of the Hamiltonian (3:24):

$$E = \langle\Psi|\hat{H}|\Psi\rangle = \sum_{i,j} h_{ij}D_{ij} + \sum_{i,j,k,l} g_{ijkl}P_{ijkl} \tag{3:25}$$

This energy expression forms the basis for the derivation of the MCSCF optimization methods. Note that the information about the molecular orbitals (the MO coefficients) is contained completely within the one- and two-electron integrals. The density matrices **D** and **P** contain the information about the CI coefficients.

3.4 Exponential operators and orbital transformations

In this section we shall go through some of the formalism needed for the coming derivation of the optimization methods. The parameters to be varied in the energy expression (3:25) are the CI coefficients and the molecular orbitals. We will consider these variations as rotations in an orthonormalized vector space. For example, variations of the MO's correspond to a unitary transformation of the original MO's into a new set:

$$\varphi' = \varphi\mathbf{U}, \tag{3:26}$$

where φ is a row vector containing the original orbitals, and φ' is the transformed orbital vector. **U** is a unitary matrix:

$$\mathbf{U}^{\dagger}\mathbf{U} = \mathbf{1}$$

The corresponding transformation of the spin-orbitals is obtained by multiplying (3:26) with an α or β spin function. When we make the transformation from one set of spin-orbitals to the other, the annihilation and creation operators will change. The following relations are easily established, by operating with the creation operator in the primed space on the vacuum state:

$$\hat{a}'_i = \sum_j \hat{a}_j U^*_{ji}$$

$$\hat{a}'^{\dagger}_i = \sum_j \hat{a}^{\dagger}_j U_{ji}$$

(3:27)

Alternatively it is possible to write the transformed annihilation and creation operators in the following form:

$$\hat{a}'_i = \exp(-\hat{T})\hat{a}_i \exp(\hat{T})$$
$$\hat{a}'^{\dagger}_i = \exp(-\hat{T})\hat{a}^{\dagger}_i \exp(\hat{T})$$

(3:28)

where $\hat{T}$ is an anti-Hermitian operator:

$$\hat{T} = \sum_{i,j} T_{ij}\hat{a}^{\dagger}_i \hat{a}_j$$

(3:29)

with the matrix **T** anti-Hermitian: $\mathbf{T}^{\dagger} = -\mathbf{T}$. The proof of relations (3:28) is obtained by expanding the operators on both sides:

$$\hat{a}'^{\dagger}_i = \hat{a}^{\dagger}_i + [\hat{a}^{\dagger}_i, \hat{T}] + \frac{1}{2}[[\hat{a}^{\dagger}_i, \hat{T}], \hat{T}] + \ldots$$

(3:30)

and a similar relation for the annihilation operator. The commutators in (3:30) are evaluated using the anti-commutator rules (3.5), for example:

$$[\hat{a}^{\dagger}_i, \hat{T}] = \sum_{k,l}(\hat{a}^{\dagger}_i \hat{a}^{\dagger}_k \hat{a}_l - \hat{a}^{\dagger}_k \hat{a}_l \hat{a}^{\dagger}_i)T_{kl} = -\sum_k \hat{a}^{\dagger}_k (\mathbf{T})_{ki}$$

$$[[\hat{a}^{\dagger}_i, \hat{T}], \hat{T}] = \ldots\ldots = \sum_k \hat{a}^{\dagger}_k (\mathbf{T}^2)_{ki}$$

(3:31)

We finally obtain by summing over all terms in the expansion (3:30):

$$\exp(-\hat{T})\hat{a}^{\dagger}_i \exp(\hat{T}) = \sum_k \hat{a}^{\dagger}_k \left(1 - \mathbf{T} + \frac{1}{2}\mathbf{T}^2 \ldots\right) = \sum_k \hat{a}^{\dagger}_k (\exp(-\mathbf{T}))_{ki}$$

(3:32)

We can now identify the unitary matrix: **U** = exp(-**T**). It is a general property of unitary matrices that they can be written as the exponential of an anti-Hermitian matrix: First it is immediately clear that exp (-**T**) is a unitary matrix, when **T** is anti-Hermitian. Secondly it is possible to show that all unitary matrices can be written in this form (see exercise 3.5).

We have thus shown the relations (3:28) for the transformation of the annihilation and creation operators to a new spin-orbital basis. We can use these relations to express an arbitrary Slater determinant in the new basis in terms of the determinants in the original basis. In order to do so, we generate the Slater determinant by applying a sequence of creation operators on the vacuum state:

$$|m'\rangle = \hat{a}'^{\dagger}_i \hat{a}'^{\dagger}_j \hat{a}'^{\dagger}_k ... |vac\rangle = \exp(-\hat{T})\hat{a}^{\dagger}_i \exp(\hat{T})\exp(-\hat{T})\hat{a}^{\dagger}_j \exp(\hat{T})... |vac\rangle =$$
$$\exp(-\hat{T})\hat{a}^{\dagger}_i \hat{a}^{\dagger}_j \hat{a}^{\dagger}_k ... |vac\rangle = \exp(-\hat{T})|m\rangle \tag{3:33}$$

This is a very important relation. It shows that the effect of an orbital transformation on a Slater determinant can be obtained simply by operating on that determinant with the exponential operator $\exp(-\hat{T})$. We shall make extensive use of this equation in the derivation of the MCSCF optimization schemes.

So far we have considered an arbitrary unitary transformation of the spin-orbitals. In practice we shall only transform the spatial part of the orbitals, the molecular orbitals. In order to see the implication of this for the operator $\hat{T}$ we order the spin orbitals by multiplying first the MO's, φ, with an α spin function, followed by the spin orbitals obtained by multiplying the same MO's with a β spin function:

$$\phi = (\varphi\alpha, \varphi\beta) \tag{3:34}$$

The matrix $\mathbf{T}$ can then be arranged as a collection of four sub-matrices corresponding to transformations within and between the two spin-orbital blocks:

$$\mathbf{T} = \begin{pmatrix} \mathbf{T}_{\alpha\alpha} & \mathbf{T}_{\alpha\beta} \\ \mathbf{T}_{\beta\alpha} & \mathbf{T}_{\beta\beta} \end{pmatrix}$$

Now, $\mathbf{T}_{\alpha\beta} = \mathbf{T}_{\beta\alpha} = \mathbf{0}$, since we are not mixing orbitals with different spin. Furthermore, the transformation matrix is the same for α and β. Thus we also have $\mathbf{T}_{\alpha\alpha} = \mathbf{T}_{\beta\beta}$. The $\hat{T}$ operator can be summed over the molecular orbitals as:

$$\hat{T} = \sum_{i,j} (T^{\alpha\alpha}_{ij} \hat{a}^{\dagger}_{i\alpha} \hat{a}_{j\alpha} + T^{\alpha\beta}_{ij} \hat{a}^{\dagger}_{i\alpha} \hat{a}_{j\beta} + T^{\beta\alpha}_{ij} \hat{a}^{\dagger}_{i\beta} \hat{a}_{j\alpha} + T^{\beta\beta}_{ij} \hat{a}^{\dagger}_{i\beta} \hat{a}_{j\beta}) \tag{3:35}$$

If we now use the relation given above for the matrix **T** we obtain:

$$\hat{T} = \sum_{i,j} T_{ij}(\hat{a}^{\dagger}_{i\alpha}\hat{a}_{j\alpha} + \hat{a}^{\dagger}_{i\beta}\hat{a}_{j\beta}) = \sum_{i,j} T_{ij}\hat{E}_{ij} \tag{3:36}$$

where we have introduced the excitation operators $\hat{E}_{ij}$, as defined in (3:7) and dropped the labels $\alpha\alpha$ ($\beta\beta$) for the matrix elements of **T**. T_{ij} is an element of the anti-Hermitian matrix **T**, which describes the unitary rotation of the MO's: **U** = exp(-**T**). We shall consider only real molecular orbitals. Thus the matrix **T** is real and anti-symmetric ($T_{ij} = -T_{ji}$). We can use this to rewrite (3:36) as

$$\hat{T} = \sum_{i>j} T_{ij}(\hat{E}_{ij} - \hat{E}_{ji}) = \sum_{i>j} T_{ij}\hat{E}^{-}_{ij} \tag{3:37}$$

An orthogonal rotation is thus described by the "replacement" operators $\hat{E}_{ij}$ - $\hat{E}_{ji}$.

3.5 Slater determinants and spin-adapted state functions

So far we have based the formalism on an N-electron basis built from Slater determinants. However, as shown above, both the Hamiltonian and the orbital rotations can be described in terms of the orbital excitation operators $\hat{E}_{ij}$. All matrix elements involving the Hamiltonian and the operator $\hat{T}$ can therefore be computed without explicit reference to the spin-orbital basis. The only thing we need is the commutator relation (3:8). We note in addition that the excitation operators commute with the spin operators $\hat{S}_z$ and $\hat{S}^2$ (see exercise). It is then possible to work entirely in a spin-adapted configurational basis. The total MCSCF wave function will be assumed to be a pure spin eigenfunction. It therefore seems favourable to be able to expand it in terms of a spin-adapted configurational basis. There are many ways in which spin-adapted configurations can be generated from a basis of Slater determinants. One of the most commonly used methods today is the Graphical Unitary Group Approach (GUGA). We shall not in these lectures discuss these methods in any detail. The subject will be covered in other parts of the course.

Most of the formalism to be developed in the coming sections of these lecture notes will be independent of the specific definition of the configurational basis, in which we expand the wave function. We therefore do not have to be very explicit about the exact nature of the basis states $|m\rangle$. They can be either Slater determinants or spin-adapted Configuration State Functions (CSF's). For a long time it was assumed that CSF's were to be preferred for MCSCF calculations, since it gives a much shorter CI expansion. Efficient methods like GUGA had also been developed for the solution of the CI problem. Recent experience has, however, shown that calculations in a Slater determinant basis can be performed very efficiently on modern computers. In fact, the largest CI calculations carried through have been performed using Slater determinants (summer 1990: full CI with more than 1 000 000 000 determinants).

The variational parameters for the CI part of the wave function could be taken to be the CI coefficients C_m in the expansion of the MCSCF wave function, which we now write as:

$$|0\rangle = \sum_m C_m |m\rangle \tag{3:38}$$

where we do not here specify the exact nature of the basis states $|m\rangle$, other than requiring orthonormality. The variation of the CI coefficients are made with the restriction that the total wave function (3:38) remains normalized:

$$\sum_m |C_m|^2 = 1 \tag{3:39}$$

We can remove the problem with this subsidiary condition by instead using as the variational space the orthogonal complement to the MCSCF state $|0\rangle$. This variational space is defined as a set of states $|K\rangle$ expanded in the same set of basis states $|m\rangle$ as $|0\rangle$:

$$|K\rangle = \sum_m C_m^K |m\rangle \tag{3:40}$$

with the property $\langle K|L\rangle = \delta_{KL}$. To each of the states $|K\rangle$ corresponds a variational parameter, describing the contribution of this state to a variation of the MCSCF state $|0\rangle$. This variation can be described as a unitary rotation between the MCSCF state and the complementary space. The operator which performs this rotation is constructed in the same way as was done for the orbital rotations. We start by defining an anti-symmetric replacement operator:

$$\hat{S} = \sum_{K \neq 0} S_{K0}(|K\rangle\langle 0| - |0\rangle\langle K|) \tag{3:41}$$

The corresponding unitary operator is $\exp(\hat{S})$. In the next chapter we shall use these operators in a derivation of the MCSCF optimization methods.

3.6 Exercises

3.1 Show explicitly that the annihilation and creation operators fulfill the anti-commutator relations (3:5).

3.2 Use (3:5) to show the commutator relation (3:8).

3.3 Calculate the matrix elements of the operators (3:11) and (3:17) in a basis of Slater determinants, and show that the normal expressions (Slater's rules) are obtained.

3.4 Show that the two-electron coupling coefficients (3:21) have the symmetry properties given by (3:22). Use the results to deduce the symmetry properties of the second-order density matrix (3:23).

3.5 Show that a unitary matrix **U** can always be written in the form **U** = exp(**T**), where **T** is an anti-Hermitian matrix.
Hint: A unitary matrix can be diagonalized by a unitary matrix.

3.6 Compute the two by two unitary matrix corresponding to the **T** matrix:

$$\mathbf{T} = \begin{pmatrix} 0 & \theta \\ -\theta & 0 \end{pmatrix}$$

3.7 Show the following formula for the unitary matrix **U** = exp(**T**):

$$\mathbf{U} = \mathbf{V}\mathbf{a}\mathbf{V}^{\dagger} + \mathbf{V}\mathbf{b}\mathbf{V}^{\dagger}\mathbf{T}$$

where $a_{ij} = \delta_{ij}\cos(\theta_i)$ and $b_{ij} = \delta_{ij}\sin(\theta_i)/\theta_i$, with $\theta_i^2 = -\varepsilon_i$.

V is the unitary matrix, which diagonalizes the Hermitian matrix $\mathbf{T}^2$ with eigenvalues $\varepsilon_i < 0$ (why?).This formula allows the construction of a unitary matrix from the matrix **T**.

3.8 Show that when $\exp(\hat{S})$ operates on $|0\rangle$, with $\hat{S}$ defined by (3:41), the following result is obtained:

$$\cos\theta|0\rangle + \frac{1}{\theta}\sin\theta\hat{S}|0\rangle, \qquad \text{where } \theta = \sqrt{\sum_{K\neq 0} S_{K0}^2}$$

3.9 Express the spin operators $\hat{S}_z$ and $\hat{S}^2$ in terms of annihilation and creation operators. Then show that the excitation operators $\hat{E}_{ij}$ commute with these spin operators.

4. The multiconfiguration self-consistent field equations

4.1 Introduction

We shall in this chapter discuss the methods employed for the optimization of the variational parameters of the MCSCF wave function. Many different methods have been used for this optimization. They are usually divided into two different classes, depending on the rate of convergence: first or second order methods. First order methods are based solely on the calculation of the energy and its first derivative (in one form or another) with respect to the variational parameters. Second order methods are based upon an expansion of the energy to second order (first and second derivatives). Third or even higher order methods can be obtained by including more terms in the expansion, but they have been of rather small practical importance.

The most prominent of these methods is probably the second order Newton-Raphson approach, where the energy is expanded as a Taylor series in the variational parameters. The expansion is truncated at second order, and updated values of the parameters are obtained by solving the Newton-Raphson linear equation system. This is the standard optimization method and most other methods can be treated as modifications of it. We shall therefore discuss the Newton-Raphson approach in more detail than the alternative methods.

The choice of optimization scheme in practical applications is usually made by considering the convergence rate versus the time needed for one iteration. It seems today that the best convergence is achieved using a properly implemented Newton-Raphson procedure, at least towards the end of the calculation. One full iteration is, on the other hand, more time-consuming in second order methods, than it is in more approximative schemes. It is therefore not easy to make the appropriate choice of optimization method, and different research groups have different opinions on the optimal choice. We shall discuss some of the more commonly implemented methods later.

4.2 The Newton-Raphson method

Before analyzing the energy expression (3:25) with respect to the variational parameters, let us briefly review the multidimensional Newton-Raphson procedure. Assume that the energy is a function of a set of parameters p_i, which we arrange as a column vector, $\mathbf{p}$. We now make a Taylor expansion of the energy $E = E(\mathbf{p})$ around a point $\mathbf{p}_0$, which we arbitrarily can put equal to zero:

$$E(\mathbf{p}) = E(0) + \sum_i \left(\frac{\partial E}{\partial p_i}\right)_0 p_i + \frac{1}{2}\sum_{i,j} p_i \left(\frac{\partial^2 E}{\partial p_i \partial p_j}\right)_0 p_j + \cdots \tag{4:1}$$

or in matrix notation:

$$E(\mathbf{p}) = E(0) + \mathbf{g}^\dagger\mathbf{p} + \tfrac{1}{2}\mathbf{p}^\dagger\mathbf{H}\mathbf{p} + \cdots \tag{4:2}$$

Here we have defined the *energy gradient* vector **g** and the *Hessian* matrix **H**, with the elements given as:

$$g_i = \left(\frac{\partial E}{\partial p_i}\right)_0 \quad \text{and} \quad H_{ij} = \left(\frac{\partial^2 E}{\partial p_i \partial p_j}\right)_0 \tag{4:3}$$

The stationary points on the energy surface (4:1) are obtained as solutions to the equations $\partial E/\partial p_i = 0$. They can be approximately solved by starting from the expansion (4:2), truncated at second order. Setting the derivatives of E in (4:2) equal to zero leads to the system of linear equations:

$$\mathbf{g} + \mathbf{H}\mathbf{p} = \mathbf{0} \qquad \text{or} \qquad \mathbf{p} = -\mathbf{H}^{-1}\mathbf{g} \tag{4:4}$$

A sequence of Newton-Raphson iterations is obtained by solving equation (4:4); redefining the zero point, $\mathbf{p}_0$, as the new set of parameters; recalculating **g** and **H** and returning to equation (4:4). Such a procedure converges quadratically, that is, the error vector in iteration n is a quadratic function of the error vector in iteration n-1. This does not necessarily mean that the NR procedure will converge fast, or even at all. However, close to the stationary point we can expect a quadratic behaviour. We shall return later to a more precise definition of what *close* means in this respect.

4.3 The MCSCF gradient and Hessian.

In this section we shall compute the energy gradient and the Hessian matrix corresponding to the energy expression (3:25). We introduce the variation of the CI coefficients by operating on the MCSCF state $|0\rangle$ with the unitary operator $\exp(\hat{S})$, with $\hat{S}$ defined by (3:41) and the orbital rotations by the operator $\exp(\hat{T})$, with $\hat{T}$ from (3:37). A variation of the MCSCF state can thus be written as:

$$|0'\rangle = \exp(\hat{T})\exp(\hat{S})|0\rangle \tag{4:5}$$

The order of the operators in (4:5) is not arbitrary, since they do not commute. The reverse order, however, leads to more complicated expressions for the Hessian matrix, and since the final result is independent of the order, we make the more simple choice given in (4:5). The energy corresponding to the varied state (4:5) will be a function of the parameters in the unitary operators, and we can calculate the first and second derivatives of this function by expanding the exponential operators to second order. The unitarity of the operators guarantees that |0'> remains normalized if the original state |0> is normalized. The energy is given as:

$$E(\mathbf{T},\mathbf{S}) = \langle 0|\exp(-\hat{S})\exp(-\hat{T})\hat{H}\exp(\hat{T})\exp(\hat{S})|0\rangle \qquad (4:6)$$

If we now expand the exponential operators to second order we obtain for the operator in (4:6) the following expression:

$$E(T,S) = \langle 0|\hat{H} + [\hat{H},\hat{T}] + [\hat{H},\hat{S}] + \frac{1}{2}[[\hat{H},\hat{T}],\hat{T}] + \frac{1}{2}[[\hat{H},\hat{S}],\hat{S}] + [[\hat{H},\hat{T}],\hat{S}] + \cdots|0\rangle \qquad (4:7)$$

The first term in this expression is the zeroth order energy $E(\mathbf{0},\mathbf{0})$. The two next terms gives the first derivatives with respect to the parameters T_{ij} (equation (3:37)) and S_{K0} (equation (3:42)). We obtain the following result for the first derivative with respect to the orbital rotation parameters:

$$\langle 0|[\hat{H},\hat{T}]|0\rangle = \sum_{i>j} T_{ij}\langle 0|[\hat{H},\hat{E}^{-}_{ij}]|0\rangle \qquad (4:8)$$

which gives the derivative as:

$$g^{(o)}_{ij} = \langle 0|[\hat{H},\hat{E}^{-}_{ij}]|0\rangle \qquad (4:9)$$

Here the superscript (o) has been used to indicate that this is the derivative with respect to the orbital rotation parameters. Note the close resemblance between (4:9) and the corresponding expression in Hartree-Fock theory - the Brillouin theorem. It is therefore sometimes called the *Extended (or generalized) Brillouin Theorem*. The Brillouin theorem states that there is no interaction between the Hartree-Fock wave function and singly excited configurations. Equation (4:9) yields the corresponding condition for an MCSCF wave function: the matrix element on the right hand side of the equation is zero for optimized orbitals. If |0> is an Hartree-Fock state, the only interesting matrix elements are those where j corresponds to an occupied and i to a virtual orbital. Setting the derivative equal to zero for these matrix elements immediately leads to the Hartree-Fock equations. Notice that, if i and j correspond to empty orbitals, or if both orbitals are doubly occupied, (4:9) is identical to zero.

The situation in the MCSCF case is obviously more complex, since orbitals may be partly occupied - they are occupied in some terms in the wave function but not in others. However, if the occupation number of both orbitals are exactly equal to two, (4:9) is again identical to zero. Like the SCF energy, the

MCSCF energy is also invariant towards rotations among the inactive (doubly occupied) orbitals. Rotations between them should thus not be included. The same is obviously true in the case where both occupation numbers are zero (virtual orbitals).

Let us now turn to the derivatives with respect to the CI parameters. Starting from:

$$<0|[\hat{H},\hat{S}]|0> = \sum_{K\neq 0} S_{K0}(<0|\hat{H}|K> + <K|\hat{H}|0>) \tag{4:10}$$

we obtain for real wave functions the derivative:

$$g_K^{(c)} = 2<0|\hat{H}|K> \tag{4:11}$$

This equation tells us that an optimized MCSCF state (for which the derivative is zero) will not interact with the orthogonal complement, that is, it is a solution to the secular problem:

$$(\mathbf{H} - E\mathbf{1})\mathbf{C} = \mathbf{0} \tag{4:12}$$

where **H** is the Hamiltonian matrix, **C** the expansion coefficients of the MCSCF wave function, and E the energy. This is the condition we expected to obtain for an optimized CI wave function.

Let us now turn to the second derivatives. The Hessian matrix is divided into three parts: the orbital - orbital part (oo); the configuration - configuration part (cc); and the so called CI coupling part (co). For the (cc) part we obtain:

$$H_{KL}^{(cc)} = 2(<K|\hat{H}|L> - \delta_{KL}<0|\hat{H}|0>) \tag{4:13}$$

which is simply the matrix elements of the reduced Hamiltonian over the complementary CI space. This is the expected result, since it is exactly the denominator which we would obtain in a second order perturbation treatment of the CI problem. For the orbital - orbital part of the Hessian we obtain after some manipulations, (which are left to the reader):

$$H_{ij,kl}^{(oo)} = <0|\hat{E}_{ij}^{-}\hat{E}_{kl}^{-}\hat{H}|0> + <0|\hat{H}\hat{E}_{ij}^{-}\hat{E}_{kl}^{-}|0> - 2<0|\hat{E}_{ij}^{-}\hat{H}\hat{E}_{kl}^{-}|0> \tag{4:14}$$

and finally for the coupling part:

$$H_{K,ij}^{(co)} = H_{ij,K}^{(oc)} = 2<K|[\hat{H},\hat{E}_{ij}^{-}]|0> \tag{4:15}$$

The formulas above give the gradient and the Hessian in terms of matrix elements of the excitation operators. They can be evaluated in terms of one- and two-electron integrals, and first and second order reduced density matrices, by inserting the Hamiltonian (3:24) into equations (4:9), (4;11), and (4:13)-(4:15). Note that transition density matrices $\mathbf{D}^{(K0)}$ and $\mathbf{P}^{(K0)}$ are needed for the evaluation of the CI coupling matrix (4:15).

We now have the expressions for the gradient and the Hessian matrix. The corresponding Newton-Raphson equations can then be written down in matrix form as:

$$\begin{pmatrix} \mathbf{a} & \mathbf{b} \\ \mathbf{b}^{\dagger} & \mathbf{c} \end{pmatrix} \begin{pmatrix} \mathbf{S} \\ \mathbf{T} \end{pmatrix} = - \begin{pmatrix} \mathbf{v} \\ \mathbf{w} \end{pmatrix} \tag{4:16}$$

where we have introduced the simplified notation:

$$\mathbf{a} = \frac{1}{2}\mathbf{H}^{(cc)},\ \mathbf{b} = \frac{1}{2}\mathbf{H}^{(co)},\ \mathbf{c} = \frac{1}{2}\mathbf{H}^{(oo)},\ \mathbf{v} = \frac{1}{2}\mathbf{g}^{(c)},\ \mathbf{w} = \frac{1}{2}\mathbf{g}^{(o)} \tag{4:17}$$

The direct solution for the rotation parameters **S** and **T** from (4:16) is not very practical if the MC expansion is large, due to the complications in computing matrix elements over the orthogonal complement space |K>, defined by equation (3:41). An M^2 transformation is needed to obtain $\mathbf{h}^{(cc)}$ from the Hamiltonian matrix elements $\langle m'|\hat{H}|m\rangle$, where M is the dimension of the MC expansion in the basis states |m>. M is often a large number (10^5 is not an unusual order of magnitude) in modern applications of the MCSCF method. It is then not feasible to perform such transformations. However, it is not necessary either.

A more attractive computational procedure can be obtained by rewriting equation (4:16) in terms of variations of the original MC expansion coefficients C_m in equation (3:38). In order to do that we introduce a matrix **C** containing the M-1 column vectors $\mathbf{C}^K$. The matrices (4:17) can now be transformed back to the original basis set |m> and we obtain:

$$\begin{pmatrix} \mathbf{C}^{\dagger}\mathbf{A}\mathbf{C} & \mathbf{C}^{\dagger}\mathbf{B} \\ \mathbf{B}^{\dagger}\mathbf{C} & \mathbf{c} \end{pmatrix} \begin{pmatrix} \mathbf{S} \\ \mathbf{T} \end{pmatrix} = - \begin{pmatrix} \mathbf{C}^{\dagger}\mathbf{G} \\ \mathbf{w} \end{pmatrix} \tag{4:18}$$

Here **A** is the CI matrix $(\mathbf{H} - E_0\mathbf{1})$ in the original basis. The other matrices are defined accordingly:

$$A_{mn} = \langle m|\hat{H} - E_0\delta_{mn}|n\rangle;\ B_{m,ij} = \langle m|[\hat{H},\hat{E}_{ij}^{-}]|0\rangle;\ G_m = \langle 0|\hat{H}|m\rangle \tag{4:19}$$

The state rotation parameters **S** are obtained from the variations of the individual CI coefficients as: $\mathbf{S} = \mathbf{C}^{\dagger}\delta\mathbf{C}_0$, but the number of parameters in $\delta\mathbf{C}_0$ is M, while the number of linearly independent parameters in **S** is only M-1. This redundancy in $\delta\mathbf{C}_0$ can be removed by adding one row to equation (4:18), where we define the redundant rotation, S_0, as $S_0 = \mathbf{C}_0{}^{\dagger}\delta\mathbf{C}_0$, and demand that $S_0 = 0$, by introducing an arbitrary, but non-zero, constant z. We thus obtain the following enlarged matrix equation:

$$\begin{pmatrix} z & 0 & 0 \\ 0 & \mathbf{C}^{\dagger}\mathbf{A}\mathbf{C} & \mathbf{C}^{\dagger}\mathbf{B} \\ 0 & \mathbf{B}^{\dagger}\mathbf{C} & \mathbf{c} \end{pmatrix} \begin{pmatrix} \mathbf{C}_0^{\dagger}\delta\mathbf{C}_0 \\ \mathbf{C}^{\dagger}\delta\mathbf{C}_0 \\ \mathbf{T} \end{pmatrix} = - \begin{pmatrix} 0 \\ \mathbf{C}^{\dagger}\mathbf{G} \\ \mathbf{w} \end{pmatrix} \tag{4:20}$$

This equation is now multiplied with the unitary rotation

$$\begin{pmatrix} \mathbf{U} & \mathbf{0} \\ \mathbf{0} & \mathbf{1} \end{pmatrix}$$

where **U** is the MxM matrix $(\mathbf{C}_0,\mathbf{C})$, which contains all the CI vectors. Introducing the projection matrices,

$$\mathbf{P} = \mathbf{C}_0\mathbf{C}_0^{\dagger} \text{ and } \mathbf{Q} = \mathbf{1} - \mathbf{C}_0\mathbf{C}_0^{\dagger} = \mathbf{C}\mathbf{C}^{\dagger} \tag{4:21}$$

we obtain the final system of equations:

$$\begin{pmatrix} z\mathbf{P} + \mathbf{QAQ} & \mathbf{QB} \\ \mathbf{B}^{\dagger}\mathbf{Q} & \mathbf{c} \end{pmatrix} \begin{pmatrix} \delta\mathbf{C}_0 \\ \mathbf{T} \end{pmatrix} = - \begin{pmatrix} \mathbf{QG} \\ \mathbf{w} \end{pmatrix} \tag{4:22}$$

The solution to this system of equations gives the parameter set **T**, from which the orbital rotations can be determined (see exercise 3.7). In addition we obtain the variations in the CI coefficients. Note that we cannot use them directly, since that does not correspond to a unitary rotation, which keeps normalization. Instead we use the definition $\mathbf{S} = \mathbf{C}^{\dagger}\delta\mathbf{C}_0$ to construct the unitary matrix exp(**S**). The rotated state is then obtained from the formula given in exercise (3.8).

4.4 Computational aspects on the Newton-Raphson procedure

When constructing methods to solve the system of linear equations (4:22) one should be aware of the dimension of the problem. It is not unusual to have CI expansion comprising 10^4 - 10^6 terms, and orbital spaces with more than two hundred orbitals. In such calculations it is obviously not possible to explicitly construct the Hessian matrix. Instead we must look for iterative algorithms where a correction vector in iteration i is obtained from the corresponding vector in iteration (i-1) by multiplication with the Hessian matrix. In such a method we could apply direct CI techniques, and construct the product directly from a list of one- and two-electron integrals together with first and second order density matrices. There are several possible iterative schemes, which can be made to work in this way. Here we shall describe one such method, which bear some resemblance to the Davidson method, which is used in the same spirit to solve large secular problems.

Let us consider a linear equation system, which we write it in the form:

$$(\mathbf{A}_0 - \mathbf{A})\mathbf{x} - \mathbf{b} = \mathbf{0}, \tag{4:23}$$

where $\mathbf{A}_0$ is a diagonal non-singular matrix (eg the diagonal part of the Hessian). Multiplying by $\mathbf{A}_0^{-1}$ we can rewrite (4.23) in the form:

$$\mathbf{x} = \mathbf{A}_0^{-1}\mathbf{A}\mathbf{x} + \mathbf{A}_0^{-1}\mathbf{b} \tag{4:24}$$

It is easy to see that the solution to this equation can be given as a power series expansion:

$$\mathbf{x} = \sum_{n=0}^{\infty}(\mathbf{A}')^n\mathbf{b} \qquad \text{where } \mathbf{A}' = \mathbf{A}\mathbf{A}_0^{-1}. \tag{4:25}$$

Direct use of this equation is not advisable, however, since it is only slowly convergent. A more efficient method is obtained by constructing a set of orthogonal vectors $\mathbf{x}_n$ as

$$\mathbf{x}_{n+1} = \mathbf{A}'\mathbf{x}_n - \sum_{l=0}^{n}\frac{\mathbf{x}_l^\dagger\mathbf{A}'\mathbf{x}_n}{\mathbf{x}_l^\dagger\mathbf{x}_l}\mathbf{x}_l \tag{4:26}$$

An improved vector **x** in iteration n is then obtained as a linear combination of the update vectors (4:26):

$$\mathbf{x} = \sum_{i=0}^{n}\alpha_i\mathbf{x}_i \tag{4:27}$$

The parameters α_i are obtained by inserting (4:27) into (4:24) and project onto each of the $\mathbf{x}_i$. The process normally converges in less than ten iterations, although, as in all iterative schemes of this kind, convergence can be slowed down considerably in near singular situations (small eigenvalues to the Hessian matrix). The fundamental operation to be performed in constructing the update vectors according to (4:26), is the multiplication of **x** with the matrix **A'**, that is, the formation of the vector **y** = **A'x** with components

$$y_i^{(n+1)} = \sum_k A_{ik} x_k^{(n)} / (A_0)_{kk} \tag{4.28}$$

Now choose A_0 to be the diagonal part of the Hessian matrix and **A** the remaining non-diagonal part. We then obtain for the CI part of the update vector **y** the following expression:

$$y_m^{(n+1)} = \sum_{n \neq m} H_{mn} \delta C_n^{(n)} / (H_{nn} - E_0) \tag{4:29}$$

which is a formula equivalent to that used in solving the CI secular problem with the Davidson iterative scheme. Equation (4:29) is of course not complete, since it should contain an additional term for the coupling of the CI part and the orbital rotations (the (co) part of the Hessian). The complete expressions for the update vectors are rather complex and will not be given explicitly here. Later we shall see how the Hessian matrix elements can be expressed in terms of integrals and density matrix elements. Using these expressions it is possible to construct the update vectors **y** through (4:28) without an explicit construction of the Hessian matrix. This is an important feature of the iterative methods, since it allows the solution of large systems of equations, corresponding to complex MCSCF wave functions, and large AO basis sets.

The second-order convergence of the Newton-Raphson procedure requires the starting vector is already in the local area, that is, rather close to the stationary point. For a calculation of the ground state of a molecule a necessary condition is that the Hessian matrix is positive definite, that is, it has only positive eigenvalues. This condition is, however, not sufficient since local minima on the energy surface may occur, which correspond to solutions with higher energy. The problem to find the area of the global minimum in an MCSCF calculation is in many cases not trivial. There is no easy way out of this problem. However, local minima often correspond to solutions of a different character than that of the global minimum. An inspection of the converged solution can therefore indicate whether the global minimum has been reached or not. For the n-th excited state, the Hessian has in well behaved cases n-1 negative eigenvalues in the local region.

When these conditions are not fulfilled the Newton-Raphson procedure may converge only slowly or even diverge. It is then necessary to introduce manual procedures, which drive the solution to the local minimum. A number of such methods have been devised, which in many cases work well. They are based on a determination of controlled step sizes and sign control, which forces the calculation to proceed in the right direction with optimal step length. Some of these methods have the disadvantage of requiring an a priori search for all negative eigenvalues of the Hessian matrix, a task which might not be easily undertaken in large calculations. However, it should be emphasized that knowledge about the eigenvalues of the Hessian gives valuable information about the stability of the calculation. Therefore a search for the n lowest roots of the Hessian at the end of a calculation on the n-th excited state is advisable.

One method, which avoids the problem with undesired negative eigenvalues of the Hessian, and which introduces an automatic damping of the rotations, is the *augmented Hessian method* (AM). To describe the properties of this method, let us again consider the Newton-Raphson equation (4:4):

$$\mathbf{H}\mathbf{p} = -\mathbf{g} \tag{4:30}$$

In the AM method one solves instead of (4:30) the secular problem:

$$\begin{pmatrix} 0 & \mathbf{g}^{\dagger} \\ \mathbf{g} & \mathbf{H} \end{pmatrix} \begin{pmatrix} 1 \\ \mathbf{p} \end{pmatrix} = E \begin{pmatrix} 1 \\ \mathbf{p} \end{pmatrix} \tag{4:31}$$

In order to relate equations (4:30) and (4:31) it is convenient to transform both equations to a basis, where **H** is diagonal. Suppose that **H** is diagonalized by the unitary matrix **U**. The corresponding diagonal is $\varepsilon = \mathbf{U}^{\dagger}\mathbf{H}\mathbf{U}$. The transformed vectors $\mathbf{p}'$ and $\mathbf{g}'$ are obtained as: $\mathbf{p}' = \mathbf{U}^{\dagger}\mathbf{p}$ and $\mathbf{g}' = \mathbf{U}^{\dagger}\mathbf{g}$. The solution of (4:30) in the new basis is trivial and yields:

$$p_i' = -g_i'/\varepsilon_i \tag{4:32}$$

while the solution of (4:31) yields:

$$p_i' = -\, g_i'/(\varepsilon_i - E_i) \tag{4:33}$$

The bracketing theorem tells us that the eigenvalues, E_i, satisfies the betweenness condition: $\varepsilon_{i-1} \leq E_i \leq \varepsilon_i$ for all i. For the lowest state we have especially: $E_1 \leq \varepsilon_1$, where ε_1 is the smallest eigenvalue of the Hessian. Equation (4:33) therefore automatically includes a sign control, since $\varepsilon_i - E_i$ is always positive even if the eigenvalue of the Hessian is negative. This feature of the AM method is especially attractive in determining MCSCF wave functions for higher roots. At convergence the two sets of rotation parameters (4:32) and (4:33) become identical, since E_i goes to zero with **g**.

Before ending this section, a few words about redundant rotation parameters. It may happen that the variation parameters in the MCSCF wave function are linearly dependent. When such a situation obtains, the Hessian will be singular and the NR equations lead to indefinite values for the rotations. A trivial example of a redundant set of parameters is the orbital rotations T_{ij}, where both indices correspond to orbitals which are doubly occupied in all configurations (inactive orbitals), or if both indices refer to empty (virtual) orbitals. The energy gradient $\partial E/\partial T_{ij}$ is in this case automatically zero, since the wave function is invariant to such rotations. Another example is given by the Complete Active Space (CAS) SCF wave function, which is complete in the CI space spanned by a set of active orbitals. The energy (but not the wave function) is here also invariant to rotations among the active orbitals. This is easily realized by considering the expression for the orbital gradient (4:9) for an orbital rotation T_{tu}, where both t and u correspond to orbitals in the active space:

$$g_{tu}^{(o)} = 2<0|\hat{H}(\hat{E}_{tu} - \hat{E}_{ut})|0> = 0 \qquad (4:34)$$

The excitation operators will generate a state which is a linear combination of the MC state $|0>$ and its orthogonal complement states $|K>$. Since we can assume that $\hat{H}$ is diagonal in this basis, the only term left is the projection of the excitation operators on the MC state itself. The zero result then follows from the relation:

$$<0|\hat{E}_{tu}|0> = <0|\hat{E}_{ut}|0> \qquad (4:35)$$

Thus the Hessian will become singular if we include rotations between the active orbitals. Redundant parameters must not be included in the Newton-Raphson procedure.They are trivial to exclude for the examples given above, but in more general cases a redundant variable may occur as a linear combination of **S** and **T** and it might be difficult to exclude them. One of the advantages of the CASSCF method is that all parameters except those given above are non-redundant.

4.5 Approximations to the general second order MCSCF procedure

The fully coupled Newton-Raphson method as it is given by equation (4:16) or (4:22) exhibits quadratic convergence behaviour in the local region, and is in this respect very attractive. Satisfactory global convergence can normally be obtained by using mode damping, step size and sign control, or the augmented Hessian method. Convergence can alternatively be improved, when far from the local region, by including cubic terms in the expansion of the energy. The full Newton-Raphson method is, however, computationally complex. Especially the calculation of the the coupling matrix **B** in (4:22) becomes time consuming for a large orbital space together with a long MC expansion. It is therefore natural to look for simplifications of the Newton-Raphson equations, which avoid some of the computational bottlenecks of the full second order procedure.

The direct solution of equation (4:18) or (4:22) is in the MCSCF language called a *one-step* procedure, since both **S** and **T** are updated simultaneously. We can formally rewrite (4:16) (or (4:22)) as a *two-step* procedure by solving for **S** first. From the first row of (4:16) we obtain:

$$\mathbf{S} = -\mathbf{a}^{-1}\mathbf{v} - \mathbf{a}^{-1}\mathbf{b}\mathbf{T} \tag{4:36}$$

which inserted into the second row gives an equation in **T** only:

$$(\mathbf{c}-\mathbf{b}^{\dagger}\mathbf{a}^{-1}\mathbf{b})\mathbf{T} = -\mathbf{w} + \mathbf{b}^{\dagger}\mathbf{a}^{-1}\mathbf{v} \tag{4:37}$$

The matrix in front of **T** is the *partitioned* orbital Hessian. Equation (4:37) is not very practical, since it involves the inverse of the CI part of the Hessian. But suppose that we work in a configuration basis (|0>,|K>), where **a** is diagonal, that is we start each iteration by solving the CI problem to all orders. The matrix **a** is then diagonal with the matrix elements $a_{KK} = (E_K - E_0)$, and the corresponding gradient vector **v** with elements $\langle 0|\hat{H}|K\rangle$ is zero. Equation (4:37) can then be written in the form:

$$\sum_{r>s}\left(c_{pq,rs} + \sum_{K}\frac{b_{pq,K}b_{K,rs}}{E_0 - E_K}\right)T_{rs} = -w_{pq} \tag{4.38}$$

This equation forms the basis for the *two-step* Newton-Raphson method. It is obviously not a very practical method to use in connection with very large MC expansions, since we cannot in this case easily obtain all the eigenvalues E_K. Attempts have been made to truncate the summation over K in (4:38) to the lowest roots, with the inherent assumption that terms with small values in the denominator will give the largest contribution to the coupling term. It turns out, however, that the convergence of the sum over coupling terms is slow in most cases. It is therefore not very useful in practice to include the coupling between the orbital and CI rotations by means of equation (4:38), except maybe in special cases with strong couplings between a few MC states, a situation which one might obtain in calculations on excited states.

An approximation, which has been used with some success, is to neglect the coupling term altogether. This leads to the *unfolded two-step* Newton-Raphson procedure, where the equations to be solved are:

$$(\mathbf{H} - E_0\mathbf{1})\mathbf{C} = \mathbf{0} \tag{4:39}$$

and

$$\mathbf{cT} = -\mathbf{w} \tag{4:40}$$

Even though this method is not quadratically convergent, it is frequently used in practical applications due to its much simpler form than the general second order approach. The secular problem (4:39) can be solved for very long MC expansions by means of efficient direct CI procedures, especially for CASSCF type wave functions.The orbital Hessian **c** can, as we shall see later, be written in terms of one- and two-electron integrals multiplied with first and second order density matrices. The approximate size (why only approximate?) of this matrix is $n_i n * n_i n$, where n_i is the number of internal (occupied) orbitals and n the total number of orbitals, and we have not assumed any reduction by symmetry. A typical calculation could have $n_i = 10$ and $n = 100$. The size of the Hessian is then 1000*1000. It is no problem to construct such a matrix and store it. Equation (4:40) or the corresponding AM secular problem can then be solved using a suitable iterative method. Alternatively the method can be programmed as a direct procedure, which computes the update vectors directly from the elements of the Hessian matrix. Such a method would be more independent of the size of the MO basis set.

4.6 The gradient and Hessian matrix elements

We shall in this section derive the explicit expressions for the elements of the gradient vector and the Hessian matrix. The derivation is a good exercise in handling the algebra of the excitation operators $\hat{E}_{ij}$ and the reader is suggested to carry out the detailed calculations, where they have been left out in the present exposition.

We start with the equation for the orbital gradient (4:9). We then have to compute the commutator between the Hamiltonian (3:24) and an excitation operator $\hat{E}_{pq}$. To do that we use the commutator relation for the excitation operators as given in equation (3:8). The result of the computation is:

$$\frac{1}{2} g_{pq}^{(o)} = F_{pq} - F_{qp} \tag{4:41}$$

where the matrix F_{pq} is defined as:

$$F_{pq} = \sum_{\alpha} h_{\alpha p} D_{\alpha q} + 2 \sum_{\alpha,\beta,\gamma} (\alpha\beta|\gamma p) P_{\alpha\beta\gamma q} \tag{4:42}$$

This is the MCSCF Fock matrix. It is defined in terms of the first and second order reduced density matrices **D** and **P** as given by (3:16) and (3:23) or with the present nomenclature:

$$D_{\alpha\beta} = <0|\hat{E}_{\alpha\beta}|0> \quad \text{and } P_{\alpha\beta\gamma\delta} = \frac{1}{2}<0|\hat{E}_{\alpha\beta}\hat{E}_{\gamma\delta} - \delta_{\beta\gamma}\hat{E}_{\alpha\delta}|0> \tag{4:43}$$

The one- and two-electron integrals occurring in (4:42) are defined as (compare (3:12) and (3:19)):

$$h_{\alpha\beta} = \int \varphi_\alpha^*(\mathbf{r})\hat{h}\varphi_\beta(\mathbf{r})dV$$

$$(\alpha\beta|\gamma\delta) = \int\int \varphi_\alpha^*(\mathbf{r}_1)\varphi_\beta(\mathbf{r}_1)(1/r_{12})\varphi_\gamma^*(\mathbf{r}_2)\varphi_\delta(\mathbf{r}_2)dV_1dV_2 \tag{4.44}$$

The Fock matrix (4:42) is in general not Hermitian for a non-converged MCSCF wave function. With optimized orbitals the gradient is zero. The MCSCF Fock matrix is thus Hermitian at this point on the energy surface. This condition has been used as a basis for optimization schemes in earlier developments of the MCSCF methodology. Convergence of such first order optimization schemes is, however, often poor, and they are not very much used today.

It is instructive to compute the gradient (4:42) for the special case of a closed shell Hartree-Fock wave function. Let us denote the occupied orbitals with labels i,j,k,l... The only non-vanishing first and second order density matrix elements are in this case:

$$D_{ij} = 2\delta_{ij}$$

$$P_{ijkl} = 2\delta_{ij}\delta_{kl} - \delta_{jk}\delta_{il} \tag{4:45}$$

The MCSCF Fock matrix then takes the form:

$$F_{pi} = 2\left(h_{pi} + \sum_j \{2(jj|pi) - (pj|ij)\}\right) \tag{4:46}$$

We recognize this as the normal Fock matrix in closed shell Hartree-Fock theory, but only for those matrix elements where the second index, i, corresponds to an occupied orbital. The first index is arbitrary. The condition for an optimum wave function is then: $F_{pi} = F_{ip}$. This is trivially fulfilled when also p corresponds to an occupied orbital, since the wave function is invariant towards rotations among these orbitals. On the other hand, when p is a virtual orbital we have $F_{ip} = 0$ (all density matrix elements in (4:42) are then zero).

Thus the Hartree-Fock condition can be written as $F_{ai} = 0$, where a is a virtual and i an occupied orbital. This condition can easily be transformed into the normal Hartree-Fock equations.

Before ending the discussion of the MCSCF Fock operator we shall transform it to a form which is more suitable in practical applications. Let us divide the occupied orbitals into two subsets: the inactive orbitals, which are doubly occupied in all configurations, and the active orbitals, which are only partially occupied. The inactive orbitals we denote with indices i,j,k,l..., and the active orbital with indices t,u,v,x... It is then possible to separate out the contribution from the inactive orbitals in the two-electron part of the Fock operator, by using the relation: $\hat{E}_{ip}|0\rangle = 2\delta_{pi}|0\rangle$. The result of this separation can be written in the following form:

$$F_{ip} = 2(F^I_{ip} + F^A_{ip})$$
$$F_{tp} = \sum_{v} D_{tv} F^I_{vp} + 2\sum_{u,v,x} P_{tuvx}(pu|vx) \tag{4.47}$$

where we have defined two matrices $\mathbf{F}^I$ and $\mathbf{F}^A$ with matrix elements:

$$F^I_{pq} = h_{pq} + \sum_{k}[2(pq|kk) - (pk|qk)]$$
$$F^A_{pq} = \sum_{v,x} D_{vx}\left[(pq|vx) - \tfrac{1}{2}(pv|qx)\right] \tag{4:48}$$

These two matrices contain all interaction with the inactive orbitals. Notice the similarity between $\mathbf{F}^I$ and the closed shell Hartree-Fock operator. The two matrices (4:48) can conveniently be calculated first in AO basis, using a list of AO two-electron integral - or, as in most SCF programs, a list of super-matrix elements - and then transformed into MO basis. This constitutes a great simplification in the calculation of the orbital gradient, since the MO two-electron integrals needed to construct the final Fock matrix (4:47) can be restricted to integrals with three of the indices in the *active* orbital space. The number of such orbitals is in most cases small, which simplifies the transformation from AO to MO basis. Normally they can be stored in the central memory of the computer, with the consequence that the two-electron term in (4:47) can be performed as a fast vector operation (a dot product). This is an important simplification, since the two-electron transformation is one of the time consuming parts of an MCSCF calculation.

Now let us turn to the different parts of the Hessian matrix. The CI section of this matrix, $\mathbf{H}^{(cc)}$, will not be discussed in detail here. It is the subject of another series of lectures. The elements of the coupling matrix, $\mathbf{H}^{(co)}$, as given by equation (4:15) have a structure very similar to the orbital gradient. The only difference is that the matrix element is now between the MC state |0> and the complementary states |K>. Performing the algebraic manipulations we obtain:

$$H^{(co)}_{K,pq} = 4(F^K_{pq} - F^K_{qp}) \tag{4:49}$$

where we have defined the transition Fock operator as:

$$F^K_{pq} = \sum_{\alpha} h_{\alpha p} D^K_{\alpha q} + 2\sum_{\alpha,\beta,\gamma} (\alpha\beta|\gamma p) P^K_{\alpha\beta\gamma q} \tag{4:50}$$

and the symmetric transition density matrices as:

$$D^K_{\alpha\beta} = \frac{1}{2}\langle 0|\hat{E}_{\alpha\beta} + \hat{E}_{\beta\alpha}|K\rangle$$
$$P^K_{\alpha\beta\gamma\delta} = \frac{1}{2}\langle 0|\hat{e}_{\alpha\beta\gamma\delta} + \hat{e}_{\delta\gamma\beta\alpha}|K\rangle \quad \text{with } \hat{e}_{\alpha\beta\gamma\delta} = \frac{1}{2}(\hat{E}_{\alpha\beta}\hat{E}_{\gamma\delta} - \delta_{\beta\gamma}\hat{E}_{\alpha\delta}) \tag{4:51}$$

We can simplify the calculation of the matrix (4:50) in the same way as we did for the MCSCF Fock matrix. The same type of integrals are thus needed here. However, the calculation is now much more complicated since we need the transition density matrices. In practical applications one does not work directly with the complementary states |K>. Instead the calculation is performed over the CI basis states |m>, where the CI coupling coefficients occur instead of the transition density matrices.
Let us finally take a closer look at the orbital Hessian matrix, $\mathbf{H}^{(oo)}$. The calculation now involves the evaluation of commutators between the Hamiltonian and products of excitation operators according to equation (4:14). In spite of the rather tiring algebra, the result takes a surprisingly simple form:

$$H^{(oo)}_{pq,rs} = (1-\hat{P}_{pq})(1-\hat{P}_{rs})\left((F_{ps}+F_{sp})\delta_{qr} - 2h_{ps}D_{qr} + 2\sum_{\alpha,\beta}[4(p\alpha|r\beta)P_{q\alpha s\beta} + 2(pr|\alpha\beta)P_{qs\alpha\beta}]\right) \tag{4:52}$$

where we have introduced permutation operators $\hat{P}_{pq}$, which permute the indices p and q. Again we can introduce simplifications in this expression by explicitly sum over the inactive orbitals. We note, however, that two-electron integrals with two arbitrary indices occur in (4:52). Thus we here need those transformed integrals, where two of the indices run over all orbitals and the other two over all occupied (inactive and active) orbitals. The two-electron transformation needed for the formation of the orbital Hessian is thus more complex than the one needed for the gradient. It is usually not possible to store these integrals in the fast memory of the computer, which makes the calculation more I/O bound. In order to see this more explicitly, let us write down one type of matrix element, corresponding to the interaction between two rotations between active and external orbitals (pq = at and rs = bu). The result is:

$$H^{(oo)}_{at,ub} = 2\left(D_{tu}F^{I}_{ab} - \delta_{ab}F_{tu} + \sum_{v,x}[P_{tuvx}(ab|vx) + (P_{txvu}+P_{txuv})(ax|bv)] \right) \quad (4:53)$$

Note that all interactions with inactive orbitals have been included into the Fock type matrices **F** and $\mathbf{F}^{I}$. The sum in (4:53) runs over active orbitals only. The two-electron integrals include two orbitals from the external space. Other type of matrix elements contain integrals where two indices are inactive. In general we need all integrals of the type $(\alpha\beta|pq)$ and$(\alpha p|\beta q)$, where α,β run over all orbitals and p,q over the occupied orbitals. The phrase 'second order integrals' is sometimes used for this subset of the total list of MO two-electron integrals. The number of operations needed for the transformation of these integrals from AO to MO basis is proportional to $n_i n^4$, where n_i is the number of occupied orbitals. This number is usually much larger than the number of active orbitals, and the second order transformation then becomes a time-consuming part of the calculation. It is therefore important to have efficient algorithms for this transformation. It is, for example, a great advantage to have the basic AO integrals computed over symmetry-adapted basis functions. The transformation can then be divided into symmetry blocks, resulting in a considerable reduction in computer time (provided there is some symmetry in the system under study). Symmetry blocks in which there are no occupied orbitals are not needed and the corresponding AO integrals can be sorted out before the transformation starts. Some of the approximate optimization methods to be discussed below avoid the second order transformation altogether by basing the algorithm entirely on the Fock matrices.

4.7 The super-CI method

Several attempts have been made to devise simpler optimization methods than the full second order Newton-Raphson approach. Some are approximations of the full method, like the unfolded two-step procedure, mentioned in the preceding section. Others avoid the construction of the Hessian in every iteration by means of update procedures. An entirely different strategy is used in the so called Super - CI method. Here the approach is to reach the optimal MCSCF wave function by annihilating the singly excited configurations (the Brillouin states) in an iterative procedure. This method will be described below and its relation to the Newton-Raphson method will be illuminated. The method will first be described in the unfolded two-step form. The extension to a folded one-step procedure will be indicated, but not carried out in detail. We therefore assume that every MCSCF iteration starts by solving the secular problem (4:39) with the consequence that the MC reference state does not interact with the complementary space. Now define a set of 'singly excited' (SX) states according to:

$$|pq\rangle = \hat{E}^{-}_{pq}|0\rangle \tag{4:54}$$

These states are called Brillouin states, since their interaction with the reference state gives the generalized Brillouin elements (4:9). A super-CI wave function is defined as a linear combination of the SX states and the reference state:

$$|SCI\rangle = |0\rangle + \sum_{p>q} t_{pq}|pq\rangle \tag{4:55}$$

The super-CI method now implies solving the corresponding secular problem and using t_{pq} as the exponential parameters for the orbital rotations. Alternatively we can construct the first order density matrix corresponding to the wave function (4:55), diagonalize it, and use the natural orbitals as the new trial orbitals in $|0\rangle$. Both methods incorporate the effects of $|pq\rangle$ into $|0\rangle$ to second order in t_{pq}. We can therefore expect t_{pq} to decrease in the next iteration. At convergence all t_{pq} will vanish, which is equivalent to the condition:

$$\langle 0|\hat{H}|pq\rangle = 0 \tag{4:56}$$

The super-CI method thus leads to the stationary point by a direct annihilation of the single excitations (4:54).

We shall now study the secular equation in some detail in order to make a comparison between super-CI and the NR method in the augmented Hessian form. The first thing to note is the non-orthogonality between the SX states:

$$S_{pq,rs} = \langle 0|\hat{E}^{-}_{qp}\hat{E}^{-}_{rs}|0\rangle \neq \delta_{pq,rs} \tag{4:57}$$

They are not even normalized ($S_{pq,pq} \neq 1$), but they are orthogonal to the reference state, due to the trivial identity $\langle 0|\hat{E}_{pq}|0\rangle = \langle 0|\hat{E}_{qp}|0\rangle$. The matrix elements of the Hamiltonian have the form:

$$\begin{aligned} H_{0,pq} &= \langle 0|\hat{H}|pq\rangle = \langle 0|\hat{H}\hat{E}^{-}_{pq}|0\rangle = w_{pq} \\ H_{pq,rs} &= \langle pq|\hat{H}|rs\rangle = \langle 0|\hat{E}^{-}_{qp}\hat{H}\hat{E}^{-}_{rs}|0\rangle = d_{pq,rs} \end{aligned} \tag{4:58}$$

where w_{pq} is a component of the gradient vector **w** in (4:17). The matrix elements $d_{pq,rs}$ are related to the orbital Hessian **c** in (4:17) as will be demonstrated below. Using the matrix elements (4:57) and (4:58), the secular equation can be written in the form:

$$\begin{pmatrix} 0 & \mathbf{w}^{\dagger} \\ \mathbf{w} & \mathbf{d}-E_0\mathbf{S} \end{pmatrix}\begin{pmatrix} 1 \\ \mathbf{t} \end{pmatrix} = \varepsilon_{SCI}\begin{pmatrix} 1 & \mathbf{0} \\ \mathbf{0} & \mathbf{S} \end{pmatrix}\begin{pmatrix} 1 \\ \mathbf{t} \end{pmatrix} \tag{4:59}$$

where $\varepsilon_{SCI} = E_{SCI} - E_0$, E_{SCI} being the super-CI energy:

$$E_{SCI} = \langle SCI|\hat{H}|SCI\rangle / \langle SCI|SCI\rangle \tag{4:60}$$

Equation (4:59) can be compared to the unfolded two-step version of the augmented Hessian method, which results in the secular equation:

$$\begin{pmatrix} 0 & \mathbf{w}^{\dagger} \\ \mathbf{w} & \mathbf{c} \end{pmatrix}\begin{pmatrix} 1 \\ \mathbf{T} \end{pmatrix} = \varepsilon_{AM}\begin{pmatrix} 1 \\ \mathbf{T} \end{pmatrix} \tag{4:61}$$

where **c** is the orbital Hessian with elements (compare eq. (4:14)):

$$c_{pq,rs} = \frac{1}{2}\langle 0|\hat{E}^{-}_{pq}\hat{E}^{-}_{rs}\hat{H}|0\rangle + \frac{1}{2}\langle 0|\hat{H}\hat{E}^{-}_{pq}\hat{E}^{-}_{rs}|0\rangle - \langle 0|\hat{E}^{-}_{pq}\hat{H}\hat{E}^{-}_{rs}|0\rangle \tag{4:62}$$

Comparing equations (4:59) and (4:61) we note two differences: the occurence of the overlap matrix in the super-CI equation, and the replacement of **c** with $\mathbf{d}-E_0\mathbf{S}$. Let us investigate the latter difference in more detail:

$$d_{pq,rs} - E_0 S_{pq,rs} = -\langle 0|\hat{E}^{-}_{pq}\hat{H}\hat{E}^{-}_{rs}|0\rangle + E_0\langle 0|\hat{E}^{-}_{pq}\hat{E}^{-}_{rs}|0\rangle \tag{4:63}$$

We can write $\hat{H}|0\rangle$ occurring in (4:62) as:

$$\hat{H}|0\rangle = E_0|0\rangle + |\perp\rangle, \tag{4:64}$$

where the second term represents some vector in the orthogonal complement to the CI space of the MCSCF wave function, ($|0\rangle$,$|K\rangle$). For a CASSCF wave function this term is a linear combination of configurations with one or two electrons in the external orbital space and/or the number of inactive electrons decreased by one or two. For an incomplete CI space, this term also includes terms from the difference between the corresponding CAS space and the current CI space. Now, inserting (4:64) into (4:62) we obtain:

$$c_{pq,rs} = d_{pq,rs} + \frac{1}{2}\langle 0|\hat{E}^{-}_{pq}\hat{E}^{-}_{rs} + \hat{E}^{-}_{rs}\hat{E}^{-}_{pq}|\perp\rangle \tag{4:65}$$

The difference between **c** and **d** thus involves the overlap between the orthogonal complement and the double excitations:

$$\hat{E}^{-}_{pq}\hat{E}^{-}_{rs}|0\rangle$$

It is natural to consider the AM method as a mode-damped version of the Newton-Raphson procedure. The super-CI method is then best regarded as an approximation to the AM method. Both AM and super-CI have the advantage of always converging to a local minimum (ε_{SCI} is always negative), in contrast to the Newton-Raphson method, which without sign and step size control can converge to a saddle point or a maximum. The AM and super-CI methods therefore often show better global convergence behaviour and is often recommended to be used in the beginning of a calculation, until the local region has been reached. The AM method also exhibits second-order convergence in the local area, if no further approximations are introduced.The AM version of the Newton-Raphson method is therefore to be preferred in comparison with the conventional method.

The super-CI method can alternatively be given in a folded form, which includes the coupling between the CI and orbital rotations. This is done by adding the complementary CI space, $|K\rangle$, to the super-CI secular problem. As in the Newton-Raphson approach, it is more efficient to transform the equations back to the original CSF space, and thus work with a super-CI consisting of the CI basis states plus the SX states. It is left to the reader as an exercise to construct the corresponding secular equation and compare it with the folded one-step Newton-Raphson equations (4:22).

Computationally the super-CI method is more complicated to work with than the Newton-Raphson approach. The major reason is that the matrix **d** is more complicated than the Hessian matrix **c**. Some of the matrix elements of **d** will contain up to fourth order density matrix elements for a general MCSCF wave function. In the CASSCF case only third order term remain, since rotations between the active orbitals can be excluded. Besides, if an unfolded procedure is used, where the CI problem is solved to convergence in each iteration, the highest order terms cancel out. In this case up to third order density matrix elements will be present in the matrix elements of **d** in the general case. Thus super-CI does not represent any simplification compared to the Newton-Raphson method.

However, what we want to achieve in an MCSCF calculation is to reach the point on the energy surface where the gradient of the energy is zero. How we get there is of less importance, as long as the process does not consume too much computer time. With this in mind one could try to simplify the Super-CI matrix **d**, with the hope that the simplified version still leads to solutions close to the exact solution, such that the iterations move the solution towards the stationary point. An approximate form of the super-CI method has been constructed, which leads to significant simplifications of the secular equation and in addition greatly reduces the transformation problem for the two-electron integrals. The idea here is to approximate the matrix elements of **d** by replacing the Hamiltonian by an effective one-electron operator constructed from a Fock type operator:

$$\hat{H}' = \sum_{p,q} f_{pq} \hat{E}_{pq} \tag{4:66}$$

This operator is used only for the construction of the **d** matrix. The gradient terms in the secular matrix are, of course, computed exactly.

The matrix **f** in (4:66) can be chosen such that the matrix elements of $\hat{H}'$ represent orbital energy differences in the spirit of Möller-Plesset perturbation theory. A suitable choice is then:

$$f_{pq} = F^{I}_{pq} + F^{A}_{pq} = h_{pq} + \sum_{r,s} D_{rs}\left[(pq|rs) - \tfrac{1}{2}(pr|qs)\right] \tag{4:67}$$

This matrix is obtained from the spin averaged matrix element over the MCSCF wave function of the operator:

$$\hat{F}_{ij\sigma} = \hat{a}_{i\sigma}[\hat{H},\hat{a}^{\dagger}_{j\sigma}] - \hat{a}^{\dagger}_{i\sigma}[\hat{H},\hat{a}_{j\sigma}] \tag{4:68}$$

A diagonal element i of (4:68) correspond to the energy difference between the system and its positive ion, when i represents an orbital which is fully occupied in $|0\rangle$. The energy difference between the system and its negative ion is obtained when spin orbital i is empty. For orbitals partly occupied in $|0\rangle$, (4:68) corresponds to an interpolation between the two extremes. Thus (4:67) gives the normal properties of the orbital energies in HF theory for orbitals that are either fully occupied or completely empty.

The operator (4:66) together with the definition (4:67) of the matrix **f** gives a surprisingly powerful approximation to the full super-CI Hamiltonian. It has been used with great success in a number of calculations of a variety of MCSCF wave functions both for ground and excited states.

We shall not carry out the calculation of the matrix elements of $\hat{H}'$ in detail. It is left as an exercise to the reader. One important computational aspect of (4:66) and (4:67) is that the matrix elements of **f** can be constructed first in AO basis and then transformed to MO basis. There is thus no need for any further two-electron transformation. As a result the computational effort is almost independent of the number of inactive orbitals. MCSCF calculations can then be performed for large molecules without the need to freeze the inactive orbitals, something which has sometimes been done in order to simplify second order calculations.

However, super-CI, especially in the approximate form described above, is only a first order optimization procedure. We can therefore not expect fast convergence to the stationary point. Especially at the end of the calculation convergence may slow down, which might make it difficult to obtain very accurately converged solutions, something that is of considerable importance when the result is to be used for the calculation of molecular gradients and Hessians (e.g. for geometry optimization).

There exists, however, a cheap procedure which can be used to improve convergence toward second order with very little extra cost. This procedure also automatically reintroduces the CI- coupling term, which is normally left out in super-CI calculations.

Only the gradient vector is calculated exactly in approximate optimization methods like the super-CI approach. This information about the exact gradients can be used to improve the convergence of the calculation via a procedure, that updates the approximate Hessian, which is implicitly used in the calculation. Suppose that we know the gradient at two consecutive points in a sequence of iterations, $p^{(n+1)}$ and $p^{(n)}$. Let us expand the gradient around the point $p^{(n+1)}$:

$$\mathbf{g}(\mathbf{p}^{(n)}) = \mathbf{g}(\mathbf{p}^{(n+1)}) + \mathbf{H}(\mathbf{p}^{(n+1)})(\mathbf{p}^{(n)} - \mathbf{p}^{(n+1)}) + O(\Delta\mathbf{p}^2) \qquad (4:69)$$

The difference between the gradients at the two points thus gives a finite difference approximation to the projection of the exact Hessian $\mathbf{H}$ on the direction $(\mathbf{p}^{(n)} - \mathbf{p}^{(n+1)})$, at the point $\mathbf{p}^{(n+1)}$. This information about the Hessian can be used to improve the convergence of the calculation. Suppose that we have computed a step on the energy surface as $\Delta\mathbf{p}^{(n+2)}$. We can always formally write this as:

$$\Delta\mathbf{p}^{(n+2)} = -\,\mathbf{H}'(\mathbf{p}^{(n+1)})^{-1}\mathbf{g}(\mathbf{p}^{(n+1)}) \qquad (4:70)$$

even if we have obtained $\Delta\mathbf{p}$ from an approximate calculation, like the super-CI method. $\mathbf{H}'$ in (4:70) is to be regarded as an approximation to the exact Hessian. Using information from the gradients calculated at preceding iterations, we can improve this step by adding a correction term, such that instead:

$$\Delta\mathbf{p}^{(n+2)} = -(\,\mathbf{H}'(\mathbf{p}^{(n+1)})^{-1} + \Delta\mathbf{H}(\mathbf{p}^{(n+1)})^{-1})\mathbf{g}(\mathbf{p}^{(n+1)}) \qquad (4:71)$$

where $\Delta\mathbf{H}^{-1}$ is a correction to the inverted Hessian, obtained from the gradients. We shall not here derive the procedures used to obtain this correction term. A number of different methods are available. All these methods are based on the fulfillment of the *quasi - Newton* condition:

$$\mathbf{H}(\mathbf{p}^{(n+1)})(\mathbf{p}^{(n)} - \mathbf{p}^{(n+1)}) = \mathbf{g}(\mathbf{p}^{(n)}) - \mathbf{g}(\mathbf{p}^{(n+1)}) \tag{4:72}$$

which follows from (4:69). Some update procedures assumes in addition that the projection of **H** on directions perpendicular to $\Delta\mathbf{p}$ do not change:

$$\mathbf{H}(\mathbf{p}^{(n+1)})\mathbf{q} = \mathbf{H}(\mathbf{p}^{(n)})\mathbf{q} \tag{4:73}$$

where **q** is perpendicular to $\Delta\mathbf{p}$.

One important aspect of these update methods is that they reintroduce the CI coupling terms in the Hessian, in cases where the original approximation is an unfolded two-step procedure without coupling. The update methods are easy to implement and turns out to be a powerful and simple way of improving convergence in an MCSCF calculation. It has recently been used with success together with the approximate super-CI method described earlier in this section. Table 1 gives an illustration of a convergence acceleration, which seems to be typical. Notice especially the fast convergence in the last iterations.

Table 1. Convergence in a CASSCF calculation on water, with a DZP basis. The approximate super-CI method was used with and without quasi-Newton update. The active space comprised 8 orbitals ($4a_1$,$2b_1$, $2b_2$ in C_{2v} symmetry), yielding 492 CSF's. The 1s orbital was inactive.

	without update			with QN update		
Iteration	Energy +76 a.u	Change	max gradient	Energy +76 a.u	Change	max gradient
1	-.10312646	$.00\ 10^{+0}$	$.23\ 10^{-01}$	-.10312646	$.00\ 10^{+0}$	$.23\ 10^{-01}$
2	-.16204065	$-.59\ 10^{-01}$	$.22\ 10^{-01}$	-.16204065	$-.59\ 10^{-01}$	$.22\ 10^{-01}$
3	-.16435412	$-.23\ 10^{-02}$	$.62\ 10^{-02}$	-.16435412	$-.23\ 10^{-02}$	$.62\ 10^{-02}$
4	-.16462534	$-.27\ 10^{-03}$	$.30\ 10^{-02}$	-.16466874	$-.31\ 10^{-03}$	$.20\ 10^{-02}$
5	-.16468364	$-.48\ 10^{-04}$	$.12\ 10^{-02}$	-.16468386	$-.15\ 10^{-04}$	$.36\ 10^{-03}$
6	-.16468633	$-.11\ 10^{-04}$	$.61\ 10^{-03}$	-.16468702	$-.32\ 10^{-05}$	$.17\ 10^{-03}$
7	-.16468708	$-.27\ 10^{-05}$	$.24\ 10^{-03}$	-.16468733	$-.30\ 10^{-06}$	$.76\ 10^{-04}$
8	-.16468730	$-.75\ 10^{-06}$	$.14\ 10^{-03}$	-.16468739	$-.60\ 10^{-07}$	$.16\ 10^{-04}$
9	-.16468737	$-.22\ 10^{-06}$	$.62\ 10^{-04}$	-.16468740	$-.10\ 10^{-07}$	$.57\ 10^{-05}$
10	-.16468739	$-.68\ 10^{-07}$	$.38\ 10^{-04}$			
11	-.16468740	$-.21\ 10^{-07}$	$.19\ 10^{-04}$			
12	-.16468740	$-.68\ 10^{-08}$	$.11\ 10^{-04}$			
13	-.16468740	$-.22\ 10^{-08}$	$.58\ 10^{-05}$			
14	-.16468740	$-.72\ 10^{-09}$	$.37\ 10^{-05}$			

4.8 Exercises

4.1 Derive the closed shell Hartree-Fock equations from the condition that the gradient (4:9) is zero for stationary orbitals. The wave function $|0>$ is assumed to be a closed shell Slater determinant. Start by deriving equation (4:41), and from there (4:46).

4.2 Illustrate the bracketing theorem mentioned in connection with the augmented Hessian method, by solving equation (4:31) for the energy E in the form: E = f(E). Plot both the functions E and f(E) and show that the crossing points (the eigenvalues E_i) satisfies the betweenness condition.

4.3 Derive the detailed expression for the orbital Hessian for the special case of a closed shell single determinant wave function. Compare with equation (4:53) to check the result. The equation can be used to construct a second order optimization scheme in Hartree-Fock theory. What are the advantages and disadvantages of such a scheme compared to the conventional first order methods?

4.4 Calculate the explicit expression, in terms of first and second order density matrices, for the overlap integral (4:57) occurring in the super-CI method.

4.5 Show that the spin averaged expectation value:

$$\frac{1}{2}\sum_{\sigma}<0|\hat{F}_{ij\sigma}|0>$$

of the operator $\hat{F}_{ij\sigma}$ given by (4:68) gives the matrix element (4:67). Show in addition that the diagonal element $<0|\hat{F}_{ii\sigma}|0>$ gives the ionization energy, when spin orbital $\phi_{i\sigma}$ is fully occupied, and the electron affinity, when it is empty.

5. Application of the MCSCF method

5.1 Introduction

Programs are available today which perform closed shell SCF (or UHF) calculations on molecules almost automatically. The user only has to provide the structural information about the molecule, and maybe suggest a basis set. This is an important development, since the user can then concentrate on his chemical problem, and does not have to know much about the program he is running. However, it is important to know what one is doing and what can be expected to be the outcome of a calculation. This means that some feeling has to be acquired for the relation between basis set and accuracy, and the size of the correlation corrections for the type of problem under study. This knowledge cannot, of course, be exact, but is rather a collection of experiences from earlier work and from calibration studies. A prerequisite for responsible studies of molecules using the SCF method is therefore a reasonably thorough knowledge of earlier work in the field.

The 'black box' situation of SCF applications has not yet been reached for the multiconfigurational SCF theory. This constitutes a major problem, since MCSCF is a much better starting point for quantum chemical calculations on many interesting chemical problems (a good example is studies of transition states for chemical reactions). A development towards more automatized procedures can consequently be expectedto take place in MCSCF theory too.

Why is it then more complicated to perform an MCSCF calculation? The basis set problem is similar in SCF and MCSCF theory, and can be solved by the construction of basis set libraries containing well documented and thoroughly tested basis sets of varying quality. The major difference lies in the construction of the wave function. In SCF theory it is given as a single determinant, where the occupied orbitals are determined by the optimization procedure. This is not so in MCSCF theory. Here a decision on the general structure of the wave function has to be made beforehand. This cannot be done without an a priori knowledge about the electronic structure. In normal bonding situations such a knowledge is not too hard to achieve. A few examples were given in chapter 2. However, in more complex situations - like the transition state for a chemical reaction - it may be difficult to make a priori judgements about the most important electronic configurations to include in the MCSCF wave function. Below we shall discuss how this problem can be, at least partly, solved by the Complete Active Space Method. Here the problem is reduced to defining a set of active orbitals, which describe the near-degeneracy effects. The choice of active orbitals requires an insight into the electronic structure, which often is rather obvious. However, not always. There are many cases where the choice is not at all clear, and several trials have to be made before the best choice has been found. This is far from black box situation, and the procedure is not easily automatized. A development towards more automatic procedures are, however, necessary if the MCSCF method is going to be widely used in chemistry, and not only by a few specialists, as the situation is today. One recent suggestion (Pulay) is to use a UHF calculation as a starting point. The natural orbitals (not the natural spin-orbitals) of a UHF wave function will have partial occupation numbers, and can be used to divide the occupied orbitals into inactive and active. The orbitals with occupation numbers different from two or zero are then chosen as active. Normally such a procedure would give a rather small active space comprising only the most important near-degeneracy effects. One problem with the method is that the active orbital space will be different on different parts of an energy surface. Anyway it is interesting, since it leads to a completely automatic way of choosing the MCSCF wave function.

5.2 The complete active space (CAS) SCF method

The CASSCF method has been mentioned at several places in these lecture notes. Here we shall give a more detailed account of this method, which is probably the most widely used MCSCF method today. It is based on a partitioning of the occupied molecular orbitals into subsets, corresponding to how they are used to build the wave function. We define for each symmetry block of MO's the following subsets:

1. Inactive orbitals
2. Active orbitals
3. External orbitals

The inactive and active orbitals are occupied in the wave function, while the external (also called secondary or virtual) orbitals span the rest of the orbital space, defined from the basis set used to build the molecular orbitals. The inactive orbitals are kept doubly occupied in all configurations that are used to build the CASSCF wave function. The number of electrons occupying these orbitals is thus twice the number of inactive orbitals. The remaining electrons (called active electrons) occupy the active orbitals.

The CASSCF method is an attempt to generalize the Hartree-Fock model to situations where near degeneracies occur, while keeping as much of the conceptual simplicity of the RHF approach as possible. Technically the CASSCF model is by necessity more complex, since it is based on a multiconfigurational wave function. The building blocks are, as in the RHF model, the occupied (inactive and active) orbitals. The number of electrons is, however, in general less than twice the number of occupied orbitals. The number of electron configurations generated by the orbital space is therefore larger than unity. The total wave function is formed as a linear combination of all the configurations, in the N-electron space, that fulfill the given space and spin symmetry requirements,and have the inactive orbitals doubly occupied. It is in this sense "complete" in the configurational space spanned by the active orbitals. The inactive orbitals represent an "SCF sea" in which the active electrons move. These orbitals have occupation numbers exactly equal to two, while the occupation numbers of the active orbitals varies between 0 and 2. The partitioning of the internal orbital space into inactive and active orbitals was illustrated in chapter 2 of these notes. It is obvious that the inactive orbitals should be chosen as the orbitals that are not expected to contribute to near-degeneracy correlation effects.

The conceptual simplicity of the CASSCF model lies in the fact that once the inactive and active orbitals are chosen, the wave function is completely specified. In addition such a model leads to certain simplifications in the computational procedures used to obtain optimized orbitals and CI coefficients, as was illustrated in the preceding chapters. The major technical difficulty inherent to the CASSCF method is the size of the complete CI expansion, N_{CAS}. It is given by the so-called Weyl formula, which gives the dimension of the irreducible representation of the unitary group U(n) associated with n active orbitals, N active electrons, and a total spin S:

$$N_{CAS} = \frac{2S+1}{n+1}\binom{n+1}{N/2-S}\binom{n+1}{N/2+S+1} \qquad (5:1)$$

Obviously N_{CAS} increases strongly as a function of the size n of the active orbital space. In practice this means that there is a rather strict limit on the size of this space. Experience shows that this limit is normally reached for n around 10-12 orbitals, except for cases with only a few active electrons or holes. The large number of CASSCF applications performed to date illustrates clearly that this limitation does not normally create any serious problem. It should be remembered that the CASSCF model is an extension of the RHF scheme. As such it is supposed to produce a good zeroth-order approximation to the wave function, when near degeneracies are present. This goal can in most cases be achieved with only a few active orbitals. The CASSCF method has not been developed for treating dynamical correlation effects, but to provide a good starting point for such studies.

However, there exist cases where it is advantageous to be able to use a larger set of active orbitals. The dimension of the CAS wave function can then become prohibitively large, and it may be of interest to look for other means of restricting the expansion length. This can be done in many ways, and results in a number of different types of MCSCF expansions. Assume that the number of inactive orbitals is n_i and the number of active orbitals n_a, the number of active electrons being N_a. We can then formally write the corresponding CAS wave function as:

$$(\varphi_1)^2(\varphi_2)^2...(\varphi_{n_i})^2(\varphi_{n_i+1},....\varphi_{n_i+n_a})^{N_a} \qquad (5:2)$$

meaning that the first n_i orbitals are doubly occupied, while the next n_a

orbitals constitute the active space, which is occupied with N_a electrons. The total number of electrons in the wave functions is $2n_i + N_a$. Let us use the N_2 molecule as an explicit example. The most simple CAS wave function which is reasonable for this molecule will have the $1\sigma_g$, $1\sigma_u$, $2\sigma_g$, and $2\sigma_u$ orbitals inactive, while the remaining valence orbitals constitute the active space. We can then write the CAS wave function for N_2 as:

$$(1\sigma_g)^2(1\sigma_u)^2(2\sigma_g)^2(2\sigma_u)^2(3\sigma_g,3\sigma_u,1\pi_u,1\pi_g)^6 \tag{5:3}$$

corresponding to the distribution of 6 active electrons in 6 active orbitals. Notice, that the active orbitals originate from the atomic 2p orbitals. The wave function (5:3) will therefore give a correct description of the dissociation of the molecule into two nitrogen atoms. It does not, however, give a very accurate description of the molecule, but can serve as a good starting point for more advanced calculations. This wave function is not very large. Using D_{2h} symmetry, the CI expansion will comprise only 32 CSF's. Even so, we shall use it as an example of how a CAS wave function can be further constrained by removing certain CSF's from the expansion. One way, which has been found useful in some applications, is to divide the active orbitals into subgroups, and fix the number of electrons in each group. In the example above we could, for example, require that there are two electrons in the 3σ orbitals and four electrons in the 1π orbitals. The corresponding constrained CAS (CCAS) wave function can formally be written as:

$$(1\sigma_g)^2(1\sigma_u)^2(2\sigma_g)^2(2\sigma_u)^2(3\sigma_g,3\sigma_u)^2(1\pi_u,1\pi_g)^4 \tag{5:4}$$

The number of CSF's is now reduced to 24. We can constrain the wave function further by requiring that there is two electrons in each of the bonding-antibonding pairs of π orbitals:

$$(1\sigma_g)^2(1\sigma_u)^2(2\sigma_g)^2(2\sigma_u)^2(3\sigma_g,3\sigma_u)^2(1\pi_{ux},1\pi_{gx})^2(1\pi_{uy},1\pi_{gy})^2 \tag{5:5}$$

which reduces the number of CSF's further to 20. Even if the wave function expansions (5:4) and (5:5) are not complete in the active space, there is no need to introduce active-active rotations in the orbital optimization procedure, since orbitals in different groups belong to different symmetries. When different groups contain orbitals with the same symmetry it is, however, necessary to introduce such rotations, since the energy is no longer invariant with respect to them.

In order to reduce further the CI expansion, it is necessary to impose restrictions to the possible spin couplings within each group. The $(3\sigma_g,3\sigma_u)$ pair, for example, contains three spin coupled orbital products:

$$(3\sigma_g)^2, (3\sigma_u)^2, (3\sigma_g,3\sigma_u)_S, \text{ and } (3\sigma_g,3\sigma_u)_T \tag{5:6}$$

We can now demand each pair of electrons in (5:5) to be singlet coupled, in which case the last term in (5:6) will not contribute to the wave function. Including only the singlet coupled terms for each pair of orbitals in (5:5) reduces the wave function further to comprise only 14 CSF's. If the open shell singlet coupled term in (5:6) is also deleted, a wave function is obtained,which spans the same CSF space as is used in the Perfect Pairing (PP) Generalized Valence Bond (GVB) method of Goddard and co-workers. It is, however, not the same wave function, since the GVB method uses a non-linear expansion in the CSF's, constructed as a antisymmetrized product of geminals.

A restriction of the full CAS wave function is not really necessary in the simple example used above, since the number of CSF's is small already when the full CAS space is used. In calculations using a larger active orbital space it may, however, be of interest to impose these constraints on the CSF space, especially since in many cases they do not lead to any serious degradation of the resulting wave function. One case where such constraints have been used successfully is in the study of the NN bond in N_2O_4, which was discussed in chapter 2 of these lecture notes. The active orbital space consisted in this case of six π orbitals together with six σ orbitals (describing the NN bond and the oxygen lone-pair electrons). The number of active electrons was 18. A full CASSCF wave function for this active space would comprise 7910 CSF's (using only the mirror plane of symmetry). Today such a CI expansion would be considered as small, and no further reduction necessary. At the time when the calculation was actually performed (1982), it was still difficult to treat wave functions comprising more than a few thousand terms. Therefore constraints where imposed by requiring that 10 electrons occupy the σ orbitals and the remaining 8 the π orbitals. The resulting wave function contains then only 976 terms, and the calculation can easily be carried out. The loss of accuracy expected from the deletion of configurations, where the number of π electrons is different from 8, can be expected to be small.

Another extension of the CASSCF model, which has recently been developed, is to introduce several active spaces, and restrict the number of electrons in each subspace in some way. The *Restricted Active Space (RAS)* model employs five orbital subspaces instead of three:

1. The inactive orbital space
2. The RAS 1 space
3. The RAS 2 space
4. The RAS 3 space
5. The external orbital space.

The inactive and external orbital spaces have the same properties as for CAS wave functions. The RAS 1 space consists of orbitals in which a certain number of holes may be created. One could for example allow single and double excitations out of this orbital space. Normally, all these orbitals would be doubly occupied in a CAS calculation. The RAS 2 space has the same properties as the active orbital space in a CAS wave function: all possible occupations and spin couplings are allowed. Finally the RAS 3 space is allowed to be occupied with up to a given number of electrons. A variety of wave functions can be created using the RAS concept. For example, by making the RAS 2 space empty, one arrives at a conventional singles and doubles (triples, quadruples,.....) CI wave function, with a single reference state. The analoge to an MR-SDCI wave function with a CAS reference space is obtained by also adding orbitals and electrons to the RAS 2 space. The RAS configuration space has the property of being closed under de-excitation, that is, moving electrons from a higher RAS space to a lower, does not generate any new configurations.

The RAS concept combines the features of the CAS wave functions with those of more advanced CI wave functions, where dynamical correlation effects are included. It is thus able to give a more accurate treatment of correlation effects in molecules. The fact that orbital optimization is included makes this method especially attractive for studies of energy surfaces, when there is a need to compute the energy gradient and Hessian with respect to the nuclear coordinates.
However, the RAS concept on the other hand complicates the orbital optimization. The wave function is no longer complete in the active orbital space. It is then necessary to introduce orbital rotations between the three RAS spaces, which may lead to convergence problems in more difficult cases.

5.3 Excited states and transition moments

Studies of excited states of molecular systems represent an interesting challenge for the MCSCF technique. Traditionally such calculations have been peformed using a common set of orbitals for all states. There are several reasons for such an approach. The different states will automatically be orthogonal to each other, and the calculation of transition properties is greatly simplified by the use of a common orbital set. Accurate results can normally also be obtained, at least for the excitation energies, if the wave function is obtained from a extended CI expansion, for example, by using the multi-reference CI method including all single and double excitations with respect to a given set of reference states. The selection of reference states can be made on the basis of an MCSCF calculation for the average energy of the electronic states under study. Simultaneously a reasonable set of orbitals is obtained.

Even if the method sketched above works well for many cases, there are several difficulties connected with it. It cannot easily be extended to larger systems, since the necessary MR-CI expansion then becomes excessively large.

Different electronic states have in many cases very differently shaped orbitals and the error introduced by using a common set cannot always be fully recovered by the MR-CI treatment. A well optimized wave function is especially important for the calculation of transition properties like the transition moments and the oscillator strength. A state specific calculation of the orbitals is more important for obtaining accurate values of the transition moments than extensive inclusion of correlation. Since excited states commonly exhibit large near-degeneracy effects in the wave function an MCSCF treatment then becomes necessary.

The major obstacle in using the MCSCF approach in studies of excited states has been the orthogonality problem. If MCSCF calculations are performed on two electronic states of the same symmetry, the resulting wave functions will not be orthogonal to each other. Independent of symmetry, the two sets of orbitals will also be different. This leads to difficulties in the calculation of the transition properties, and the interpretation is obscured by the non-orthogonality of the two wave functions. Ideally one would like to perform the calculation on the excited states including the condition that it remains orthogonal to the lower states of the same symmetry. Such a procedure, however, leads to a formidable computational problem, which has so far not been satisfactorily solved.

There exists today an alternative approach which has made MCSCF calculations on excited states feasible, also for rather large systems. A method has been developed which makes it easy to obtain orthogonal wave functions and transition densities from CASSCF wave functions optimized independently for a number of excited states of different or the same symmetry as the ground state. The method has been called the *CAS State Interaction (CASSI) method*. It will be briefly described below.

The possibility to use the CASSCF method for calculations of excitation energies and transition probabilities makes it suitable for studying problems in spectroscopy, photochemical reactions, and other areas of chemistry where excited states are involved.

MCSCF calculations on excited states are, however, more difficult than calculations on the ground state, especially when the excited state is not the lowest in its symmetry. First of all the selection of the state to study is not always obvious. Consider for example the calculation of a valence excited state in an aromatic system. Note that we are not asking for the second, third, etc excited state, but for a specific *valence* excited state. There could exist excited states of a different origin with lower energy, as for example Rydberg states, a situation which is not uncommon. In an unknown situation the safest procedure is then to start by optimizing the orbitals for the average energy of a number of excited states, large enough to include also the state of interest. Such calculations are not more difficult than calculations on a specific state. We start with the energy functional for the average energy of some states {I, I=1,M}:

$$E_{av} = \sum_{I} \omega_I E_I \tag{5:7}$$

where ω_I are weight factors. If we insert the general expression (3:25) for the energy into (5:7) the only change which takes place is the replacement of the single state density matrices with average matrices. Thus the whole formalism for the orbital and state optimization remains valid, with only small modifications to account for the fact that we now need to optimise more than one CI vector. In an unfolded two-step procedure, like the super-CI method, this is easily achieved by simply extracting the M lowest eigenvalues from the CI secular equation (M being the number of roots used in the averaging).

The result of such a calculation will be one set of orbitals and M CI wave functions. Usually it is then possible to identify the root of interest and start an MCSCF calculation for that specific root. The average orbitals are normally good starting orbitals. Now one of two things will happen. If we are lucky, the root will stay in the same position relative to the other roots. The result of the calculation is then a set of orbitals optimized specifically for this root and a wave function whith optimized CI coefficients. There are, however, cases where optimization of the orbitals for an excited state leads to a reordering of the solutions to the CI secular problem. Say that we are looking for the third root. When we follow the MCSCF iterations we notice that the energy of the third root drops more than that of the second root, and at a certain iteration they switch places. A *root flipping* has occurred. Normally the calculation will not converge, if we have instructed the program to optimize the third root, since that root is now the second. It is sometimes possible to solve the problem by instructing the program not to pick a specific root, but a root, where the wave function has certain well defined characteristics. One way is to guess the most important CI coefficients and ask for the root that has the largest overlap with the guessed structure. The preceding average calculation can be used to obtain good approximations to the CI coefficients. This procedure will not always work, since the orbital rotations can change the values of the CI coefficients considerably. The best solution in such situations seems to be to perform instead an average calculation for the second and the third root, with weights chosen to emphasize the third root as much as is possible without causing root flipping.

A typical case where root flipping frequently occurs is in studies of excited states for transition metal compunds. The metal atom often has an occupation of the d- shell which is different in the ground and the excited state. For example let the ground state occupation be d^{n+1} and the excited state occupation d^n. The d orbitals of the excited state will then be very different from those in the ground state. A calculation which optimizes the orbitals for the excited state then often increases the energy of the ground state so much that it becomes higher than that of the excited state. Root flipping has occurred, as illustrated below:

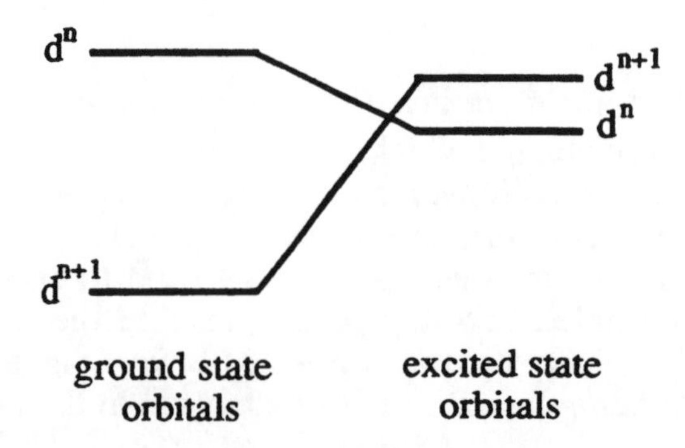

Wave functions obtained from a calculation where root flipping has occurred often has a large overlap with wave functions for the lower roots, and it is not always certain that they will give a good representation of the electronic state under study. It is probably better in most cases to use a wave function obtained from an average calculation.

The choice of active orbitals is often more difficult in excited state calculations and it is sometimes necessary to perform several trial calculations with different choices. Suppose, for example, that you want to compute the π excited states of an aromatic molecule which has one or several σ lone-pairs If you want a specific example, think about an azabenzene, like pyrazine or pyrimidine. Obviously the active orbital space has to include the six π orbitals, one for each heavy atom. But should it also include some sigma orbitals? When lone-pairs are present we might expect to find low lying $\sigma - \sigma*$ excitations that can interact with the $\pi - \pi*$ states (note that the important $\sigma - \pi*$ excitations have a different symmetry and do not interact). It might therefore be wise to include also one pair of σ (σ and $\sigma*$) orbitals for each of the lone-pairs. Actual calculations have in fact shown that these type of interactions have a profound effect on the transition moments, especially for the higher states. In calculations on π excited states one also has to be aware of the interaction between the valence states and the Rydberg states. Since the Rydberg states have quite differently shaped orbitals, it is in such cases necessary to include a second shell of π orbitals into the active space. As you notice, these considerations easily lead to active spaces that are too large. One then has to make compromises, or use for example the RASSCF method instead of CASSCF.

One should be aware of the fact that the MCSCF results will not by themselves yield very accurate excitation energies. Dynamical correlation effects are often different for different excited states, and if they are not accounted for, errors

as large as 1 - 2 eV can easily occur. On the other hand we can expect to obtain rather accurate values for the transition densities, since the most important terms have been included into the MCSCF wave functions for the two states, and the orbitals have been optimised. The limited experience which we have today strongly supports this presumption. The correlation error can be accounted for by means of MR-CI calculations in cases where they are feasible.
However, for larger molecules this approach might not be possible. The treatment of the differential dynamical correlation effects is for such cases not an easy task, which still awaits a satisfactory solution.

Now to the problem of calculating the transition densities. We need these quantities in order to be able to compute transition properties like the transition dipole moment. When we use a common orthonormal set of molecular orbitals for both the electronic states, the formalism developed in chapter 3 can be applied. For a one-electron operator $\hat{A}$ the transition matrix element is obtained from the simple formula:

$$<i|\hat{A}|j> = \sum_{p,q} D^{ij}_{pq} A_{pq} \tag{5:8}$$

for two electronic states i and j. A_{pq} are the matrix elements of the one-electron operator $\hat{A}$ and the sum is over all pairs of occupied molecular orbitals. The transition density matrix elements are obtained as

$$D^{ij}_{pq} = <i|\hat{E}_{pq}|j> \tag{5:9}$$

As was shown in chapter three we can compute the transition densities from the CI coefficients of the two states and the CI coupling coefficients. Matrix elements of two-electron operators can be obtained using similar expresssions involving the second order transition density matrix. This is the simple formalism we use when the two electronic states are given in terms of a common orthonormal MO basis. But what happens if the two states are represented in two different MO bases, which are then in general not orthonormal? We can understand that if we realize that equation (5:8) can be derived from the Slater-Löwdin rules for matrix elements between Slater determinants. In order to be a little more specific we expand the states i and j :

$$<i|\hat{A}|j> = \sum_{m,n} \sum_{p,q} D^{mn}_{pq} C_{im} C_{jn} A_{pq} \tag{5:10}$$

Assume for a moment that the expansion was made in a Slater determinant basis, and ignore the spin summation.The coupling coefficients are then given by the Slater-Löwdin rules, which in general are complicated expressions, but they depend only upon determinants of overlap integrals. Therefore, if the overlap matrix is a unit matrix, these expressions are simplified to the normal values given by the Slater rules. In this case, we can also sum over spin, and use CSF's instead of Slater determinants. The coupling coefficients are then the same as those used in chapter 3.Thus we recover the simple expressions for the matrix element when the two sets of MO's are *bi-orthonormal*. Call the two sets φ_A and φ_B. The bi-orthonormality condition then reads:

$$<\varphi_{iA}|\varphi_{jB}> = \delta_{ij} \tag{5:11}$$

If we transform the MO's such that condition (5:11) is fulfilled, the resulting transition density matrix will be obtained in a mixed basis, and can subsequently be transformed to any preferred basis The generators $\hat{E}_{pq}$ of course have to be redefined in terms of the bi-orthonormal basis, but this is a technical detail which we do not have to worry about as long as we understand the relation between (5:9) and the Slater rules. How can a transformation to a bi-orthonormal basis be carried out? We assume that the two sets of MO's are expanded in the same AO basis set. We also assume that the two CASSCF wave functions have been obtained with the same number of inactive and active orbitals, that is, the same configurational space is used. Let us call the two matrices that transform the original non-orthonormal MO's φ_X and φ_Y for $\mathbf{C}_{AX}$ and $\mathbf{C}_{BY}$,respectively:

$$\varphi_A = \mathbf{C}_{AX}\varphi_X \qquad \varphi_B = \mathbf{C}_{BY}\varphi_Y \tag{5:12}$$

The orthonormality condition (5:11) results in the following condition for the tranformation matrices:

$$\mathbf{S}_{XY}^{-1} = \mathbf{C}_{BY}^{\dagger}\mathbf{C}_{AX} \tag{5:13}$$

There are many choices of the transformation matrices that fulfill condition (5:13), but not all of them are allowed here. The reason is that it must be possible to express the transformed CASSCF wave functions in the new orbitals, using the same type of configurations as was originally used. The CI space must be *closed* under the transformation. If this condition is not fulfilled we cannot use the procedure sketched above in the calculation of the transition densities. We know that the CI space is invariant towards rotations among the inactive or among the active orbitals. We thus have total freedom in choosing these blocks of the transformation matrices. What about tranformations between active and inactive orbitals? The condition is here that the number of active electrons may not change. It is thus not allowed to mix in active orbitals among the inactive orbitals. That block of the transformation matrices must be zero. On the other hand nothing prevent us from mixing inactive orbitals into the active orbitals. The inactive orbitals are already doubly occupied and the

Pauli principle will ensure that no extra terms occur due to such a mixing. The transformation matrix for orbital set A then has the following blocked form:

$$C_{AX} = \begin{pmatrix} C_{AX}^{ii} & \mathbf{0} \\ C_{AX}^{ai} & C_{AX}^{aa} \end{pmatrix} \tag{5:14}$$

where i and a stand for inactive and active orbitals, respectively.The transformation matrix for orbital set B has a similar strucuture. There are many ways in which we can construct these matrices such that condition (5:13) is fulfilled. It is left as an exercise to the reader to suggest a specific procedure.
An important feature is the non-unitarity of the transformation (5:14). The transformed orbital sets will thus not be orthonormal, but this property is not needed since they will only be used to compute the transition densities where bi-orthonormality is the only condition which has to be fulfilled.

The next step in the calculation is to transform the CI coefficients to the new orbital basis. This can be done by a sequence of one-electron transformations, which we shall not describe in detail here. We note only that it is an essential part of the theory that such a transformation is possible. It is obviously not possible to obtain the transformed CI vector by performing one extra CI calculation. Such a calculation is virtually impossible to perform in a basis of non-orthonormal MO's.

Once the transformation to the bi-orthonormal basis has been performed we can calculate the first- and second order transition density matrices. From them we can obtain the desired transition properties. But we can do one more thing: Having the transition densities at hand we can compute the matrix elements of the Hamiltonian between the CASSCF states. The only extra thing needed in order to perform such a calculation is a transformation of the one- and two-electron integrals to the bi-orthonormal basis. It is then possible to set up and solve the secular equation for the M interacting CASSCF states. The non-orthogonality problem in excited state CASSCF calculations has then been solved. The states obtained from a CASSI calculation will be both orthogonal and non-interacting. Up till now CASSI calculations have been performed where as many as 20 CASSCF states have been allowed to interact. The method can easily be extended to handle MCSCF wave functions of the RASSCF type. The property of the RAS CI space to be closed under de-excitation here becomes essential, since it allows a blocking of the orbital transformation matrices in the same way as illustrated above (equation (5:13).

5.4 A test example, The CCCN radical

As a final example in these lectures we shall go through in some detail an MCSCF study of the C_3N radical. The method chosen will be CASSCF and the calculations will be performed using the Lund quantum chemistry software

package MOLCAS-1.
The C_3N radical is an important astro-chemical compound. Its chemical properties are not well understood. Not even the structure of the system is known. An experimental group is investigating the spectral properties of the radical. They have difficulties in locating the lowest electronically excited states and would like to be aided by theoretical calculations. The quantum chemical problem is then to compute the electronic structure for the two lowest states of the radical.
The first problem is to choose an adequate AO basis set for the calculation. As always in this situation we have to make a compromise between accuracy (large basis set) and economy (small basis set). The system is not so big, so we should be able to afford to use a reasonably good basis set, at least for the final calculations. On the other hand, it is clear that we have to optimize the geometry of both electronic states of the system. This we may have to do using a smaller basis set.
The basis set will be of the contracted gaussian type. There are many different possibilities for the specific choice of basis, and some programs have libraries containing a variety of basis sets. Experiences during the last couple of years have shown that the so called Atomic Natural Orbitals (ANO) in many respects are superior to earlier basis sets. They are constructed using a *general contraction*, where each contracted function (CGTF) is given as a linear combination of all primitive gaussian functions. The ANO's are obtained from studies of the individual atoms, as the natural orbitals from calculations including a large fraction of the atomic correlation energy (using SDCI or CPF methods). Normally additional atomic calculations with the atom in a homogeneous electric field are performed in order to determine the polarization functions.

High quality basis sets of the ANO type have recently been constructed in Lund and are easily available together with the programs we are using. For the first row atoms they are based on a primitive set consisting of 14 s-type, 9 p-type, 4 d-type, and 3 f-type gaussian functions. The occupation numbers obtained in the atomic calculations with this basis set indicate that the largest sensible ANO basis, that should be used in connection with this primitive set, contains 5 s-type, 4 p-type, 3 d-type, and 2 f-type functions. This is, however, too large for our preliminary study. It would result in a basis set comprising 220 CGTF's. The symmetry of the molecule is low (if it has any symmetry at all), and the number of two-electron integrals would become excessively large. For the final calculations we might try a basis set of the size 5s,4p,2d,1f, resulting in 156 CGTF's, but for the geometry optimizations it is advisable to use a somewhat smaller basis. It can be anticipated that the f-type functions will only have a minor effect on the geometry, and they can rather safely be excluded. Also the s and p part of the basis set is a little bit too large. We therefore decide to reduce the basis set down to 4s, 3p, and 2d. The total number of basis functions is now 100, which is small enough to make the calculations cheap.

Notice that we want to use 2 d-type functions in the basis. This is because for correlated wave functions the d-type functions serve two purposes, which cannot easily be accomplished by one basis function: First they serve as polarization functions, polarizing the electron density in the bonding region, a feature that is especially important in π-bonded systems. Secondly, they give important contributions to the angular correlation effects, which is very important in linear molecules. Studies of the ANO's shows that the two properties of the d-type functions cannot be incorporated into one function. The polarization d is rather diffuse, while the correlating d is a much more contracted function. We therefore prefer to include two d-type functions in the basis set.

Having decided about the basis set, we now turn to the problem of choosing an appropriate form of the wave function. Since we are going to calculate an energy difference between two electronic states, it is clear that we have to include as much correlation effects as possible. We shall do that in two steps: In the first step we try to account for the near degeneracy effects by using the CASSCF method with an appropriate choice of the active space. On top of that we shall have to perform multi-reference CI calculations to account for the dynamical correlation energy.

However, here we shall concentrate on the MCSCF part of the calculation. Since we are going to use the CASSCF method our problem is the choice of active orbitals. Let us try to make some guesses about the electronic structure of the system. The ground state is probably a sigma radical with the odd electron at the end carbon atom. This structure follows naturally, since CCCN is obtained by abstracting a hydrogen atom from the linear molecule HCCCN. The electronic structure can then be written as:

$$\cdot C \equiv C - C \equiv N$$

This structure strongly indicates that the radical is linear in its ground state in accordance with the molecular structure of acetylene and RCN. We can expect some conjugation between the triple bonds, making each of them weaker and the central CC bond stronger and shorter. It is not immediately clear which of the π orbitals will be directly involved in the conjugation. A more detailed analysis would probably tell us, but that is not really necessary, since the most safe choice of active orbitals is all the π-type orbitals together with the singly occupied σ-type orbital. In this case we will cover all possible near degeneracy effects in the π system. It is not a very large set of active orbitals, so there should no problem to perform the calculation.

The lowest excited state is most probably generated by exciting a π electron into the singly occupied σ orbital. The electron will come from the highest occupied π orbital, which is one of the CC triple bond electrons. The electronic structure of the lowest excited state can then be written as:

$$:C\overset{...}{\equiv}C—C\equiv N$$

This structure has a weakened CC triple bond (longer bond length) which can be expected to increase the near degeneracy effects in the π system. With all the π orbitals active, this effect is included in our wave function. We cannot immediately conclude that this state is linear. The CCC angle might become non-linear, since the end CC bond is now intermediate between a triple and a double bond. The degenerate $^2\Pi$ state splits into a $^2A'$ and a $^2A''$ state when the molecule bends. The ground state is for such geometries also of $^2A'$ symmetry. Thus we can expect some interaction. If the excitation energy is low, which is not unlikely, this could lead to a non-linear structure for the ground state too. However, the CN group is probably going to be rather inert. It might therefore be sufficient to optimize the two CC bond lengths and the CCC angle, using the experimental value for the CN bond length, for example, taken from the known bond length in HCCCN, and a CCN angle of 180°. Such a restriction in the number of geometry parameters may not be necessary if we have access to an energy gradient program for geometry optimizations.

We have thus established the basis set to use and the active space for the CASSCF calculations. To perform the CASSCF calculations we need in addition a set of starting orbitals. These are most easily obtained from an SCF calculation. If we have access only to a closed shell SCF program we can perform the calculation on the negative ion $CCCN^-$. We also need a starting geometry. A reasonable guess would be to use the geometry of HCCCN. This is a linear molecule with the bond distances: C_1C_2 = 1.20 Å, C_2C_3 = 1.38 Å, and C_3N = 1.16 Å (the atoms are labelled as $C_1C_2C_3N$). So, let us perform CASSCF calculations for the lowest $^2\Sigma^+$ and $^2\Pi$ state of CCCN in this geometry.

We first compute the integrals, order them, and store them to disk. We then generate a super- matrix file from the integrals and save it (this step is not performed in later version of the MOLCAS system, since all Fock matrices are generated directly from the integrals). These two job steps will not be described in detail, since they are outside the scope of these lectures. We then perform an SCF calculation on $CCCN^-$ and save the resulting MO's for use as trial orbitals in the MCSCF calculation. Now, let us take a closer look at the input for the CASSCF calculation, as it is organized in MOLCAS-1. The calculations will be performed in C_{2v} symmetry. This point group has four irreducible representations, a_1, b_1, b_2, and a_2. The σ orbitals transform as a_1, and the π orbitals as b_1 and b_2. The integral package of the MOLCAS-1 system (program MOLECULE) computes the one- and two-electron integrals directly in a symmetry adapted basis. We can therefore directly represent our molecular orbitals in this basis.

Here is the input for the CASSCF calculation on the ground state. An explanation is given after each of the input cards. The input is organized as a sequence of code words followed by appropriate data, if needed.

01. &CASSCF &END

This is a namelist input, which is only used to locate the CASSCF input in a more extended input stream.

02. TITLE
03. C3N (CC=1.19 Å, C-C=1.38 Å, CN=1.15 Å), ANO(4,3,2) C2v(z)
04. 2SIGMA state: CASSCF: (0000/8000/1440) 0+16+9 electrons

This is the title of the calculation. Up to ten title cards can be used.

05. SYMMETRY
06. 1

These cards give the symmetry information for the calculation. The dimension of the point group is 4 (C_{2v}). This information is transferred to the CASSCF program from the integral generation step. The irreducible representation for the state we want to calculate is 1. In the MOLCAS system it is possible to work with point groups which are subgroups of D_{2h}. This restriction is governed by the integral code .

07. SPIN
08. 2

The spin multiplicity (2S+1) of the wave function is given here, where S is the total spin quantum number.

09. ELECTRONS
10. 9

This input specifies how many electrons are occupying the active orbitals. The number of electrons in inactive orbitals is of course twice the number of inactive orbitals.

11. FROZEN ORBITALS
12. 0 0 0 0

We may want to freeze some of the inactive orbitals, that is, not optimize them, but leave them as they are given by input. There are several reasons why one would like to do that. If there are many core orbitals (like in a transition metal) freezing them minimizes the basis set superposition error (BSSE). Moreover, in second order MCSCF optimization procedures freezing orbitals simplifies the optimization. This is, however, not a problem in the super-CI procedure which is used in the MOLCAS CASSCF program. In this case we therefore prefer not to freeze any orbitals. It should be noted that the ANO basis sets result in very small BSSE's.

```
13. INACTIVE ORBITALS
14.    8  0  0  0
```

The number of inactive orbitals in each symmetry is given here. In our case they are all the doubly occupied σ orbitals, which all belong to the first irreducible representation, a_1, of C_{2v}.

```
15. ACTIVE ORBITALS
16.    1  4  4  0
```

Here we give the active orbitals: one σ orbital (the lone-pair at the end carbon atom) and 4+4 π orbitals, belonging to symmetries 2 and 3 (b_1 and b_2). Each atom thus contributes one π orbital to the active space.

```
17. DELETED ORBITALS
18.    8  0  0  0
```

Deleted orbitals are those which are not to be used in the orbital optimization process. MOLECULE computes integrals over Cartesian gaussian functions. Thus the six 3d functions include a 3s component. These 3s orbitals are orthogonalized to the remaining orbitals and placed at the end of the SCF orbital set. In order to avoid linear dependencies, and to have a clear understanding of the 3d contributions in the MO's, they are deleted from the calculation, meaning that these orbitals are not to be rotated into the occupied orbital space. There are eight of them, all belonging to the first symmetry, since the basis set contains two 3d functions on each centre. Later versions of MOLCAS use a basis set of (real) angular momentum functions. Thus 3s and higher contaminants do not enter the basis set at all.

```
19. CIROOT
20.    1  1
21.    1
```

This input specifies the root(s) of the CI wave function which we want. The program has the possibility to compute orbitals for a weighted average of a number of electronic states. This is of interest when studying excited states of the same symmetry as the ground state. It may then be necessary to have the same orbitals for all the states. This can be achieved by optimizing orbitals for the average energy. In the MCSCF formalism this is done by replacing the single state density matrices with average density matrices. The first number in the input gives the number of states for which the energy is to be averaged. The second number gives the minimal size of the CI matrix which has to be used in the Davidson diagonalization of the CI wave function. It should be at least as large as the first number. The input on the second row (card 21) gives the states which are to be used in the energy averaging. The number of states equals the first item on card 20. If this number is larger than one, a third card follows, which gives the weights of the different states in the averaging. In our case the input is simple, since we want only the lowest root.

22. THRSHD
23. 0.0000001 0.000010 0.000010

This is the input for convergence thresholds. Default values can be used, so it is actually not needed. The first number is the energy threshold, the second the threshold for the gradient (the largest component of the gradient vector must be smaller than this value at convergence), and the third value is the threshold for the orbital rotation matrix (the largest non-diagonal element of exp(T) must be smaller than the threshold). The calculation is considered to be converged when all three thresholds are met.

24. LEVSHF
25. 0.5

This is input for level shift, which is also default. A level shift is used to raise the diagonal elements of the super-CI matrix (if needed), such that the smallest value equals the input value. This is to avoid too large rotations, especially in the beginning of a calculation. The approximate super-CI method used in MOLCAS sometimes diverges, when no level shift is employed.

26. LUMORB

This input card indicates that the input orbitals are given in formatted form on the orbital input unit. The alternative is the code word JOBIPH, which indicates that the orbitals are to be taken from an output file produced by an earlier CASSCF calculation (for example at a nearby geometry).

27. ITER
28. 50

Max number of CASSCF iterations. The calculation stops after 50 iterations, if it has not converged earlier (which it normally has).

29. EOD

End of data card.

That concludes the description of the input. To do a calculation on the lowest state of $^2\Pi$ symmetry we can use the same input, except that the symmetry of the wave function is now 2 (or 3). We can then include an extra input which forces the π orbitals of symmetry 2 and 3 to be identical. If we do not do that the wave function may not have linear symmetry. However, since we are also going to do calculations for non-linear geometries, we might prefer not to force linear symmetry, since that leads to an energy that does not vary in a continuous way as a function of the bending angle.

Let us now assume that we have performed the calculation for the $^2\Sigma^+$ state, and take a look at the result. It turns out that the number of configurations in the wave function is 2068. The calculation converges (starting with SCF vectors from $CCCN^-$) in 20 iterations. The energy is then converged to better than 10^{-7} a.u., the gradient to $4x10^{-6}$, and the rotation matrix to $6x10^{-6}$. This can be considered as satisfactory. One CASSCF iteration takes around 15 seconds, and the whole calculation then takes 5 minutes (on an IBM 3090-17S computer).

The SCF configuration: $(1\pi)^4(2\pi)^4(9\sigma)^1$

has a coefficient of 0.9178 in the final wave function, thus contributing 84.2 % to the CASSCF wave function. The most important of the other configurations correspond to double excitations to anti-bonding π orbitals, as could be anticipated, with CI coefficients as large as 0.11. We shall not analyze these numbers in details, but look only at the occupation numbers of the active natural orbitals. They come out as follows:

orbital	occupation number
1π (mainly C_3N bonding orbital)	3.8876
2π (mainly C_1C_2 bonding orbital)	3.8124
3π (C_1C_2 anti-bonding, C_3N anti-bonding, C_2C_3 bonding)	0.2032
4π (C_1C_2, C_2C_3, and C_3N anti-bonding)	0.0970

The σ orbital has of course the occupation number 1.00, since any excitation to or from this orbital leads to configurations of the wrong symmetry. We notice especially the strong occupation of the orbital 3π which is bonding in the region between the acetylene and the cyano bond. This occupation results in an increased conjugation of the two triple bonds.
The calculation automatically gives a population analysis, yielding the following gross atomic charges for the four atoms:

$q(C_1) = -0.01$, $q(C_2) = +0.05$, $q(C_3) = +0.17$, $q(N) = -0.22$

We notice that the CC triple bond is almost non-polar, while the CN bond is polarized towards nitrogen, as expected. The computed dipole moment for the radical in the chosen geometry is -2.91 Debye.

Now let us turn to the lowest state of $^2\Pi$ symmetry. The number of CSF's for this symmetry is 2352, with the active space we have chosen. The CASSCF calculation now converges in 18 iterations, with about the same residual values for the energy, gradient, and rotation matrix. The leading configuration is for this state:

$$(1\pi)^4(2\pi)^3(9\sigma)^2$$

It has a coefficient of 0.9360 in the CASSCF wave function, corresponding to a weight of 87.6%. We notice that this is larger than the weight of the SCF configuration in the ground state. Our presumption that the near-degeneracy effects in the π system would be larger for the $^2\Pi$ state was thus not correct. Let us take a look at the occupation numbers and see if we can understand the difference (the orbitals have the same qualitative structure as for the ground state):

orbital	occupation number
1π	3.8556
2π	2.9540
3π	0.1560
4π	0.0820
9σ	1.9524

Comparing with the ground state we find that the 1π orbital has about the same occupation as in the ground state. The occupation of the 2π orbital has decreased with almost one electron, as expected. As a result there is less occupation in the 3π orbital, which was the orbital leading to increased conjugation between the triple bonds. In retrospect this is rather obvious, since there are now fewer electrons to excite into this orbital. Double excitation from 2π to 3π is now possible only for one of the components. As a result the C_2C_3 bond ought to be longer in the $^2\Pi$ state than it is in the $^2\Sigma^+$ state. The geometry optimization will show if this is really the case. We notice that the total number of π electrons is slightly larger than 7.00, while the occupation number of the active 9σ orbital is smaller than 2.00. This is due to the appearance of configurations with zero electrons in 9σ and 9 electrons in the π shell.

Moving one electron to the σ lone-pair orbital changes the polarity of the radical. The calculated gross atomic charges are now:

$$q(C_1) = -0.15,\ q(C_2) = +0.20,\ q(C_3) = +0.13,\ q(N) = -0.18$$

We notice that both the triple bonds are now polar leading to a small overall polarity of the system. The computed dipole moment is also small: 0.27 Debye (notice the reversed polarity compared to the ground state).

The most interesting result of this preliminary calculation is probably the computed energy difference between the two states. The total energy computed for the $^2\Sigma^+$ ground state is -168.077552 a.u.. The corresponding energy for the $^2\Pi$ state is -168.040109 a.u. The energy difference is only 0.037443 a.u. or 1.02 eV. When the radical bends one component of the $^2\Pi$ state transforms into the same symmetry as the ground state ($^2A'$). Thus the two states can interact. The small energy difference indicates that the interaction could lead to a rather strong mixing, resulting in a stabilization of a bent structure even for the ground state. A CASSCF study of the energy as a function of the bending angle will show if this is going to happen.In fact, preliminary results indicate that this is not the case. Both state are linear.
These results and also the results for the excitation energies will of course be modified by dynamical correlation effects, which will be accounted for by means of MR-CI calculations based on the CASSCF orbitals for each state. The transition moments will be computed using the CASSI method, where the two A' are allowed to interact. The final results can be expected to be accurate to within 0.1 eV for the excitation energies and 10% for the transition moments. The study of the CCCN system is not yet finished. Hopefully some result can be presented in the lectures.
Note added 1991: The work discussed above has now been finished and the results can be found in Chem. Phys. Letters **180**, 81 (1991).

5.5 Exercises

5.1 Show that the gradient elements given by equation (4:9) are identically zero if the orbital indices i,j both belong to active orbitals, and |0> is a CAS wave function.
In CASSCF calculations only the rotation parameters T_{it}, T_{ia}, and T_{ta} have to be used, where i represents an inactive orbital, t an active orbital, and a an external orbital. Motivate!

5.2 You want to calculate an accurate potential curve for the $^2\Pi$ ground state of the OH radical. The method to be used is CASSCF plus MR-CI. Suggest a basis set for the calculation, an active orbital space for the CASSCF calculation, and reference states for the MR-CI calculation (the last part cannot be done in detail without knowledge of the CASSCF wave function).

5.3 The same task, but now for the dissociation process:

$$H_2O \rightarrow H_2 + O(^1D)$$

along a dissociation path of C_{2v} symmetry.

5.4 You want to find the transitions state for the Diels-Alder reaction:

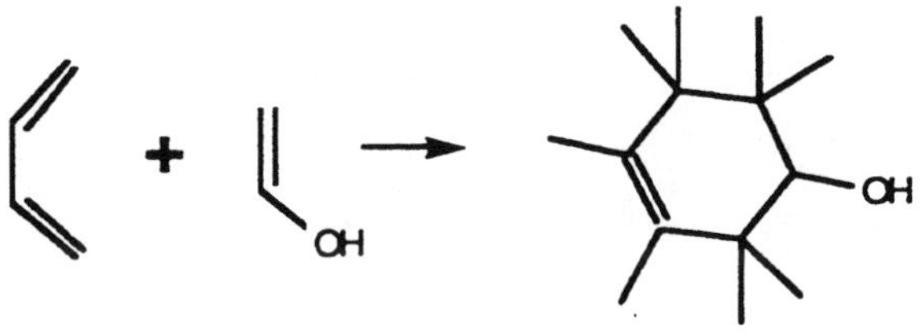

OH OH

Suggest a computational procedure with which you believe the problem can be solved. Consider the basis set problem, the wave function, and the method you would like to use to reach the transition state geometry. This is a rather large computational problem. Try to set up a procedure that fits into the computer you are using. Try to estimate the computer time you would need to use on your installation.

6. Further reading

Below are listed some recent review articles, which are recommended as further reading. They give details on subjects, which have merely been touched upon in the present notes. Much of the material used here has been drawn from these articles. They contain a large number of references to original articles, which have laid the foundation to contemporary MCSCF theory.

B.O.Roos, *The Multiconfigurational (MC) SCF Method,* in Methods in Computational Physics (G.H.F Diercksen and S.Wilson, eds), D.Reidel Publishing Company, Dordrecht (1983).

Three articles in: *Ab Initio Methods in Quantum Chemistry, Part II* (K.P. Lawley,ed), John Wiley & Sons Ltd, Chichester (1987):
B.O.Roos, *The Complete Active Space Self-Consistent Field Method and its Application in Electronic Structure Calculations.*
R.Shepard, *The Multiconfiguration Self-Consistent Field method.*
H.J.Werner, *Matrix-Formulated Direct Multiconfigurational Self Consistent Field and Multiconfiguration Reference Configuration-Interaction Methods.*

THE CONFIGURATION INTERACTION METHOD.

Per E.M. Siegbahn
Institute of Theoretical Physics, University of Stockholm.
Vanadisvägen 9, S-11346 Stockholm, Sweden.

I. Introduction.

The simplest *ab initio* quantum chemical method is the Hartree-Fock method which has been outlined in detail in a preceding chapter. Other names for this method are the Self Consistent Field (SCF) method and the independent particle method. Due to its simplicity the Hartree-Fock method has become a very useful tool for studying stable molecules around their equilibrium geometries. The Hartree-Fock total energy differs from the exact non-relativistic energy by an energy, which for historical reasons is called the correlation energy. There are essentially two shortcomings of the Hartree-Fock method which sometimes make the correlation energy large. The most important of these shortcomings is connected with how the Hartree-Fock method is normally used, namely with the restriction that two electrons are placed in each spatial orbital, the Restricted Hartree-Fock (RHF) procedure. This restriction is responsible for, for example, the very large error of several eV obtained at the asymptotic limit when the hydrogen molecule is dissociated. Other examples are given in the chapter on the Multi- Configuration SCF (MCSCF) method. As mentioned in that chapter the problem with improper dissociation is corrected by the use of a few additional determinants (or configurations). For this reason this error is said to be due to near-degeneracy effects.

The other error in the Hartree-Fock method is of a completely different physical origin. In the independent particle model the electrons are moving in the average field of the other electrons, which is obviously an approximation. In reality the motion of the electrons depends on the instantaneous positions of all the other electrons, the motion of the electrons is "correlated". The lack of correlation of the motion of the electrons in the Hartree-Fock method is the reason the energy error using this method is called the correlation energy. This error is always present in the Hartree-Fock method even for stable molecules at their equilibrium positions and is responsible for the lack of quantitative accuracy obtained at the Hartree-Fock level in these cases. The error explicitly caused by the independent particle approximation is normally called the dynamical correlation energy in contrast to the error caused by the RHF approximation which is then called the non-dynamical correlation energy. This division of the error in the Hartree-Fock approximation is physically motivated and has been extremely useful in devising methods to correct for this error, but is not easily defined mathematically or methodologically. Normally there is a floating boarder-line between dynamical and non-dynamical correlation effects as will become more evident later on in this chapter. We will, however, return with a suggestion of a more methodological and less ambiguous definition of the different parts of the correlation energy.

The simplest and most straightforward method to deal with the correlation energy error is the Configuration Interaction (CI) method. In this method the single determinant Hartree-Fock wavefunction is extended to a wavefunction composed of a linear combination of many determinants in which the coefficients are variationally optimized. These determinants are set up by selecting occupied and virtual one particle orbitals generated by a preceding Hartree-Fock or MCSCF calculation. This method is in principle an extremely simple extension of the Hartree-Fock method and it was therefore suggested early in the 1930's. Owing to the lack of sufficiently powerful computers the applications of the CI method on molecules did not appear (with a few exceptions) until the end of the 1960's. One problem with the CI method is that even in a rather small basis set the possible number of determinants which can be generated is extremely large. It depends in an n-factorial way on the number of electrons and on the number of orbitals. The type of CI where all these determinants are used, complete CI or full CI, is therefore normally not a tractable method for attacking the correlation problem. Instead, the complete CI method has found its use as a benchmark test for approximate schemes where the configurations are selected in one way or another. The simplest selection scheme starts out with the Hartree-Fock configuration and adds all configurations where one or two of the occupied orbitals have been replaced by virtual orbitals. This is the single-reference state CI method. To make this type of configuration selection scheme open ended, the configuration list can be extended by single and double replacements out of the most important configurations. This is the multi-reference CI (MR-CI) method, to which most of this chapter will be devoted. Since the list of configurations selected in the MR-CI method can be quite large the number of configurations is sometimes reduced further by the use of perturbation theory. If the list of configurations is reduced sufficiently (at present to the order of 10^4 configurations) a workable scheme to solve the CI problem was suggested already in the early 1950's. This method is used even today and has been termed the conventional CI method. Much of the work on the CI method, mainly during the 1970's, was devoted to devising methods which could handle the full list of configurations generated in the MR-CI method without any further selections. In this context the "direct" CI method has become the generally accepted strategy. The details of the direct CI method will be outlined in this chapter. But before the various methods are described a section will be devoted to examples of the type of problems that can be studied and the accuracy which can be obtained at present by applications of the CI method.

II. Examples of CI calculations.

In this section examples will be given of what can presently be done using the MR-CI method. There are several reasons for bringing this subject up at this early stage in a chapter about the CI method. First of all the idea is to show what kind of effects can be expected from dynamical correlation of the electronic motion. The MR-CI method is the most accurate quantum chemical tool that exists but it is also one of the most expensive tools. There is no reason to always go to this level of approximation in the calculations. A knowledge of the size of the correlation effects is therefore important to have in order to make the decision whether a CI calculation is necessary or not. It is possible that this degree of accuracy is not needed for the particular problem in question. Or worse, the demand of accuracy is much greater than what can presently be achieved by any quantum chemical method.

There is no textbook in quantum chemistry which does not discuss the bonding in the hydrogen molecule in detail. Most qualitative aspects of the quantum chemical methods we have today can be understood from an application on this simple system. We have already mentioned the fact that H_2 does not dissociate correctly in the RHF approximation and its important consequences for the division of the correlation energy into a dynamical and a non-dynamical part. Therefore, if the RHF energy for H_2 at its equlibrium geometry is compared to the closed shell RHF energy for H_2 at long distance between the hydrogen atoms the binding energy D_e will actually be larger than 10 eV which is much larger than the experimental value for D_e which is only 4.75 eV. If, on the other hand, the RHF energy for H_2 is compared to the energy for two separately calculated hydrogen atoms (which are exactly described at the RHF level) the binding energy is 3.64 eV. In the rest of this chapter this way of calculating D_e at the RHF level is what will be referred to as the RHF D_e. The error at the RHF level, the dynamical correlation energy contribution to D_e is thus 1.11 eV, which is what one has to expect for the breaking of a single covalent bond in any molecule. As discussed in a previous chapter on the MCSCF method, the incorrect dissociation of H_2 is corrected by adding to the Hartree-Fock configuration σ_g^2 the configuration σ_u^2. This MCSCF calculation will describe the two hydrogen atoms at long distance from each other exactly and thus remove the error from non-dynamical correlation. The addition of the σ_u^2 configuration will, however, also give an energy lowering at the equilibrium geometry for H_2 and will thus contribute to the description of what could be regarded as the dynamical part of the correlation energy. This illustrates the floating boundary between the non-dynamical and the dynamical parts of the correlation energy. It is therefore customary to relate the energy lowering from all configurations required for a proper dissociation to the non-dynamical correlation energy. An even more technical definition which is usually adopted is to relate the energy lowering obtained from all configurations with coefficients larger than say 0.05 to near-degeneracy or non-dynamical correlation effects. A more strict definition is possible and will be addressed later. Going back to H_2, the dissociation energy obtained from the two configuration MCSCF wavefunction is 4.13 eV, which is a substantial improvement compared to the RHF value but it is still 0.62 eV lower than experiment. For H_2 it is possible to continue adding important configurations in an MCSCF treatment and in this way obtain a compact and accurate treatment. This was an important development at a time when computations were much more expensive than they are nowadays but it is not the way the development has continued. It is, however, still interesting that it is possible to get a dissociation energy for H_2 of 4.63 eV using a so called Optimized Valence Configuration (OVC) wavefunction which is an MCSCF wavefunction containing less than 10 configurations. Today, a "brute force" CI treatment of H_2 giving a dissociation energy above 4.70 eV represents a trivial calculation. Since there are only two electrons, a calculation including all single and double replacements out of the Hartree-Fock configuration gives the exact non-relativistic result within the chosen basis set. With a single *d*-function on hydrogen "chemical accuracy" is obtained for D_e, usually defined as 1 kcal/mol (0.05 eV). With two *d*-functions the error is half of that, 0.025 eV, and the calculation is still very inexpensive.

It is interesting to see how the contribution to D_e for a molecule with a single bond varies as the number of electrons increases. The next examples we will discuss are therefore Li_2 and F_2. For Li_2 the experimental dissociation energy is 1.05 eV, which is much smaller than the value for H_2. This is simply a consequence of the much more diffuse binding orbital for Li, which leads to a smaller nuclear attraction to the other nucleus for the electrons forming the bond than in the case of H_2. The RHF value for D_e is 0.17 eV which means that dynamical correlation contributes 0.88 eV to D_e. The influence of the 1s core orbitals of Li has thus not changed the qualitative picture obtained for H_2. In fact, correlation of the 1s electrons in Li_2, contributes less than 0.02 eV to D_e. Core correlation in Li_2 has larger effects on other properties as we will discuss below.The simplest improvement of the RHF value for D_e is obtained in the two-configuration MCSCF treatment leading to proper dissociation for Li_2. This treatment gives a D_e of 0.46 eV. The error at this level of treatment, 0.59 eV, is thus very nearly the same as for H_2. Since, to a good approximation, we can neglect the core correlation effect on D_e a high accuracy CI treatment of Li_2 is nearly as trivial as for H_2. Again, a basis set including a single *d*-function is sufficient for obtaining chemical accuracy (1 kcal/mol).

The binding in the F_2 molecule is more complicated than in H_2 and Li_2, although the qualitative picture of the bonding with a single covalent bond is the same. Since the bond in F_2 is formed with a 2p orbital and there are other non-bonding 2p electrons present, a considerable change of the dynamical correlation of the 2p shell upon bond-formation is expected. This leads to a similar (but less severe) problem as when *d*-electrons form bonds in transition metal complexes, which we will return to below. The experimental value of D_e for F_2 is 1.68 eV. This rather small value is a consequence of the repulsion from the non-bonding 2p electrons and is not due to the diffuseness of the binding orbital as in Li_2. A consequence of the small binding energy and the large changes in the dynamical correlation energy is that the RHF value for D_e actually becomes negative by 1.37 eV. This should not be interpreted as if the F_2 molecule would fall apart in the RHF approximation. In fact, if the bond is optimized at the RHF level a reasonable (although not very accurate) bond distance is obtained. This means that at a qualitative level the RHF wave-function is still a reasonable representation of the correct wave-function. It is in our opinion an exaggeration of the low quality of the RHF wave-function to state that the F_2 molecule solely stays together because the RHF picture leads to incorrect dissociation, as if the value for the bond distance was obtained purely by chance. At the two-configuration MCSCF level there is a large improvement of the dissociation energy to a positive value of 0.54 eV. One might perhaps expect that the qualitative picture of the bonding in F_2 should change dramatically as the binding energy changes from negative at the RHF level to positive at the MCSCF level but this is not so. In fact, it is hardly possible to see any change in the orbitals or in the density. Since the correlation problem is much more complicated for F_2 than for H_2 and Li_2 a considerably more advanced CI treatment is needed. Single and double replacements out of the Hartree-Fock configuration is no longer sufficient for high accuracy. In the most accurate CI calculation performed so far for F_2 eight reference states were used and a basis set including *g*-functions. The error for D_e at this level is less than 0.02 eV and the result is thus of chemical accuracy. If only one reference state is used the error is 0.5 eV and even if the two configurations needed for proper dissociation are used as reference states the error is still 0.3 eV. It should be added that there exist ways in which in particular the single-reference state CI result can be improved and we will describe these so called "size consistency" corrections later on in this chapter.

The diatomic molecule formed from two first row atoms which is the most difficult to treat quantum chemically is the N_2 molecule (possibly with the exception of CO). The reason for this is obviously the formation of a triple bond in this molecule. This causes a considerable redistribution of the electrons and thus a large change of the correlation energy. For N_2 quantum chemists have long struggled in vain to obtain chemical accuracy (1 kcal/mol) for the dissociation energy. The experimental value for D_e is as large as 9.90 eV. At the RHF level only slightly more than half of this value is obtained, 5.27 eV. If the configurations needed for proper dissociation are used in an MCSCF treatment the value for D_e is improved to 7.27 eV. This is for a basis set including two *d*-functions on each nitrogen atom, and this may still be a few tenths of an eV away from the basis set limit. The remaining 2.6 eV dynamical correlation energy is very difficult to obtain, requiring large basis sets and large reference spaces. It is a general experience that basis set convergence is much slower at the CI level than at the RHF and MCSCF levels. In the most accurate MR-CI calculations performed so far a basis set including *h*-functions on each nitrogen atom was used and the effect of adding *i*-functions was not considered completely negligible (0.02 eV). With this basis set and using 32 reference states a D_e value of 9.80 eV is obtained, which includes an effect of 0.03 eV from correlating the core electrons. The conclusion from this work was that the major part of the remaining error comes from deficiencies in the chosen configuration space, which means that other, more generally chosen reference configurations need to be selected to reach chemical accuracy. However, in a recent study of N_2 it was suggested that a large part of the remaining error in the above mentioned study still originates from limitations in the basis set. At the moment this is still an unsettled question.

A very important aspect of the results described above for D_e is that the error obtained at a certain level of approximation is systematic. This fact combined with the fact that the results improve as the method improves are aspects of ***ab initio*** methods which are at least as important as the final accuracy of the results. So far the only property discussed is D_e. It is clear that the most important chemical information, such as reaction pathways and thermochemistry, is obtained from relative energies, but the accuracy of other properties is also of interest. If we look at the equilibrium bond distance R_e and the harmonic vibrational frequency ω_e, these properties also display a systematic behaviour depending on the method chosen and this systematic behaviour is easy to understand. Since the RHF method dissociates incorrectly, the potential curves tend to rise too fast as the bond distance is increased. At the RHF level this leads to too short equilibrium bond distances and vibrational frequencies that are too high. When proper dissociation is included at the MCSCF level, the opposite trend appears. Since the dissociation energies are too small at this level the potential curves rise too slowly as the bond distance increases. This leads to too long bond distances and too low frequencies. These systematic trends are nicely illustrated by the results for three of the previously discussed diatomic molecules. For H_2 the experimental value for R_e is 1.40 a_0 and for ω_e it is 4400 cm^{-1}. At the RHF level R_e becomes too short, 1.39 a_0, and ω_e becomes too high, 4561 cm^{-1}. At the two configuration MCSCF level R_e becomes too long, 1.42 a_0, and ω_e becomes too low, 4214 cm^{-1}. For F_2 the experimental value for R_e is 2.668 a_0 and for ω_e it is 892 cm^{-1}. At the RHF level R_e becomes much too short, 2.50 a_0, and ω_e much too high, 1257 cm^{-1}. At the two configuration MCSCF level the corresponding results are 2.74 a_0 and 678 cm^{-1}, which are too long and too low respectively as expected. For N_2 finally the experimental values are 2.074 a_0 and 2358 cm^{-1}, the RHF values 2.013 a_0 and 2729 cm^{-1} and the MCSCF values 2.077 a_0 and 2359 cm^{-1}, respectively. Incidentally, due to cancellation of errors these latter MCSCF values are much too close to the experimental values to be typical. A factor which slightly complicates the simple trends are angular correlation effects which tend to make the bonds shorter. These effects are quite large for the weak bond in Li_2 due to the near degeneracy between the 2s and 2p orbitals. Therefore the simple trends do not hold for this molecule. Here the experimental values are 5.05 a_0 and 351 cm^{-1} and the RHF values 5.26 a_0 and 326 cm^{-1}, which are contrary to the expected trends. At least for molecules composed of first and second row elements, Li_2 is a rare exception. At the MR-CI level reasonably high quantitative accuracy has been obtained for R_e and ω_e for all these molecules in the most recent calculations. For the two most difficult molecules to treat, F_2 and N_2, the above mentioned MR-CI calculations give the following results. For F_2 the calculated R_e is 2.672 a_0 and for ω_e it is 932 cm^{-1}, i.e. errors of 0.004 a_0 and of 15 cm^{-1} respectively. For N_2 the corresponding results are 2.081 a_0 and 2343 cm^{-1} with errors of 0.006 a_0 and 15 cm^{-1}. Since these latter calculations are by normal standards very large, this size of errors is what one has to live with for at least a couple of more years.

We mentioned earlier the fact that core correlation has a larger effect for other properties than for D_e. Of the molecules we have looked at, Li_2 is expected to show the largest effect of core correlation, since the 1s core orbitals are fairly close to the valence orbitals. In particular for the bond distance the effect, which shortens the bond by 0.06 a_0, is not negligible. This effect is fairly difficult to obtain in a straightforward MR-CI treatment since very large basis sets are required. It turns out that the core correlation effect can be regarded as an instantaneous polarization of the core in the field of the valence electrons. By introducing a simple core polarization operator the effect of core correlation is easy to model and this is presently the most promising route to follow for obtaining these difficult correlation effects.

One of the greatest challenges for quantum chemists today is to be able to accurately treat molecules containing transition metals. From the MR-CI point of view there are several reasons why transition metals are difficult to treat. First, for transition metals there are always three low lying states with different numbers of *d*-electrons. For example, for nickel the $^3F(d^8s^2)$ state and the $^3D(d^9s^1)$ state are almost degenerate and the $^1S(d^{10})$ state is only 1.74 eV higher in energy. The transition metal bonds are often formed from a hybridized state composed of a mixture of these states and it is therefore very important to describe the energy differences between the states accurately. Since the number of 3*d* electrons varies from state to state there will be a large difference in the 3*d* shell correlation energy between the states and this correlation energy is very difficult to obtain accurately, requiring basis functions with high angular quantum numbers. A second reason for the difficulties to treat transition metals is that covalent bonds are often formed with both the *d*-orbitals and the *s*-orbitals and these orbitals have very different spatial distribution. For nickel $< r >$ for the $3d$ orbital is about 1.0 a_0 whereas $< r >$ is as large as 3.0 a_0 for the $4s$ orbital. This means that for at least one of these orbitals the bond to a ligand will have to be formed with non-optimal overlap. The problem with incorrect dissociation for these systems therefore enters already at the equilibrium geometry. The third reason for the difficulties is that bonds are always formed in the presence of a large number of electrons. This situation is similar to the bond formation in F_2 but the problem is even more severe. These problems added up means that to accurately treat transition metal complexes a large number of electrons have to be correlated using both large basis sets and a large number of reference configurations.

To illustrate some of these problems we will take three examples, the Ni atom, the NiH molecule and the $Ni(CO)_n$ molecules where n=1 and 4. The experimental value for the splitting between the d^9 and the d^{10} states of the Ni atom is 1.74 eV. At the non-relativistic Hartree-Fock limit the value for this splitting is as poor as 4.20 eV, even though there are no significant near degeneracy effects in any of these states. Adding relativistic effects (by perturbation theory) leads to an even worse result of 4.41 eV. In the following numbers this relativistic effect is included. The (essentially dynamic) correlation effect on the splitting is thus 2.67 eV. With one reference state, size consistency corrections and very large basis sets including h-functions the splitting is improved to 2.27 eV, which is still 0.53 eV away from the experimental value. This was actually the best computational result obtained until a few years ago. At this time a multireference treatment was performed where the most important $3d$ to $4d$ excitations were included in the reference space. Above a threshold of 0.01 there are about 20 such configurations for both states, but these configurations turn out to be much more important for the d^{10} state. At this MR-CI level the splitting is 1.87 eV using a basis set including g-functions, which is a quite acceptable value with an error of only 0.13 eV.

For NiH we will concentrate our interest on the bond distance R_e and the frequency ω_e. The experimental values are 2.75 a_0 and 1993 cm^{-1}, respectively. At the Hartree-Fock level the results are 2.88 a_0 and 1726 cm$^-$1, which are contrary to the normal trends for molecules containing first row atoms. The main reason for these trends is the high polarizability of the $3d$ orbitals, which in a sense belong to the core orbitals of the Ni atom since they are far inside the $4s$ orbital. The instantaneous influence of this polarizability leads to an attractive term in the interaction between the $3d$ electrons and the valence electrons, which will shorten the bond. This is a typical dynamical correlation effect and is thus missing at the Hartree-Fock level but appears at the MR-CI level through configurations in which one electron is excited from the valence orbitals and the other is excited from the $3d$ (core) orbitals. These excitations are therefore normally called core-valence excitations. In a multi-reference CI treatment with a threshold selection of 0.05 the bond distance is decreased to 2.69 a_0 and the frequency is increased to 1971 cm^{-1}, which are acceptable values. To obtain high quantitative accuracy a lower selection threshold for the reference configurations of 0.02 is needed, just as for the splitting of the Ni atom. For NiH this leads to 29 reference configurations and the results are 2.76 a_0 and 1997 cm^{-1} with a basis set including several f-functions.

The above examples have illustrated the need to go to very long MR-CI expansions to reach high accuracy for transition metal systems. The linear molecule NiCO will instead serve to illustrate the necessity to correlate many electrons in these systems. The D_e value for this molecule for dissociation into Ni and CO is experimentally uncertain but should be in the interval 30-40 kcal/mol. At the Hartree-Fock level a value of -67 kcal/mol is obtained. We have earlier seen an example where the dissociation energy is negative at the Hartree-Fock level, namely for F_2, but NiCO is thus much worse. Since there is no actual standard bond formation in this molecule one might suspect that the main reason for the poor value lies in the treatment of the $3d$ electrons. This turns out to be the case and an MR-CI calculation (reference selection threshold 0.05) correlating only the nickel $3d$ and $4s$ electrons leads to a positive value for D_e of 16 kcal/mol, which is still less than half of the actual dissociation energy. The only electrons on CO which are close to the nickel atom are the carbon lone pair electrons. Correlating also these electrons improves D_e to 24 kcal/mol. Continuing to account for the correlation effects of the remaining CO valence electrons gradually improves D_e to a value of 33 kcal/mol, with even the carbon and oxygen $2s$ electrons contributing a few kcal/mol to the dissociation energy. If high quantitative accuracy should be reached it is thus necessary to correlate all valence electrons in a rather large surrounding of the transition metal atoms.

The final example we will discuss here is $Ni(CO)_4$, for which the experimental dissociation energy into a Ni atom and four CO's is 140 kcal/mol. Based on the above experience for NiCO a CI calculation was set up in which all 50 valence electrons were correlated. It could be of interest to know that such a calculation is actually feasible even using a rather large basis set. When this calculation was performed it was only possible to use one reference state in the CI expansion and the resulting value for the dissociation energy including size consistency corrections was 120 kcal/mol. To account for some of the remaining error compared to experiment of 20 kcal/mol a multireference treatment including some $Ni(3d)$ to $CO(\pi^*)$ excitations in the reference will be necessary. An alternative procedure would be to include the effect of higher than double excitations by a perturbational treatment of the triple excitations, and such calculations are presently being performed.

III. The configuration interaction method.

There are essentially two different quantum mechanical approaches to approximately solve the Schrödinger equation. One approach is perturbation theory, which will be described in a different set of lectures, and the other is the variational method. The configuration interaction equations are derived using the variational method. Here, one starts out by writing the energy as a functional F of the approximate wavefunction ψ,

$$F = \frac{< \psi \mid \hat{H} \mid \psi >}{< \psi \mid \psi >} \tag{3.1}$$

(Later on, in section X, we will show that certain important and desirable features are obtained if the energy functional is written in a somewhat different form.) The value of the functional has to be larger than or equal to the lowest exact eigenvalue of the Hamiltonian. This is trivial to see if the approximate wavefunction ψ is expanded in terms of the complete set of exact eigenfunctions to the Hamiltonian operator, as obtained from the Schrödinger equation. The variational method is said to yield an "upper bound" to the true energy. The variational principle, which in other contexts is postulated and considered a very basic law of nature, is thus contained in the Schrödinger equation. By expressing the approximate wavefunction in terms of parameters and minimizing the functional with respect to these parameters one can thus obtain progressively better approximations to the energy and the eigenfunctions of the Hamiltonian. The simplest example of an application of the variational method is when the variational parameters appear linearly in the approximate expression for the wavefunction. This is the case for the derivation of the CI equations. The wavefunction is here written simply as

$$\psi = \sum_i C_i \Phi_i \tag{3.2}$$

where Φ_i are the configurations which are chosen as described in the next section. Minimization of the functional (3.1) with respect to the parameters C_i leads to the secular equations which are identical to the CI equations,

$$\sum_j (H_{ij} - E S_{ij}) C_j = 0 \tag{3.3}$$

where E=F is the energy and H_{ij}, S_{ij} are the Hamiltonian and overlap matrix elements between the configurations. Unlike the MCSCF method the form of the configurations, i.e. the orbitals, are not varied in the CI method.

When the variational method is applied to the functional (3.1) the convergence of the energy E is particularly efficient and much faster than the convergence of other properties which can be derived from the same wavefunction. This can be seen by the following set of operations. Say that the approximate wavefunction ψ has a small error $\Delta\Phi$ which can be chosen orthogonal to the exact wavefunction Φ. The energy functional (3.1) can then be written,

$$F = \frac{< \Phi - \Delta\Phi \mid \hat{H} \mid \Phi - \Delta\Phi >}{< \Phi - \Delta\Phi \mid \Phi - \Delta\Phi >} \tag{3.4}$$

which can be simplified by using the fact that the Hamiltonian operator can act both to the left and to the right on the exact eigenfunction Φ to give $E_0\Phi$ where E_0 is the exact energy. Since Φ and $\Delta\Phi$ are orthogonal all linear terms in $\Delta\Phi$ will vanish and the error in the energy will thus only contain quadratic terms in the error of the wavefunction. In perturbation theory there is a corresponding result saying that the energy of order 2n (actually 2n+1) can be obtained from a wavefunction of order n. This means, among other things, that if an iterative method is used to converge the energy, more iterations are needed to converge another property (such as the dipole moment) than is needed for the same accuracy for the energy. A related result is that when the variational method is used there are seldom convergence problems for the energy. Straightforward iterative methods usually give 5 decimal places for the energy in around 10 iterations. Such iterative methods have been discussed in the mathematical lectures and will also be briefly described in section VI.

IV. Configurations and the configuration expansion.

The configurations, which constitute the basis set in the CI method, are normally chosen to fulfill certain symmetry conditions. It is important to note that it is not necessary in principle to apply these symmetry conditions. If all the symmetry components are included in the wavefunction expansion, correct symmetry combinations will automatically be obtained anyway. The main reason to apply these conditions is instead to save computation time. The first of these conditions is a symmetry which is always imposed and this is the exchange symmetry. This symmetry means that a wavefunction composed of fermions should be antisymmetric with respect to permutations of the fermions. As has already been discussed in previous chapters this symmetry is most easily fulfilled by requiring that the configurations are (linear combinations of) Slater determinants, i.e. antisymmetrized products of orbitals. In practically all existing CI methods the orbitals from which the Slater determinants are built are furthermore required to be orthogonal. Due to the simplicity of the matrix elements between single Slater determinants the most efficient implementation of some CI methods is actually obtained when each configuration is a single Slater determinant without any further requirements. This is, for example, the case at present for complete CI expansions.

a. Spatial symmetry.

There are very few existing CI methods which are able to make optimal use of spatial symmetry in a general way. This does not mean that there is a fundamental problem to obtain correctly symmetry adapted configurations of any symmetry. On the contrary, several simple algorithms exist for projecting out any space symmetry in a configuration. However, all of these schemes destroy the important simple structure of the CI expansion which is essential, in particular, for efficiently implementing the direct CI strategy. However, in the case of CI methods for atoms the gain in using the full atomic symmetry can overpower the loss of efficiency in abandoning the direct CI strategy. For molecules the dominating strategy, to at least partly utilize spatial symmetry, is to restrict the use of symmetry to Abelian point groups, which are characterized by one-dimensional irreducible representations. For such point groups the product of two irreducible representations is another irreducible representation. If the orbitals that make up the configurations can be characterized as belonging to irreducible representations, then the total spatial symmetry of the configuration is obtained as a simple product between the symmetries of the orbitals that build up that configuration. This means that symmetry adaptation to Abelian point group symmetry does not require any linear combination of determinants, each determinant is automatically symmetry adapted.

A simple example of the use of symmetry can be taken from the water molecule. The water molecule at its equilibrium geometry has C_{2v} symmetry and this point group is

Abelian with 4 irreducible representations: A_1, B_2, B_1 and A_2. Say that a configuration can be written as

$$[1a_1^2\ 2a_1^2\ 3a_1^1\ 1b_2^1\ 1b_1^2 2b_1^1\ 1a_2^1]$$

which is a double excitation from the Hartree Fock determinant. The total symmetry is then obtained by a simple table look up procedure from the C_{2v} multiplication table which is at hand in the computer core storage. First, the doubly occupied orbitals do not have to be considered since their symmetries will always be totally symmetric, i.e. of A_1 symmetry. For the singly occupied orbitals in this configuration, multiplying A_1 with B_2 leads to B_2, continuing to multiply this result with B_1 leads to A_2 and finally multiplying this result with A_2 leads to A_1, which is then the total symmetry of this configuration. Therefore, if a CI calculation is to be performed for a state of H_2O which is not of A_1 symmetry, this configuration can be disregarded.

b. Spin symmetry.

The final symmetry which is accounted for in most CI methods is spin symmetry. The handling of spin is in general much more complicated than the normal handling of spatial symmetry as described above. The most essential difference is that spin adaptation of configurations leads to configurations which are no longer single but linear combinations of Slater determinants with fixed, spin adaptation, coefficients. The brute force evaluation of matrix elements between configurations will therefore be to first evaluate the matrix elements between the individual Slater determinants building up the configurations followed by a summation of these matrix elements using the spin adaptation coefficients. This is an inefficient procedure and a large part of the development of modern CI methodologies has been devoted to the problem of obtaining matrix elements directly over spin adapted configurations. Here the language of the unitary group has been particularly useful and we will return to how this formalism has been used for the evaluation of matrix elements between spin adapted configurations later on in this chapter.

To understand the spin properties of Slater determinants one first has to have an expression for the $\hat{S}^2$ operator. From the definition of $\hat{S}^2$ and by using the step up and step down spin operators it is easy to show that

$$\hat{S}^2 = \hat{S}_z(\hat{S}_z + 1) + \hat{S}_-\hat{S}_+ \tag{4.1}$$

where for a many-electron system

$$\hat{S}_z = \sum_i \hat{s}_{zi} \tag{4.2}$$

and similarly for $\hat{S}_+$ and $\hat{S}_-$. The sum in (4.2) is over the electrons in the system. In all practical implementations of the CI method the orbitals forming the configurations are chosen as eigenfunctions of $\hat{s}_{zi}$. The action of the above one-electron spin operators on the orbitals are then well defined in the normal way. It is left as an exercise to show that when $\hat{S}^2$ acts on a determinant the result can be written as,

$$\hat{S}^2\Phi = \{\sum_P \hat{P}_{\alpha\beta} + \frac{1}{4}[(n_\alpha - n_\beta)^2 + 2n_\alpha + 2n_\beta]\}\Phi \tag{4.3}$$

where $\hat{P}_{\alpha\beta}$ is an operator which exchanges α and β spins in Φ. The sum is over all possible such interchanges. It is easily realized from the expression for $\hat{S}^2$ that the contribution from doubly occupied orbitals will be zero to the spin expectation value, which means that these orbitals can be disregarded when the electrons of a configuration are spin coupled. For a two-electron system (equivalently, a many-electron system with two open shells) one can form two different determinants with M_s=0, and it is easily verifiable that the sum of the two determinants,

$$[\varphi_i\overline{\varphi}_j] + [\overline{\varphi}_i\varphi_j] \tag{4.4}$$

has S=1 and the corresponding difference has S=0, by the application of the $\hat{S}^2$ operator. Each one of these determinants will not have a well-defined value for S. With more than two open shells the situation becomes more complicated and it is for these cases generally possible to form more than one linearly independent combination of Slater determinants which will have the same value for S. For three open shells and M_s=1/2 one finds the two orthogonal doublets (S=1/2),

$$[\varphi_i\varphi_j\overline{\varphi}_k] - [\varphi_i\overline{\varphi}_j\varphi_k] \tag{4.5}$$

and

$$2[\overline{\varphi}_i\varphi_j\varphi_k] - [\varphi_i\varphi_j\overline{\varphi}_k] - [\varphi_i\overline{\varphi}_j\varphi_k] \tag{4.6}$$

and for four open shells and M_s=0 two orthogonal singlets (S=0) can be written as,

$$[\varphi_i\overline{\varphi}_j\varphi_k\overline{\varphi}_l] - [\varphi_i\overline{\varphi}_j\overline{\varphi}_k\varphi_l] - [\overline{\varphi}_i\varphi_j\varphi_k\overline{\varphi}_l] + [\overline{\varphi}_i\varphi_j\overline{\varphi}_k\varphi_l] \tag{4.7}$$

and,

$$2[\varphi_i\varphi_j\overline{\varphi}_k\overline{\varphi}_l] - [\varphi_i\overline{\varphi}_j\varphi_k\overline{\varphi}_l] - [\varphi_i\overline{\varphi}_j\overline{\varphi}_k\varphi_l] - [\overline{\varphi}_i\varphi_j\varphi_k\overline{\varphi}_l] - [\overline{\varphi}_i\varphi_j\overline{\varphi}_k\varphi_l] + 2[\overline{\varphi}_i\overline{\varphi}_j\varphi_k\varphi_l] \tag{4.8}$$

It is left as exercises to verify that the two latter examples are actually eigenfunctions of $\hat{S}^2$. It should be noted that whenever there is more than one eigenfunction of $\hat{S}^2$ with the same eigenvalue, any linear combination of these eigenfunctions will also be an eigenfunction of $\hat{S}^2$ so that the spin eigenfunctions are not uniquely defined.

We have already mentioned that several ways exist for constructing proper linear combinations of determinants which will be eigenfunctions of $\hat{S}^2$. We will here sketch three of these methods. The first of these is one which is used in several CI programs. One starts out by, for a given number of open shells listing all possible determinants which have the same expectation value of $\hat{S}_z$. For the four open shell case with M_s=0 given above there are six such determinants. In this basis one sets up the $\hat{S}^2$ matrix using relation (4.3) for the action of $\hat{S}^2$ on a general determinant. A diagonalization of this matrix, using standard techniques gives the desired eigenfunctions and these will be automatically orthogonal. This is a simple, brute force, method for obtaining proper spin functions. A drawback is that the spin-adaptation coefficients will not be given by analytical expressions which may be useful sometimes. The second method we will mention starts out by setting up the determinant with the highest possible S and the highest possible M_s=M_s(max). For the four open shell case this will be the quintet determinant where all orbitals have α spin. The method proceeds by operating with $\hat{S}_-$ on this determinant which gives a configuration with the same value for S but with M_s=M_s(max)-1. By orthogonalization the configurations (in general many) with S=S(max)-1 and M_s=M_s(max)-1 will be formed etc. The third method starts out with an arbitrary trial determinant with the desired M_s value. On this determinant one applies a projection operator,

$$\hat{O}_K = \frac{\hat{S}^2 - K(K+1)}{S(S+1) - K(K+1)} \tag{4.9}$$

which annihilates the component with S=K. One proceeds by annihilating all undesired spin-components until a proper spin state is found (or 0 in which case one has to start out with another trial determinant). This method will not automatically lead to orthogonal spin-adapted configurations, but has to be followed by an orthogonalization procedure if this is desired. This method can also be used in more complicated situations. For example, for the case of an MP2 (second order perturbation theory) wavefunction based on an UHF determinant this method has recently been used to annihilate the worst spin-contaminants.

We will finish this section by pointing out a property which will be useful later on in this chapter. In the mathematical lectures the generators of the unitary group have been defined as,

$$\hat{E}_{ij} = \hat{a}^+_{i\alpha}\hat{a}_{j\alpha} + \hat{a}^+_{i\beta}\hat{a}_{j\beta} \tag{4.10}$$

The usefulness of this operator is partly based on the fact that when it operates on a properly spin-adapted configuration it will generate another spin-adapted configuration with the same spin. This result may be considered trivial since the operator is a sum (trace) over all possible M$_s$ eigenvalues. As an exercise it is useful to check this property out by, for example, applying the product operator $\hat{E}_{ai}\hat{E}_{bj}$ on a closed shell determinant $\mid 0 >$ and verify that the resulting linear combination of determinants is a spin-adapted configuration with S=0. The orbitals i and j are occupied in $\mid 0 >$ but not the orbitals a and b. This is thus a properly spin-adapted doubly excited configuration which can be directly used in a CI calculation. For future purposes it is also useful to show that the spin-adapted configuration obtained by applying $\hat{E}_{aj}\hat{E}_{bi}$ on $\mid 0 >$ is linearly independent, but not orthogonal, to the previously defined doubly excited state.

c. The MR-CI configuration expansion.

So far in this section we have only discussed the form of the configurations, not the choice of configurations in the CI expansion (3.2). To do this, we will start out by the simplest example, where already the Hartree-Fock configuration is a qualitatively good approximation to the true wave-function, and can therefore by itself be selected as a zeroth order approximation to the wave-function. Using the normal vocabulary, the Hartree-Fock configuration is selected as a single reference state. From the rules for the matrix elements between determinants (which will be given later) we know that only determinants which differ by at most two orbitals will have non-zero matrix elements with each other. This results from the simple fact that the Hamiltonian operator does not contain more than two-electron operators. From perturbation theory we further know that the configurations which have non-zero matrix elements with the zeroth order wave-function should be the dominant corrections to the wave-function. Therefore, the simplest choice of CI expansion, which is actually often a very good approximation to the true wave-function, can be written as,

$$\Psi_{SD} = C_0\Phi_0 + \sum_{ia} C_i^a \Phi_i^a + \sum_{ijab} C_{ij}^{ab}\Phi_{ij}^{ab} \tag{4.11}$$

where Φ_0 is the Hartree-Fock configuration and Φ_i^a and Φ_{ij}^{ab} are single and double replacements (excitations) out of Φ_0. The occupied (internal) orbitals i,j are replaced by (excited to) the unoccupied (external or virtual) orbitals a,b. This wave-function is usually called the SD-CI (singles and doubles CI) wave-function. Including size-consistency corrections (see section X in this chapter) this type of wave-function has been shown to yield results of very high quantitative accuracy for cases where the Hartree-Fock wave-function is a good zeroth order wave-function, i.e. normally for molecules around their equilibrium geometries. The size-consistency corrected SD-CI wave-function is in practice of the same quality as the more elaborate coupled cluster SD wave-function (see a different chapter).

Today there exist two fundamentally different ways to extend the SD-CI wave-function for situations where the Hartree-Fock wave-function is a less adequate zeroth order starting point. In the first approach the wave-function is simply extended by adding triply and even quadruply excited configurations,

$$\Psi_{SDTQ} = \Psi_{SD} + \sum_{ijkabc} C_{ijk}^{abc}\Phi_{ijk}^{abc} + \sum_{ijklabcd} C_{ijkl}^{abcd}\Phi_{ijkl}^{abcd} \tag{4.12}$$

The problem with this type of extension is clearly that the number of configurations grows very rapidly with the level of excitation. Even the SD-CI wave-function may contain a large number of terms but can today using the direct CI strategy (see section VI in this chapter) be handled in essentially all cases for which the two-electron integrals can be stored on disk. If not more than 50 electrons are correlated and 200 basis functions are used (which is at present a normal limit for storing the integrals) there will be at most a few million configurations, which is expensive but feasible to handle. The number of triple excitations will be at least two orders of magnitude larger than the number of double excitations and can clearly not be handled without some sort of approximation in the normal case. To treat the triple excitations at the lowest order of perturbation theory is still a rather expensive but possible alternative (see different chapter). In the normal CI formalism the inclusion of higher than double excitations in a general way is not a realistic possibility unless very few orbitals are included or only a few (at most 5) electrons are correlated. The extension of the SD-CI wave-function has therefore taken a different route and is connected with the development of the MCSCF method.

In section II we have already discussed how the qualitative problems with the single-configuration Hartree-Fock approximation can normally be corrected by adding a few other configurations. The simple idea to extend the SD-CI wavefunction is then to add also the single and double replacements from these other important configurations, which in the CI formalism will be termed reference states. The wave-function can then be written,

$$\Psi_{MR-CI} = \sum_I \Psi_{SD}(I) = \sum_I \{C(I)\Phi(I) + \sum_{ix} C_i^x(I)\Phi_i^x(I) + \sum_{ijxy} C_{ij}^{xy}(I)\Phi_{ij}^{xy}(I)\} \quad (4.13)$$

where the sum over I runs over all selected reference states. This is the MR-CI (multi-reference CI) wave-function. It should be noted that since a single or double replacement from one reference state can also be a single or double replacement from another reference state, a unique list of configurations has to be constructed by deleting doubly counted configurations. We have used the letters x and y, rather than a and b for the orbitals to which electrons are excited, to emphasize that an orbital which is unoccupied (external) for one reference state may be occupied (internal) for another reference state. It turned out that the extension of the direct CI method to the MR-CI expansion became very difficult. One reason for this was actually that the way the wave-function was written seemed to imply that there were very many special cases. In one attempt to generalize direct CI to the case of many closed shell reference states, over 100 different cases of integral contributions were identified, and that method was still far from the desirable general MR-CI method. Perhaps surprisingly, a major step towards the generalization of the direct CI method was taken by rewriting the MR-CI wave-function in the following way,

$$\Psi_{MR-CI} = \sum_{\mu} C_{\mu}\Phi_{\mu} + \sum_{\mu a} C_{\mu}^{a}\Phi_{\mu}^{a} + \sum_{\mu ab} C_{\mu}^{ab}\Phi_{\mu}^{ab} \quad (4.14)$$

where the letters a and b have been used for orbitals which are unoccupied in all reference states and μ is used as a compound index for the occupied orbitals. This can be exactly the same wave-function as (4.13) but the number of external orbitals rather than the excitation level for the configurations has been emphasized. It turned out that after this reformulation, the number of special integral cases in the direct CI method was reduced to less than 20 even for the most general MR-CI case. The three groups of configurations in (4.14) will in the following be denoted as valence, singly external and doubly external configurations. The relation between the groups of configurations in (4.13) and (4.14) is the following. The configurations in the first group in (4.13), the reference configurations, is included in the valence configurations in (4.14). The single excitations in (4.13) can be either valence or singly external configurations, whereas the double excitations in (4.13) can belong to any of the three groups in (4.14). One should also note that all the doubly external configurations in (4.14) are double excitations from the reference states but this group does not contain the full list of double excitations.

d. The occupation graph and the branching diagram.

One of the first things that is done in an MR-CI program is to generate a unique set of configurations. For this purpose the occupation graph and the branching diagram are very useful. An example of an occupation graph is given in Fig.1.

The idea is that all possible occupations in the MR-CI expansion can be generated by following a path in the diagram from the bottom to the top of the graph. The simplest organization is to leave out the external orbitals and instead have three different graphs for the internal orbitals, one for each group of configurations in (4.14). The graph consists of points called vertices and lines called arcs. The vertices are arranged along horizontal lines, one horizontal line for every orbital. The inclination of the arcs gives the orbital occupation. The arc with the smallest angle with the horizontal lines denotes double occupation, the one with a somewhat larger angle denotes single occupation and the vertical arc denotes zero occupation. At each vertex in the graph a number is given which is equal to the total number of electrons at that vertex summed along any path to that vertex from the bottom of the graph. This number must consequently not exceed the total number of electrons that should be distributed in the internal orbitals, which means that there can be no vertices to the left of the top vertex. In the construction of the graph one must also exclude vertices which can not be reached from the top vertex even if all the top orbitals are doubly occupied. By having the orbitals which are doubly occupied in all reference states at the bottom of the graph it is furthermore easy to exclude vertices with more than two holes in these orbitals. Any further reduction of the vertices and arcs in the graph is not generally possible for a normal MR-CI case. Therefore the occupations which are generated by following all possible paths from the bottom to the top of the graph will normally exceed the number which is actually going to be used in the calculation. These unwanted occupations have to be deleted by a comparison to each individual reference state.

Having constructed all the desired occupations in the MR-CI expansion, there remains to construct all possible spin-couplings for each occupation, before a full list of configurations is obtained. The total number of spin-couplings for a given number of singly occupied orbitals is conveniently given by the branching diagram which is shown in Fig.2.

The x-axis in this diagram gives the number of electrons N in singly occupied orbitals and the y-axis gives the total spin quantum number S. The number given at each vertex in the graph is equal to the number of spin-couplings for given values of N and S. The construction of this diagram starts from the bottom left part. The number at this vertex gives the number of spin-couplings when there are no singly occupied orbitals. This number clearly has to be one. Arcs are then drawn to connect the vertices with a certain number of singly occupied orbitals to the vertices with one more singly occupied orbital. The numbers at the vertices are obtained by adding the numbers of the vertices which are connected to that vertex by the arcs coming from the left, i.e. in most cases two numbers. The fact that there are two connecting vertices is a simple consequence of the fact that the spin of an electron can couple in two ways, up or down, to generate $S=S+1/2$ and $S=S-1/2$ respectively. One notable fact in this diagram is that the number of possible spin couplings grows rapidly with the number of singly occupied orbitals. Since for many methods, prototype matrix elements between different spin-couplings have to be stored in core-storage, a calculation using these methods can be limited by the maximum allowed number of singly occupied orbitals in the program.

We will end this section by pointing out that there also exists a graphical representation where both the occupations and spin-couplings are given for each configuration.

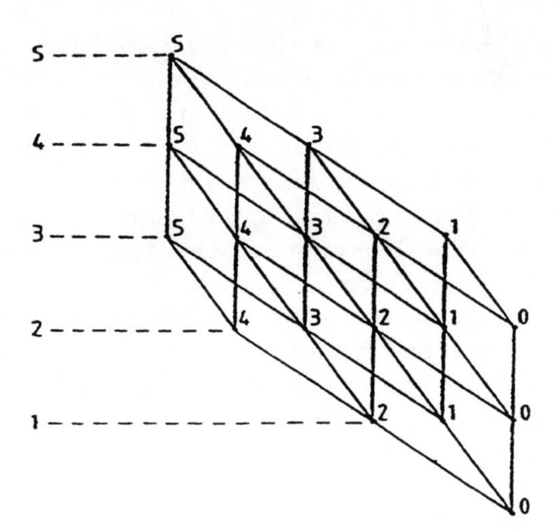

Fig.1 Occupation graph with 5 electrons in 5 orbitals

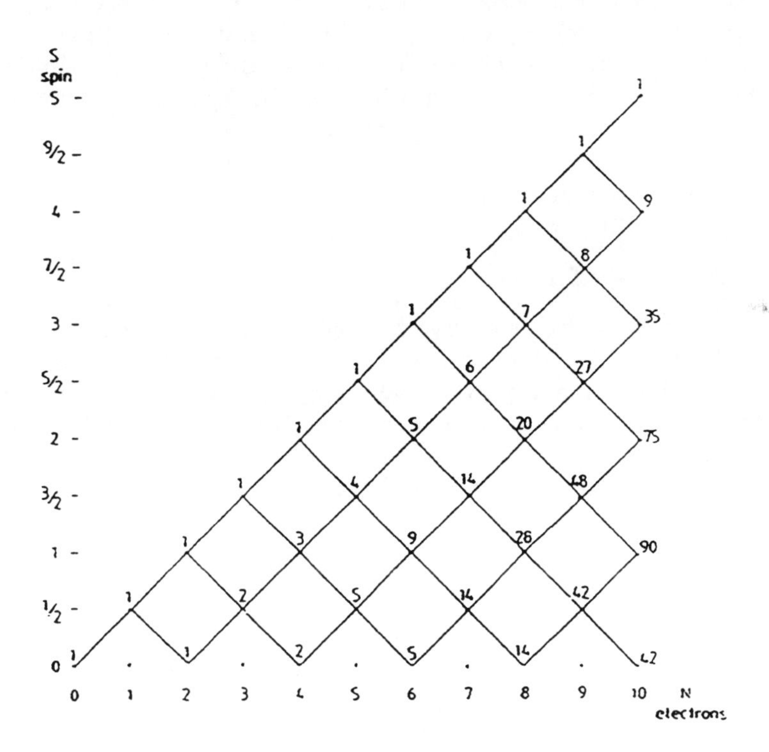

Fig.2 Branching diagram

This graph was first used by Shavitt and we will therefore call it the "Shavitt graph", which was actually prior in time to the occupation graph described above. The main difference between these two graphs is that in the Shavitt graph there are four different types of arcs rather than three in the occupation graph. The arcs that denote zero and double occupation of the orbitals have the same meaning in the two graphs. These two occupancies will not affect the spin coupling in contrast to the case of single occupancy. In the latter case there are two different types of arcs in the Shavitt graph, one denoting spin-coupling from S to S+1/2 and the other from S to S-1/2. To keep track of both occupation and spin-coupling each vertex has two different values, one for the number of electrons and the other for the total spin both counted from the bottom of the graph. An organisation where occupation and spin-coupling are treated on an equal footing, as in the Shavitt graph, is usually referred to as the "unitary group" approach to CI. The previously described organisation based on a separate organisation for the occupations, as in the occupation graph, and of the spin-couplings, as in the branching diagram, is instead referred to as the "symmetric group" approach to CI. There was originally a much larger difference between the symmetric and unitary group approaches, particularly concerning the evaluation of Hamiltonian matrix elements, but as the development has proceeded the two approaches have become nearly identical apart from the organisation of the configurations. At present for the MR-CI method, the symmetric group organisation is the one that has the most advantages.

In an MR-CI program it is also necessary to be able to assign a unique number to each configuration. A simple procedure to do this has been outlined by Shavitt. In this procedure one assigns a number to each arc in the graph. By summing these numbers for the path describing a configuration one obtains a unique number for that configuration. To describe how the numbers for the arcs are defined would carry us into too many details and we will here instead refer to the original papers by Shavitt. It is, however, useful to know that such a procedure exists.

e. The coupled cluster wave-function.

In section c above, two different ways of writing the wave-function have been described by (4.12) and (4.13), which in principle can be extended to the exact wave-function. There is a third alternative which has many advantages and is connected with one of the presently dominating ways of solving the Schrödinger equation, and this is the coupled cluster wave-function. To define this wave-function it is convenient to define certain excitation operators. A general n-tuple excitation operator T_n is defined as

$$\hat{T}_n = \sum_{i,j..,a,b..} T^{a,b..}_{i,j..} \hat{a}^+_a \hat{a}_i \hat{a}^+_b \hat{a}_j ... \tag{4.15}$$

where the cluster amplitudes $T^{a,b..}_{i,j..}$ are related to the CI-coefficients $C^{a,b..}_{i,j..}$ as given, for example, in (4.12). The exact coupled cluster wave-function can then be written as

$$\Psi_{CC} = e^{\hat{T}} \Phi_0 \tag{4.16}$$

with $\hat{T} = \hat{T}_1+\hat{T}_2+...$ (number of electrons). From the definition of $\hat{T}_n$ it also follows that the Ψ_{SD} wave-function defined in (4.11) can be conveniently written as

$$\Psi_{SD} = (1 + \hat{T}_1 + \hat{T}_2)\Phi_0$$

The corresponding coupled cluster wave-function is instead

$$\Psi_{CCSD} = e^{(\hat{T}_1+\hat{T}_2)}\Phi_0$$

The main advantage of writing the wave-function in an exponential form follows easiest by considering non-interacting systems. When a method is applied to a case of non-interacting systems the energy obtained should ideally be the sum of the energies obtained using the same method for the separate systems. It follows from the Schrödinger equation that if this should hold the wave-function should be the product of the wave-functions for the separate systems. (Another possibility is to change the energy expression, see section X).

To rationalize the use of the exponential form of the wave-function in (4.16), the case of non-interacting He-atoms will first be considered. The exact wave-function for a He atom can be written in terms of the operators $\hat{T}_1$ and $\hat{T}_2$ only, since it is a two-electron system. For simplicity we will only keep the $\hat{T}_2$ operator in this derivation. The wave-function for He atom A is then

$$\Psi_A = (1 + \hat{T}_{2A})\Phi_{0A}$$

With a similar wave-function for He atom B the product wave-function will be

$$\Psi_{AB} = \Psi_A\Psi_B = (1 + \hat{T}_{2A} + \hat{T}_{2B} + \hat{T}_{2A}\hat{T}_{2B})\Phi_{0A}\Phi_{0B}$$

The $\hat{T}_2$ operator for the composite system is equal to $\hat{T}_{2A} + \hat{T}_{2B}$. Since $\hat{T}_{2A}^2$ acting on Φ_{0A} must be zero it then follows that Ψ_{AB} can be written as

$$\Psi_{AB} = (1 + \hat{T}_2 + \frac{1}{2!}\hat{T}_2^2)\Phi_{0A}\Phi_{0B}$$

It is easy to derive similar expressions for the case of three and four He-atoms as exercises, and continuing this way it is realized that the exact wave-function Ψ for an infinite number of non-interacting He-atoms can be written as

$$\Psi = (1 + \hat{T}_2 + \frac{1}{2!}\hat{T}_2^2 + \frac{1}{3!}\hat{T}_2^3 + ...)\Phi_0 = e^{\hat{T}_2}\Phi_0$$

It should be noted that this expression is, of course, also valid for a limited number of He-atoms.

Another rationalization of the exponential form of the wave-function is obtained by considering two non-interacting N-electron systems. If a wave-function of type $(1+\hat{T}_2)\Phi_0$ is tried, the product wave-function is

$$\Psi_A\Psi_B = (1 + \hat{T}_{2A} + \hat{T}_{2B} + \hat{T}_{2A}\hat{T}_{2B})\Phi_{0A}\Phi_{0B}$$

If the same type of wave-function, $(1+\hat{T}_2)\Phi_0$ with $\hat{T}_2 = \hat{T}_{2A} + \hat{T}_{2B}$, is written directly for the composite system one obtains instead

$$\Psi_{AB} = (1 + \hat{T}_{2A} + \hat{T}_{2B})\Phi_{0A}\Phi_{0B}$$

which is different from the product wave-function. To obtain a better agreement between $\Psi_A\Psi_B$ and Ψ_{AB} it is easily realized that a wave-function of the type $(1+\hat{T}_2+\frac{1}{2!}\hat{T}_2^2)\Phi_0$ is needed. This type of wave-function leads to differences between $\Psi_A\Psi_B$ and Ψ_{AB} which appear when operators of the order of $\hat{T}_2^3$ and $\hat{T}_2^4$ act on Φ_0. To eliminate also these differences, wave-functions of the type $(1+\hat{T}_2+\frac{1}{2!}\hat{T}_2^2+\frac{1}{3!}\hat{T}_2^3)\Phi_0$ are needed etc., and it is realized that full agreement between $\Psi_A\Psi_B$ and Ψ_{AB} is not reached until the full exponential form of the wave-function is used.

This short sub-section on the coupled cluster method can clearly not cover everything of this important area, but a few more aspects should be pointed out here. First, a procedure is needed to obtain the cluster amplitudes $T_{i,j..}^{a,b..}$. For the CCSD wave-function described above, this is normally done by simply taking enough equations out of the full CI secular problem. Since the number of unknown amplitudes in this case is equal to the number of single and double replacements the number of equations is equal to this number and also includes the first equation, which defines the energy. Another way of describing this procedure is to say that the Schrödinger equation for the CCSD wave-function is projected on to the space of single and double replacements. The equations of the CCSD method will be similar to the corresponding CISD equations but will also contain products of coefficients for double replacements, for single replacements and for single and double replacements. The CCSD equations are thus a set of coupled *non-linear* equations.

An interesting point of the coupled cluster method concerns the treatment of quadruple excitations. If the CCD method is considered, in which only the $\hat{T}_2$ operator is retained in the exponent, the amplitudes for these excitations are given as products of amplitudes for double excitations according to the term $\hat{T}_2^2$. In fact T_{ijkl}^{abcd} is a sum of the 18 products of type $T_{ij}^{ab}T_{kl}^{cd}$ (with phase factors) which can be formed when the indices i,j,k,l and a,b,c,d are permuted in all different ways. It turns out that this is exactly the same expression as that obtained in 4th order perturbation theory. The treatment of quadruple excitations is for this reason expected to be quite accurate in the CCD (and the CCSD) method. There is no corresponding product form of the triple excitations in perturbation theory. The major correction to the CCSD method therefore appears for these excitations and has to explicitly take the operator $\hat{T}_3$ into account.

In the coupled cluster equations the energy as expressed through the first equation is cancelled in all other equations. In the CCD method the energy is cancelled by matrix elements between double excitations and quadruple excitations. This cancellation is an important property and is the origin of the size consistency (see section X) and size extensivity (see other lectures) properties of the coupled cluster method. Since this cancellation is the main condition for size consistency it means that some simpler methods of the same type as the coupled cluster method can also have this property as long as the energy is cancelled. A method of this type is the QCISD (quadratic CISD) method, where terms coming from $\hat{T}_1^3$, $\hat{T}_1^4$, $\hat{T}_1^2\hat{T}_2$ etc. have been neglected.

V. Matrix elements and the conventional CI approach.

In this section we will briefly outline the different steps in a CI calculation. Before we do this we first need to recapitulate the main results for the Hamiltonian matrix elements between determinants. In this course it is assumed that the students have already seen a detailed derivation of the matrix elements and we will therefore only sketch this derivation here.

We start out by writing a determinant in terms of the antisymmetrizer $\hat{A}$ as,

$$D = \hat{A}\Omega = \hat{A}(\varphi_1\varphi_2...\varphi_N) \tag{5.1}$$

where Ω on which $\hat{A}$ acts is a simple product of the N occupied spin-orbitals entering the determinant. The antisymmetrizer $\hat{A}$ is chosen so that its operation on the spin-orbital product leads to a normalized Slater determinant. With this definition $\hat{A}$ can be written,

$$\hat{A} = (N!)^{-\frac{1}{2}} \sum_P \epsilon(P)\hat{P} \tag{5.2}$$

where $\hat{P}$ is a permutation operator and $\epsilon(P)$ is the normal phase factor. The sum runs over all possible permutations of the electrons. It is then relatively easy to show that,

$$\hat{A}^2 = (N!)^{\frac{1}{2}}\hat{A} \tag{5.3}$$

which means that $\hat{A}$ apart from a constant factor fulfills the idempotency condition of a projection operator. Since $\hat{A}$ is a sum of simple permutation operators and the Hamiltonian operator is independent of the numbering of the electrons it is clear that $\hat{A}$ commutes with $\hat{H}$. Therefore, a Hamiltonian matrix element between two determinants can be immediately simplified as,

$$< D \mid \hat{H} \mid D' >=< \hat{A}\Omega \mid \hat{H} \mid \hat{A}\Omega' >= (N!)^{\frac{1}{2}} < \Omega \mid \hat{H} \mid \hat{A}\Omega' >=$$

$$= \sum_P \epsilon(P) < \Omega \mid \hat{H} \mid \hat{P}\Omega' > \tag{5.4}$$

With the further condition that the spin-orbitals are orthogonal the special cases, Slater's rules, for matrix elements between determinants are obtained from this formula by inspection. The general formula can be written

$$< D \mid \hat{H} \mid D' >= \sum_i n_i h_{ii'} + \sum_{ij} n_{ij}[(\varphi_i\varphi_i' \mid \varphi_j\varphi_j') - (\varphi_i\varphi_j' \mid \varphi_i\varphi_j')] \tag{5.5}$$

where

$$n_{ij} = 1 \text{ if } \varphi_k = \varphi_k' \text{ for all } k \neq i,j,.. \tag{5.6}$$

and zero otherwise. In summary, if the determinants are identical this expression will be a single sum over one-electron and a double sum over two-electron integrals. If the determinants differ in one spin-orbital there will be a single one-electron integral and a single sum over two-electron integrals and finally if the determinants differ in two

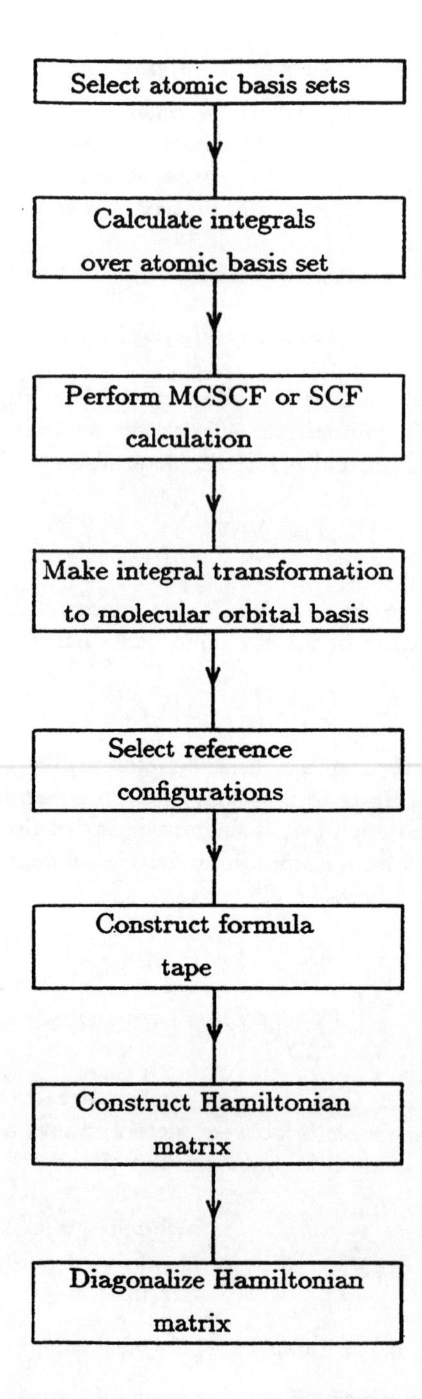

Fig.3. Flow chart for a conventional CI calculation.

spin-orbitals there will be just two two-electron integrals in the expression. All other matrix elements will be zero.

Based on the above formulas and knowledge from this and previous chapters it is easy to sketch the different steps in an MR-CI calculation. First, an atomic basis set is selected and all one- and two-electron integrals over this basis are constructed and stored. As a second step an MCSCF (or sometimes simply an SCF) calculation is performed. Since the matrix elements between determinants (configurations) are expressed in terms of integrals over molecular rather than atomic orbitals, in the third step the atomic integrals have to be transformed to the molecular orbital basis as obtained from the MCSCF step. As a start of the actual MR-CI calculation the reference states are selected based on the MCSCF calculation and the MR-CI expansion can be set up. In the next step the matrix elements over the Hamiltonian are constructed from the general expressions described above and using the transformed molecular integrals generated in the third step. In many MR-CI programs the construction of the Hamiltonian matrix is preceeded by a construction of a "formula tape". This formula tape contains all the formal knowledge required for constructing the Hamiltonian matrix, such as the sequence numbers of the interacting configurations for a particular matrix element, the sequence numbers for the integrals entering this matrix element and finally the coefficients for the integrals in the matrix element. When the Hamiltonian matrix has been constructed and stored all that remains is to diagonalize this matrix for the root of interest. This procedure is illustrated in Fig.3.

The above sequence of steps is what enters a so called "conventional" MR-CI calculation. This is still a widely used method to solve the MR-CI equations and during the past decades many tricks have been developed to circumvent some of the inherent problems with this approach. These problems are mainly due to the storage of large data sets, in particular the storage of the Hamiltonian matrix elements or even worse the storage of the formula tape. Therefore, if this approach is used the MR-CI expansion has to be drastically truncated. The maximum number of configurations which can be handled is in practice 10 to 20 thousand terms. The Hamiltonian matrix will then contain on the order of a few million non-zero terms. Since an MR-CI expansion without truncation in normal applications is 10^5 to 10^7 configurations the adopted truncation scheme has to be extremely efficient if the final result should still be accurate. In the next section we will discuss an alternative approach by which it is possible to handle the non-truncated MR-CI expansion without approximations.

VI. Diagonalization and the direct CI equations.

The direct CI equations are obtained by combining the normal CI equations (3.3) with an iterative diagonalization procedure. (The same direct CI equations can also be obtained within a perturbation theory approach). Since diagonalization procedures have been described in another set of lectures we will here only repeat the most essential results. The simplest iterative procedure is obtained by moving everything but the diagonal terms in the CI equations over to the right hand side and assume that this side of the equations can be obtained from the CI vector of the previous iteration $\mathbf{C}^k$.

An improved CI vector $\mathbf{C}^{k+1}$ is then obtained as,

$$C_\mu^{k+1} = C_\mu^k + \frac{1}{E^k - H_{\mu\mu}}(\sum_\nu H_{\mu\nu}C_\nu^k - E^k C_\mu^k) \tag{6.1}$$

where $\mathbf{E}^k$ is the variational energy evaluated using the vector $\mathbf{C}^k$. The start, $\mathbf{C}^0$, of this iterative procedure is normally obtained by a diagonalization of the small reference space, which can usually be done using the Jacobi method. A straightforward application of this iterative procedure will not be very efficient and actually diverge in many cases. A very efficient scheme is instead obtained by setting up a small CI equation system where the basis vectors are the $\mathbf{C}^k$ vectors. The Hamiltonian matrix elements over these vectors are very easily obtained as a byproduct in constructing $\mathbf{C}^{k+1}$. The dimension of this secular equation system is thus equal to the number of CI iterations and the equation system is therefore always so small that it can easily be solved. Convergence to six decimal places in the energy is normally obtained in about 10 iterations.

The most important point in the above outlined procedure is that essentially all the work in a CI iteration is performed in generating the vector σ, which can be written as,

$$\sigma_\mu = \sum_\nu H_{\mu\nu}C_\nu^k \tag{6.2}$$

In the perturbation theory approach a similar vector is obtained by operating with the perturbation operator $\hat{V}$ on the vector of the previous order (=iteration). Since the matrix H and the matrix V differ only in trivial diagonal terms the work in the two approaches is nearly identical in each iteration.

In order to be able to write out all the terms of the direct CI equations explicitly, the Hamiltonian operator is needed in a form where the integrals appear. This is done using the language of second quantization, which has been reviewed in the mathematical lectures. Since, in the MR-CI method, we will generally work with spin-adapted configurations a particularly useful form of the Hamiltonian is obtained in terms of the generators of the unitary group. The Hamiltonian in terms of these operators is written,

$$\hat{H} = \sum_{pq} h_{pq}\hat{E}_{pq} + \frac{1}{2}\sum_{pqrs}(pq \mid rs)(\hat{E}_{pq}\hat{E}_{rs} - \delta_{qr}\hat{E}_{ps}) \tag{6.3}$$

Combining this expression for the Hamiltonian with expression (6.2), the final direct CI equations for the σ vector can be written as,

$$\sigma_\mu = \sum_\nu \{\sum_{pq} h_{pq}A_{pq}^{\mu\nu} + \frac{1}{2}\sum_{pqrs}(pq \mid rs)A_{pqrs}^{\mu\nu}\}C_\nu \tag{6.4}$$

where,

$$A_{pq}^{\mu\nu} = < \mu \mid \hat{E}_{pq} \mid \nu > \tag{6.5}$$

and

$$A_{pqrs}^{\mu\nu} = < \mu \mid \hat{E}_{pq}\hat{E}_{rs} - \delta_{qr}\hat{E}_{ps} \mid \nu > \tag{6.6}$$

We will in the following call these coefficients "the direct CI coupling coefficients" or simply the coupling coefficients.

The main point in this reformulation is that the explicit reference to the Hamiltonian matrix is avoided. The so called "residual vector" σ is obtained directly from the one- and two-electron integrals, the previous CI vector and the coupling coefficients. The sizes of the different vectors and matrices in a normal MR-CI calculation illustrate the great advantage in the direct CI reformulation. The CI vector and the residual vector have the same size which is equal to the total MR-CI expansion which can be of the order of 10^6 terms and can be held in core storage. The number of integrals depends on the chosen atomic basis set and for up to 200 basis functions this list contains on the order of 10^7 terms and is held on peripheral storage. The Hamiltonian matrix, on the other hand, contains in this case on the order of at least 10^{10} non-zero terms, which can not be held even on the largest available storage devices. The number of coupling coefficients in the direct CI equations is, on the other hand, as large as the number of Hamiltonian matrix elements and they can thus not be stored either. Since these coefficients do not depend on the actual values of the integrals but only on the structure of the interacting configurations they can in principle be calculated as they are needed given the proper algorithms. Of key importance for the efficiency of the direct CI method is thus that extremely efficient such algorithms for the direct CI coupling coefficients can be obtained. We will address this question in the next section.

VII. Evaluation of the direct CI coupling coefficients.

It is possible to write a section on the direct CI coupling coefficients which is full of detailed formulas which can be directly put into a computer program. We have, however, for this course found it more appropriate to just give a general orientation of how these coefficients can be evaluated, which means that there will be much more words than formulas in the text. One reason for this is of course that very few of the normal users of direct CI programs will ever write their own code. Another reason is that the present strategy for evaluating these coefficients is by no means going to be permanently fixed, but could easily change the next year. We have also chosen to describe the development in a historical perspective. If the reader is interested in the explicit details we refer instead to the references at the end of this chapter.

The first application of the direct CI method was for the case of an SD-CI wavefunction based on a closed shell reference determinant. In this case it was possible to identify all the different possible values the coupling coefficients can have and program these values directly into the code. The contributions to double replacement elements μ in the residual vector σ from double replacement elements ν in the CI vector $\mathbf{C}^k$ (in the following this will be called the interactions between μ and ν) were divided into 5 different groups. A group was identified by the number of external (= unoccupied orbital in the reference state Φ_0) indices in the integral $(pq \mid rs)$ in the interaction (6.4). Group 1 contained integrals with 0 external indices and group 5 contained integrals with 4 external indices. In each group there was also a Coulomb and exchange type contribution. To define every possible interaction between the double replacements μ and ν these configurations were further divided into 5 different classes, generating a 5x5 table of different coupling coefficients for the interaction between the different classes. In summary, the integral group in a particular interaction was first determined. Then the integral was identified as being either Coulomb or exchange. Finally, the class of the interacting double replacements was identified and the coupling coefficient could then in principle be fetched in a table stored in the core memory. These coupling coefficients typically have values such as $\sqrt{3}/2$ and $-\sqrt{2}$ etc. and were determined by the programmer once and for all by using Slater's rules for prototype configurations. Technically, this procedure was made more efficient by grouping together all integrals of a particular group and treating them at the same time in one subroutine. In this subroutine there was then a loop over the 5x5 different types of interactions as described above. As a general strategy, this is actually not very different from how the general direct MR-CI case is handled today. The main difference is, of course, that in the general case every possible coupling coefficient for every possible type of interaction can not be obtained in such a simple way as for the closed shell SD-CI case.

The first attempts to generalize direct CI to other cases than the closed shell SD-CI wave-function were made in the same spirit as the original scheme. The case of a single reference determinant with one open shell (a doublet) is illustrative of the problems that were encountered. The introduction of an open shell orbital in the reference increases the number of types of orbitals from two (internal and external) to three. This may not seem drastic, but the number of integral groups, as determined by the number of orbital indices from the different groups of orbitals in the integral $(pq \mid rs)$ has now increased from 5 to 15. The number of classes of configurations (single and double replacements) increases in the same way from 6 to at least 11 depending on the actual organization. The total number of cases thus goes from 5x6x6=180 to 15x11x11=1815, which is a drastic increase. Already this slight generalization of the direct CI method was thus a rather difficult programming task. The situation obviously becomes even more complicated for more general wave-functions. As has already been mentioned, for the case of a reference composed of a set of closed shell determinants, over 100 different groups of integrals were identified, each one of them requiring special formulas for different classes of interactions.

It is clear that a full generalization of the direct CI method required a different approach with a more general formalism, which would allow a reduction of the number of special cases. The first step in this direction was the introduction of the unitary group formalism which, actually for the first time, lead to explicit formulas for the direct CI coupling coefficients. These formulas have already been given above in (6.5) and (6.6), but historically they appeared many years after the first direct CI programs. The next important, but seemingly trivial, step in the generalization of the direct CI method has also been given above and this was the rewriting of the MR-CI expansion according to equation (4.14). The significance of this reformulation is that it takes away all the special cases as to how the reference states are selected. The structure of the MR-CI expansion is always the same.

The final step in the generalization of the direct CI method was the realization that the full list of coupling coefficients $A^{\mu\nu}_{pqrs}$ can be generated from a small set of prototype coupling coefficients $B^{\mu\nu}_{pqrs}$, initially termed "internal" coupling coefficients.

The relation between these coefficients can simply be written as,

$$A^{\mu\nu}_{pqrs} = B^{\mu\nu}_{pqrs} C^{\mu\nu}_{pqrs} \tag{7.1}$$

where $C^{\mu\nu}_{pqrs}$ were initially termed "external" coupling coefficients. The idea is to determine the internal coupling coefficients within an orbital space containing the full list of internal orbitals but replacing the long list of external orbitals by a few prototype external orbitals. The function of the external coupling coefficients $C^{\mu\nu}_{pqrs}$ would then be to relate a particular prototype internal coupling coefficient $B^{\mu\nu}_{pqrs}$ to a general coupling coefficient $A^{\mu\nu}_{pqrs}$ with the same internal orbitals involved as in $B^{\mu\nu}_{pqrs}$, but where the external orbitals involved will be completely general. The key to the efficiency of this reformulation is of course first that the number of internal coupling coefficients is so small that they can be stored on a peripheral device and also that the external coupling coefficients are extremely easy to determine from the structure of the interaction. The simple form of the external coupling coefficients is best seen if a general doubly external configuration is written as,

$$\Phi_\mu = \Phi^{ab}_\kappa = [\Phi_\kappa(N-2) * (2+2\delta_{ab})^{-\frac{1}{2}}(\varphi_a\overline{\varphi}_b + p\varphi_b\overline{\varphi}_a)] \tag{7.2}$$

where $\Phi_\kappa(N-2)$ is a function of the N-2 electrons in the internal orbitals and p is +1 for singlet coupling and -1 for triplet coupling of the external orbitals. What is of interest here is to see how the form of this type of configuration changes as a function of a and b. We first note that when a and b are interchanged there will be no change in the configuration if a and b are singlet coupled, whereas the configuration will change sign if a and b are triplet coupled. The only other case of interest is when a=b. In this case the triplet coupling of the external indices will lead to a vanishing configuration. For singlet coupling and a=b the normalization coefficient will introduce a factor $+\sqrt{2}$. This simple structure of the external part of the configurations is what leads to the simple form of the external coupling coefficients.

To illustrate the action of the external coupling coefficients we will take an explicit example. Let us look at the interaction between the configurations ϕ^{ac}_μ and ϕ^{bc}_ν where μ and ν stand for the occupations of the internal orbitals in the respective configurations. If the occupations of μ and ν will lead to an interaction, these two configurations will interact with an integral of the general form $(ai \mid bj)$ where i and j are internal orbitals. The question is which coupling coefficient $A^{\mu\nu}_{aibj}$ that enters into this interaction. To determine this coupling coefficient we first evaluate the interaction when a,b and c take prototype values with a>c and b>c. This gives us the prototype coupling coefficient $B^{\mu\nu}_{aibj}$ for the interaction. In the computer program where the interaction is determined there will then be a triple loop over the external orbitals a,b and c. In this loop, whenever a>c and b>c the actual coupling coeffcient $A^{\mu\nu}_{aibj}$ will be equal to $A^{\mu\nu}_{aibj}$. If the configurations are triplet coupled there will be a factor -1 entering the interaction when a<c and another factor -1 when b<c. The product of these phase factors is the external coupling coefficient. Similarly if.the configurations are singlet coupled there will be a factor $+\sqrt{2}$ entering the interaction when a=c and another factor $+\sqrt{2}$ when b=c, and so on. It is clear that these various situations are very easily identified in the loop system in the program.

It is easy to utilize the simple structure of the external coupling coefficients for an efficient implementation on a vector computer. This is done by modifying the form of the configurations. A set of configurations of type (7.2) differing only in the external orbitals will be viewed as a matrix in a and b. The CI coefficients with singlet coupling of a and b will then be stored as a symmetric matrix where the diagonal elements are multiplied with a factor $+\sqrt{2}$. When a and b are triplet coupled the CI coefficients will instead be stored as an anti-symmetric matrix which thus has a vanishing diagonal. In this formulation the external coupling coefficients are no longer needed and the actual coupling coefficients $A^{\mu\nu}_{pqrs}$ can be directly identified with the corresponding internal prototype coupling coefficients $B^{\mu\nu}_{pqrs}$. In the remainder of this chapter we will therefore not make a distinction between these types of coefficients. The matrix structure of the direct CI equations is best seen in an explicit example. For the case of integrals with two external indices (6.4) can be written,

$$\sigma^{ac}_{\kappa} = \sum_{\lambda b}\{\delta_{\kappa\lambda}h_{ab} + \frac{1}{2}\sum_{ij}[A^{\kappa\lambda}_{abij}(ab \mid ij) + A^{\kappa\lambda}_{aibj}(ai \mid bj)]\}C^{bc}_{\lambda} \tag{7.3}$$

With the introduction of a matrix $\mathbf{F}$, with matrix elements $F^{ab}_{\kappa\lambda}$ equal to the expression inside {}, (7.3) can be written in matrix form as,

$$\sigma_{\kappa} = \mathbf{F}_{\kappa\lambda} * \mathbf{C}_{\lambda} \tag{7.4}$$

The dimension of these matrices is thus equal to the number of virtual orbitals (per symmetry) which is normally large enough to make vectorization efficient on most computers.

The above formalism has thus reduced the very large number of actual coupling coefficients where the virtual orbitals take all possible values, to a much smaller number of coupling coefficients where the external indices take only a few prototype values. The question is now how these coefficients should be evaluated in the most efficient way. In the first general direct CI program these coefficients were evaluated using a factorization algorithm. If we concentrate on the one-electron coupling coefficients these can be written as,

$$A^{\mu\nu}_{pq} = \prod_{t=p}^{q} W_t \tag{7.5}$$

where the factorization is only over the occupied orbitals of the interacting prototype configurations. For each orbital the factor, or "segment value" W_t can be identified depending on the type of interaction. These segment values can best be identified from the "shape" and position of the interaction in the Shavitt graph and can then be fetched from predetermined tables stored in the memory. Since each of the interacting configurations generates a path in the graph the two configurations will together generate a characteristic shape in the graph. These are the principles behind the first workable schemes for evaluating the direct CI coupling coefficients. This description has been deliberately vague since the actual evaluation contains a large number of details and since this formalism is not in use anymore, at least not in the most recent and efficient programs.

The presently most efficient method for evaluating the direct CI coupling coefficients is based on another type of factorization, the details of which has been developed over the past few years. The steps in this factorization are easy and are based on the use of projection operators. In an infinite configuration basis the following identity holds,

$$\sum_{\kappa} | \kappa >< \kappa | = 1 \tag{7.6}$$

This expression is often called "the resolution of the identity". The expression (6.6) for the two-electron coupling coefficient can now be rewritten by introducing expression (7.6) between the generators of the unitary group. The result is,

$$A^{\mu\nu}_{pqrs} = \sum_{\kappa} A^{\mu\kappa}_{pq} A^{\kappa\nu}_{rs} - \delta_{qr} A^{\mu\nu}_{ps} \tag{7.7}$$

At first sight this reformulation only seems to complicate matters since an infinite sum over configurations is introduced. It should, however, be noted that the sum is only infinite in principle. First, it is clear that since the generators of the unitary group replace one of the orbitals in the selected orbital set by another orbital in the same set, there will be no matrix elements for these operators outside the space spanned by the chosen orbitals. This restriction still makes the summation very long, in principle over the complete CI list in the selected orbital basis. If we now look at each individual matrix element, with fixed values of μ,ν,p,q,r and s, we find that the summation in (7.7) is cut down to a quite tractable length. It is clear that for this coupling coefficient to be different from zero the occupation of κ must simultaneously be a single excitation from p to q in μ and s to r in ν. The in principle infinite sum over configurations in (7.7) is thus for a particular matrix element reduced to a sum where κ has at most one particular orbital occupation. In fact, most matrix elements over the product $E_{pq}E_{rs}$ for different p,q,r,s and between fixed μ and ν will be zero, which is of course a consequence of the normal selection rules between determinants. The reduction of the sum to a single occupation for a particular matrix element does not mean that there is only one term in the sum. For a particular occupation of κ there are still normally several different possible spin couplings which should enter into the summation. The final form of the two-electron coupling coefficient can consequently be written,

$$A^{\mu\nu}_{pqrs} = \sum_{\kappa_s} A^{\mu\kappa_s}_{pq} A^{\kappa_s\nu}_{rs} - \delta_{qr} A^{\mu\nu}_{ps} \tag{7.8}$$

where κ_s only runs over the spin couplings of the fixed occupation in κ. The expression for a two-electron coupling coefficient is thus reduced to a rather simple expression of products of one-electron coupling coefficients.

The expression (7.8) formed the basis for the first vectorized direct CI formulation of complete CI (see section VIII) where the one-electron coupling coefficients were precomputed, reordered and stored on peripheral device. It is, however, possible to make a formulation of direct CI in which it is not necessary to precompute all the one-electron coupling coefficients. In this formulation it is necessary to study the structure of the one-electron coupling coefficients in detail. The idea is to set up tables over prototype one-electron coupling coefficients, which can be precomputed and held in memory, from which the actual coefficients can easily be obtained. This is somewhat in the spirit of the factorization of the direct CI coupling coefficients into prototype (internal) and external parts as in expression (7.1). To be able to set up prototype one-electron coefficients it is necessary to fix the spin coupling in some way: For a certain number of singly occupied orbitals in a configuration the spin couplings of the different components always have to be the same. This is the basis for the symmetric group organisation of the CI expansion. It is then easy to see that the possible values of the one-electron coupling coefficient $A_{pq}^{\mu\nu}$ only depends on the position of p and q in relation to the positions of the singly occupied orbitals in μ and ν. This follows since it is first obvious that these values will not depend on unoccupied orbitals, and also since the position of a doubly occupied orbital can always be changed without modifying the phase or the spin coupling.

The evaluation of prototype one-electron coupling coefficients starts out by setting up prototype configurations with predetermined spin couplings of the singly occupied electrons. If a configuration has n singly occupied orbitals these are placed at positions 2,4,6 etc up to 2n. A loop over prototype generators $\hat{E}_{pq}$ acting on the configurations is then set up where p and q run over the sequence values 1,2,3, etc up to 2n+1, which allows the generator indices to take all possible positions in relation to the singly occupied orbitals. For the orbitals which are not singly occupied it is only necessary to define the occupation of p and q and it is clear that p has to be unoccupied and q doubly occupied if the action of the generator should yield a non-zero result. After this, the matrix element of the generator between this configuration and any other prototype configuration is completely defined and the value of the prototype coupling coefficients can be determined using any normal method. The simplest way is to write out the prototype configuration in terms of determinants and spin coupling coefficients and use Slater's rules. It is not necessary to have a very efficient method for evaluating the prototype coefficients since the time required to calculate them is always very small compared to the total time in a direct CI calculation. The resulting prototype coupling coefficients are finally stored in tables organized after the type of interaction. As an actual one-electron coupling coefficient is needed later on in the calculation all that is required is that the proper table should be located. This is done by counting the number of singly occupied orbitals in the interacting configurations and the position of the generator indices among these orbitals. All this information can conveniently be obtained using the computer bit algebra or it can also be stored in tables.

There is one severe problem with the scheme described above and this is a storage problem. Say that one-electron coupling coefficients are required for a singlet state with at most 12 singly occupied orbitals, which is far from an abnormal situation in an MR-CI calculation. From the branching diagram we see that the number of spin coupling possibilities is as large as 132. For each prototype generator a table of 132x132 elements thus has to be set up and stored. The number of prototype orbitals is 2n+1 which is thus 25 in this case, and the number of prototype generators $\hat{E}_{pq}$ is thus 25x25. This case would thus require the storage of on the order of 10^7 terms in fast core which is not practical on most computer installations. This storage problem can be solved, at least to a large extent, by using the algebra of the generators of the unitary group and the introduction of what we will call "ghost" orbitals. A ghost orbital x is defined by requiring that this orbital should be outside the chosen orbital space. If μ and ν belong to the selected configuration space the following relation is then obtained from the commutation relation between the generators of the unitary group,

$$< \mu \mid \hat{E}_{pq} \mid \nu >= \sum_{\kappa_s} < \mu \mid \hat{E}_{px} \mid \kappa_s >< \kappa_s \mid \hat{E}_{xq} \mid \nu > \tag{7.9}$$

which means that a single generator can be replaced by a product of two generators. (It should be noted that the introduction of ghost orbitals has *nothing* to do with the resolution of the identity (7.6) and there should thus obviously *not* be any summation over x in (7.9)). The resulting expression for a two-electron coupling coefficient will then be,

$$A^{\mu\nu}_{pqrs} = \sum_{\kappa_s\lambda_s\tau_s} A^{\mu\kappa_s}_{px} A^{\kappa_s\lambda_s}_{xq} A^{\lambda_s\tau_s}_{rx} A^{\tau_s\nu}_{xs} - \delta_{qr} \sum_{\kappa_s} A^{\mu\kappa_s}_{px} A^{\kappa_s\nu}_{xs} \tag{7.10}$$

which is clearly a more complicated expression to evaluate than the original expression (7.8). This prize is, however, well worth paying since the core storage demand is much smaller. The reason for this is that to evaluate (7.10) prototype matrix elements are only needed where one of the generator indices is the ghost orbital x, which could for example be placed last of all the orbitals in the prototype configuration. This means that there will only be one running index in the prototype generator. In the above example with 12 open shells the storage requirement for the prototype one-electron coupling coefficients will thus be reduced by a factor of 25, which is what is required to make this calculation possible.

In summary, the evaluation of the large number of coupling coefficients has been reduced to the calculation of a manageable number of prototype coupling coefficients through the relation (7.1). The calculation of these prototype coefficients is performed by a sequence of matrix multiplications according to (7.10), where each matrix is a matrix over spin-couplings only. These latter matrices can easily be identified as equal to prototype matrices which can be held in core storage through the introduction of ghost orbitals as in relation (7.9).

VIII. Complete CI expansions.

Complete CI, or full CI, is configuration interaction with a configuration list which includes all possible configurations of proper spin and space symmetry in the chosen orbital space. As has been mentioned previously, the number of configurations in complete CI will depend in an n-factorial way on the number of electrons and the number of orbitals and it will therefore quickly become too large to be handled. This method is therefore not very well suited as a standard model to solve quantum chemical problems. There are, however, two situations where an efficient complete CI method is useful to have. The first of these is in connection with the CASSCF method which has been described in another chapter. The other is in connection with bench mark tests. Since any other CI method selects configurations after some principle, a comparison to complete CI is the way to check these principles out. We will therefore in this section briefly outline the main steps in the complete CI method as it is carried out today.

The formalism in the direct CI implementation of complete CI starts out with the normal direct CI equations for the residual vector σ according to equations (6.4)-(6.6). In these equations the resolution of the identity (7.6) is introduced as in (7.7) and the resulting equation is,

$$\sigma_\mu = \sum_\nu \{\frac{1}{2} \sum_{pqrs} \sum_\kappa (pq \mid rs) A^{\mu\kappa}_{pq} A^{\kappa\nu}_{rs}\} C_\nu \tag{8.1}$$

where only terms depending on products of generators have been kept. After some reorganisation equation (8.1) can then be rewritten as,

$$\sigma_\mu = \frac{1}{2} \sum_\kappa \sum_{pq} A^\mu_{\kappa,pq} \sum_{rs} I_{pq,rs} D_{rs,\kappa} \tag{8.2}$$

which has been written to emphasize the matrix structure with

$$A^\mu_{\kappa,pq} = A^{\mu\kappa}_{pq} \ , \ I_{pq,rs} = (pq \mid rs) \ , \ D_{rs,\kappa} = \sum_\nu A^{\kappa\nu}_{rs} C_\nu \tag{8.3}$$

Equation (8.2) in matrix notation is

$$\sigma_\mu = \frac{1}{2} \mathrm{Tr}(\mathbf{A}^\mu * \mathbf{I} * \mathbf{D}) \tag{8.4}$$

The gain in this reformulation is twofold. The storage of the long list of two-electron coupling coefficients is avoided and the calculation of the σ vector according to (8.4) is likely to work well on a vector computer. In the first implementation of this algorithm the one-electron coefficients $A^{\mu\kappa}_{pq}$ were first calculated and stored. As the complete CI expansion is increased the storage of the one-electron coefficients becomes another bottle-neck. It was then suggested that rather than using the rather complicated spin-adapted configurations, one should use simple determinants. In this case the non-zero one-electron coupling coefficients are either +1 or -1, and can be evaluated as they are needed and thus need not to be stored. This variant of using equation (8.4) is the most efficient method to perform complete CI at present.

IX. Approximate CI methods.

Since the solution of the CI problem with a long CI expansion is time consuming there is often a need for faster approximate methods. In this section we will describe two such methods both based on contractions of the CI expansion.

In the so called "externally contracted CI" (CCI) method the general MR-CI expansion (4.14) is rewritten as,

$$\Psi_{MR-CI} = \sum_{\mu(N)} C_\mu \Phi_\mu + \sum_{\mu(N-1)} C_\mu \sum_a \tilde{C}^a_\mu \Phi^a_\mu + \sum_{\mu(N-2)} C_\mu \sum_{ab} \tilde{C}^{ab}_\mu \Phi^{ab}_\mu =$$

$$= \sum_{\mu} C_{\mu} \varphi_{\mu} \tag{9.1}$$

where the summations have been written to emphasize that the internal part of the configurations contain different number of electrons in each sum. The C_{μ} are obtained variationally but the contraction coefficients $\tilde{C}_{\mu}^{a}$ and $\tilde{C}_{\mu}^{ab}$ are obtained by perturbation theory as,

$$\tilde{C}_{\mu}^{ab} = \frac{< O \mid \hat{H} \mid \Phi_{\mu}^{ab} >}{E_0 - < \Phi_{\mu}^{ab} \mid \hat{H} \mid \Phi_{\mu}^{ab} >} \tag{9.2}$$

The number of variational parameters in (9.1) is therefore much smaller than it is in (4.14) to the price that some accuracy is lost and that each function φ_{μ} is much more complicated than the original configurations Φ_{μ}^{ab}. A typical reduction is from 10^6 to 10^3 variational parameters. Since the number of functions φ_{μ} is so small a CCI calculation can be performed in a way which is a compromise between the conventional CI and the direct CI strategies. The matrix elements over φ_{μ} are determined using the formalism of direct CI. The matrix elements are then written on a peripheral device and the Hamiltonian matrix is diagonalized as in a conventional CI calculation. The former step takes in the limit of large basis sets about half an iteration of an uncontracted CI calculation, and the second step is usually fast. Altogether, the time for a CCI calculation is one order of magnitude faster than the corresponding uncontracted calculation. The loss of correlation energy by the contraction is usually on the order of 1-3 %. It should be added that the CCI approximation becomes better the larger the reference space is, and it is not to be recommended if configurations with coefficients larger than 0.10 are not included in the reference space.

The second approximate scheme we will discuss here is the internally contracted CI (ICCI) method. In this method correlating configurations are formed by applying excitation operators (the generators of the unitary group) directly on the full reference CI vector. The four types of configurations thus formed can be written as,

$$\Psi_0 = \mid O >= \sum_{\mu} C_{\mu} \Phi_{\mu} \tag{9.3}$$

$$\Psi_i^a = \hat{E}_{ai} \mid O >= \sum_{\mu} d_{\mu} \Phi_{\mu}^a \tag{9.4}$$

$$\Psi_{ij}^{ak} = \hat{E}_{ai} \hat{E}_{kj} \mid O >= \sum_{\mu} d_{\mu} \Phi_{\mu}^a \tag{9.5}$$

$$\Psi_{ij}^{ab} = (\hat{E}_{ai} \hat{E}_{bj} + p \hat{E}_{aj} \hat{E}_{bi}) \mid O >= \sum_{\mu} d_{\mu} \Phi_{\mu}^{ab} \tag{9.6}$$

where p is +1 for singlet coupling and -1 for triplet coupling of the external orbitals a and b. The configuration given in (9.5) has not been explicitly discussed in this chapter before. It is usually called a "semi-internal" configuration. The contraction coefficients

d_μ will be linear combinations of the coefficients C_μ which define the reference CI vector of interest. The total ICCI wavefunction is then written as,

$$\Psi = C_0\Psi_0 + \sum_{ia} C_i^a\Psi_i^a + \sum_{ijka} C_{ij}^{ak}\Psi_{ij}^{ak} + \sum_{ijab} C_{ij}^{ab}\Psi_{ij}^{ab} \tag{9.7}$$

In the single-reference case the internally contracted CI method is identical to the uncontracted case, but it is easy to see that in the multi-reference case the number of variational parameters can be drastically reduced. In fact, the number of variational coefficients C in (9.7) is almost independent of the number of reference states used. This method has therefore its main strength in cases with very many reference states. The largest multi-reference CI calculation performed to this date has been done using this method for the molecule Cr_2 for which 3088 reference configurations were used. Since the contraction error in the ICCI method is usually quite small, on the order of 0.1-0.2 % of relative correlation energy contributions, this method is very promising. At present, however, there is no existing program where the full internal contraction according to (9.3)-(9.6) is implemented. The calculation on the Cr_2 molecule was done with the singly excited (9.4) and semi-internal configurations (9.5) left uncontracted. This calculation was therefore quite expensive. To construct a program where the internal contraction is fully utilized therefore still remains a challenge. It is likely that such a program will be made the coming 1-2 years.

It is clear that matrix elements over the Hamiltonian between the internally contracted configurations (9.3)-(9.6) will be rather complicated. Since both the configurations and the Hamiltonian are and can be written in terms of the generators of the unitary group, the formal expressions for these matrix elements can be obtained through the commutation relations between these generators. For the most complicated matrix element, that between the semi-internal configurations, the expression for the matrix element will contain products of up to 6 generators and the indices for these will contain 10 internal orbitals. The generators containing external orbitals can be made to vanish by commuting the generators to positions where, $\hat{E}_{pa}$ acts on $\mid 0 >$. The result of this action is by definition zero since the external orbital a is not occupied in $\mid 0 >$ and can therefore not be annihilated. The final expression for this matrix element will thus contain products of up to 5 generators with indices that are all occupied (internal). With proper symmetrization of the indices entities of this type are called the 5th order reduced density matrix. We will not here give the general definition of this density matrix but the definition of some lower order density matrices could be of some interest. The first order density matrix is defined as,

$$D_{pq}^{(1)} = < 0 \mid \hat{E}_{pq} \mid 0 > \tag{9.8}$$

the second order matrix element is defined as,

$$D_{pq,rs}^{(2)} = < 0 \mid \hat{E}_{pq}\hat{E}_{rs} - \delta_{qr}\hat{E}_{ps} \mid 0 > = < 0 \mid \hat{E}_{pq,rs} \mid 0 > \tag{9.9}$$

where the last equality is used to define a symbol which is needed for the higher order density matrices. From the definitions (9.8) and (9.9) it is clear that we can write the variational energy of a normalized wavefunction $| 0 >$ as,

$$E_0 = \sum_{pq} D^{(1)}_{pq} h_{pq} + \frac{1}{2} \sum_{pqrs} D^{(2)}_{pq,rs} (pq \mid rs) \tag{9.10}$$

This expression is actually sometimes used to define the first and second order density matrix. It is anyway useful to know that the first order density matrix elements are equal to the coefficients for the corresponding one-electron integrals in the energy expression and similarly for the second order density matrix elements. The definition of the third order density matrix is,

$$D^{(3)}_{pq,rs,tu} = < 0 \mid \hat{E}_{pq,rs} \hat{E}_{tu} - \delta_{st} \hat{E}_{pq,ru} - \delta_{qt} \hat{E}_{pu,rs} \mid 0 > \tag{9.11}$$

etc. These density matrix elements are all needed in the ICCI method and they need to be generated with high efficiency if this method should be applied to long MR-CI expansions. The best way at present to calculate these matrix elements is essentially the same as the one described above for the calculation of the direct CI coupling coefficients, i.e. using product formulas similar to (7.10). The main difference is that there should also be a summation over the reference configurations, which means that there will be matrix times vector operations rather than matrix times matrix operations as in (7.10). Both these types of operations will, however, vectorize well on most vector computers.

X. Size consistency.

The final topic we will discuss in this chapter is size-consistency, which has been mentioned several times already. A method is said to be size-consistent if the computed energy of the composite system A + B, with A and B at infinite distance from each other, yields the same energy as if the method is applied to A and B separately and the energies are added, i.e. E(A+B)=E(A)+E(B). Some of the methods we have discussed are automatically size-consistent. This is true, for example, for the Hartree-Fock method and the complete CI method, and it is also true for the methods discussed in the chapter on perturbation theory, such as the coupled cluster method. It is, however, not true for the SD-CI or the MR-CI method. We will in this section show that it is possible, by a slight modification of the formalism, to correct these CI methods to be approximately size-consistent. The experience gathered over the past two decades on size-consistency corrections indicates that the calculated results are much improved at the SD-CI level, whereas relative energies are improved at the MR-CI level but the situation for geometries is less clear at this level.

The reformulation of the CI method which leads to approximate size-consistency starts out by writing the correlation energy functional in a more general and slightly different form than is obtained from (3.1) as,

$$F_c = \frac{< \psi_0 + \psi_c \mid \hat{H} - E_0 \mid \psi_0 + \psi_c >}{< \psi_0 \mid \psi_0 > + g < \psi_c \mid \psi_c >} \equiv \frac{Nu}{De} \tag{10.1}$$

where E_0 is the reference energy, ψ_0 is the reference part of the CI vector and ψ_c is the remaining part of the vector. It can be seen that with g=1, F_c in (10.1) reduces to the normal expression for the correlation energy as obtained from the functional F in (3.1). By inspection of (10.1) it can be realized that the problem of size-inconsistency appears in the expression for the denominator De. This is clear since for infinitely separated systems A and B the numerator must be Nu(A+B)=Nu(A)+Nu(B), which is easily realized since $< A \mid \hat{H} \mid B >$ must be zero. With a constant denominator independent of the size of the system, F_c must also be size-consistent. The simplest way to achieve this property for the denominator is to put g=0. When this is done we have a size-consistent method which has many names, simply because it has been rediscovered several times. Some of these names are CEPA-0 (coupled electron pair approximation 0), L-CPMET (linearized coupled pair many-electron theory) and DE-MBPT (doubly excited many-body perturbation theory). A related entity is the so called "Davidson correction" which is frequently used to correct a CI calculation for size-inconsistency. The expression for this correction to the CI energy is simply obtained by taking the difference between the size-consistent functional with g=0 and the normal CI functional with g=1. The result for the single-reference SD-CI case is,

$$E_{Dav} = \frac{1 - C_0^2}{C_0^2} E_c \tag{10.2}$$

where C_0 is the coefficient for the reference configuration and E_c is the correlation energy. The detailed derivation of (10.2) is left as an exercise. A completely analogous result is obtained in the multi-reference case. It is worth noting that adding the Davidson correction to the CI energy will not make the energy exactly size-consistent even though the use of the energy functional with g=0 is exactly size-consistent. This is so since the Davidson correction is evaluated using CI coefficients which have been optimized using a functional which is not size-consistent.

The above description is one way to realize that the SD-CI method is not size-consistent. Another way is to look in detail at what happens when this method is used on the composite and on the separated systems. It is clear that if the energy of A and B should be additive the corresponding wavefunction for (A+B) should be equal to the wavefunction of A times the wavefunction of B. This means that since in the calculations on the separated systems there are local double excitations on both A and B, the product wavefunction will contain certain quadruple excitations. In the SD-CI calculation on the composite system these quadruple excitations are clearly missing and this is the reason for the size-inconsistency. It is also clear that for the SD-CI method E(A)+E(B) must be lower in energy than E(A+B).

Another point worth making is that since the SD-CI method is exact within the chosen basis set for a two-electron system, it must be size-consistent in this particular case. Nevertheless, when Davidson's correction is applied to an SD-CI wave-function for a two-electron system it will give a non-zero contribution, which is thus an artefact of this correction. (The same error appears also when the functional (10.1) is used with g=0.) This artefact can be simply removed and this is done in the Averaged Coupled Pair Functional (ACPF) method. In this method the factor g is considered to be a function of the number of electrons N, g=g(N), and one considers the special case of n separated He atoms. If the denominator De in (10.1) for one He atom is

$$1 + g(2) < \psi_c^0 \mid \psi_c^0 > \tag{10.3}$$

where it has been assumed that the wavefunction is normalized with $< \psi_0 \mid \psi_0 >= 1$, then the denominator for n He atoms will be,

$$1 + g(2n) < \psi_c \mid \psi_c >= 1 + g(2n) * n < \psi_c^0 \mid \psi_c^0 > \tag{10.4}$$

where a corresponding normalization has been used. If we now use the information that for a two-electron system as in (10.3) the normal CI expression is exact we will require that g(2) should be equal to one. Since the requirement for size-consistency is that the denominator should be constant, independent of the number of electrons, then g(2n)*n should also be equal to one. With $n = \frac{N}{2}$ these two conditions can be fulfilled by requiring that

$$g = \frac{2}{N} \tag{10.4}$$

Since in a normal CI calculation g is equal to one, the effect of using the ACPF g-value will be particularly important when there are many electrons. The ACPF g-value will in fact correct an important deficiency of the SD-CI method, namely that the correlation energy will not grow linearly with the number of electrons N in the system as it should, but rather as $\sqrt{N}$. It should also be noted that the ACPF method is not exactly size-consistent in the general case. However, in test calculations the energy E(A+B) has been calculated for a variety of systems and been found to be very nearly equal to E(A)+E(B). Since the ACPF method is easily generalized to the MR-CI case, this method appears very promising for MR-CI calculations on systems with a large number of electrons.

XI. Summary.

We will finish this chapter by going back to one of the fundamental points on the correlation energy mentioned already in the introduction. There, it was mentioned that the correlation energy is usually divided into a non-dynamical (near-degeneracy) part and a dynamical part. This division is not easily defined, but has anyway been extremely useful for designing methods and has therefore also influenced the division of the lectures given at this course. Recently there has been some work which could turn out to be quite important in the context of defining the different parts of the correlation energy. From this work it has been shown that in order to partition the correlation energy, a useful starting point is the Unrestricted HF (UHF) method. In this method the restriction that the electrons of α and β spin should have the same spatial orbital is released, which removes the main origin for the near-degeneracy error in the RHF method. To define the non-dynamical correlation energy the following procedure has been suggested. First, perform a UHF calculation and set up the first order charge density matrix and obtain the natural orbitals by diagonalization. Four types of occupations will appear, those very nearly equal to two, those somewhat smaller than two, those a bit larger than zero and finally those close to zero. Second, perform a CASSCF (see different chapter) calculation with the active orbitals chosen as the ones belonging to the second and third class of the natural orbitals. The difference between this CASSCF energy and the RHF energy will then be defined as the non-dynamical correlation energy. The remaining part of the total correlation energy would be the dynamical correlation energy. Why is a methodological definition of the non-dynamical correlation energy important ? The answer is that this definition also suggests a procedure for carrying out calculations that is well defined, which has so far been missing for calculations involving the MCSCF method. Among expert quantum chemists it has long been known that calculations based on the MCSCF and MR-CI methods have a higher general accuracy than methods based on Hartree-Fock and single-reference procedures. Nevertheless, these latter methods have been the dominating tools for chemists when they perform quantum chemical calculations. With a well-defined partitioning of the correlation energy, and thereby a well defined "black box" procedure to do MR-CI calculations, the situation may change in the future.

In this chapter we have tried to give a general introduction to the structure and concepts of the configuration interaction method, which is one of the most widely used methods to calculate correlation energies for molecular systems. One of the objectives has been to carry the reader up to the most modern parts of the recent developments. This has meant that in a course of this size the presentation has sometimes been rather vague and many of the details have been left out. It is, however, hoped that with this chapter as a background the reader should be able to go directly to some of the original papers given at the end of this chapter, in order to find more details of a particular subject of interest. Another hope is also that those students, who are only going to apply the CI method have been motivated to learn more about this method. It turns out in general that the best calculations are performed by those who best know the methods.

XII. Further reading.

I. Shavitt, The Method of Configuration Interaction, in Methods of Electronic Structure Theory (H.F. Schaefer III, ed.), Plenum Press, New York (1977).

B.O. Roos and P.E.M. Siegbahn, The Direct Configuration Interaction Method from Molecular Integrals, in Methods of Electronic Structure Theory (H.F. Schaefer III, ed.), Plenum Press, New York (1977).

H.-J. Werner, Matrix-Formulated Direct Multiconfiguration Self-Consistent Field and Multiconfiguration Reference Configuration-Interaction Methods, in Ab Initio Methods in Quantum Chemistry - II (K.P. Lawley, ed.), John Wiley & Sons Ltd, Chichester (1987).

P.E.M. Siegbahn, The Direct CI Method, in Methods in Computational Molecular Physics, (G.H.F. Diercksen and S. Wilson, eds.) D. Reidel Publishing Company, Dordrecht (1983).

P.E.M. Siegbahn, The Direct Configuration Interaction Method with a Contracted Configuration Expansion, Chem. Phys. 25, 197 (1977).

P.E.M. Siegbahn, Multiple Substitution Effects in Configuration Interaction Calculations, Chem. Phys. Letters 55, 386 (1978).

V.R. Saunders and J.H. van Lenthe, The Direct CI Method- A Detailed Analysis, Mol. Phys. 48, 923 (1983).

P.E.M. Siegbahn, A New Direct CI Method for Large CI Expansions in a Small Orbital Space, Chem. Phys. Letters 109, 417 (1984).

P.E.M. Siegbahn, Current Status of the Multiconfiguration-Configuration Interaction (MC-CI) Method as Applied to Molecules Containing Transition-metal Atoms, Faraday Symp. Chem. Soc. 19, 97 (1984).

R.J. Gdanitz and R. Ahlrichs, The Averaged Coupled-Pair Functional (ACPF): A Size-Extensive Modification of MR CI(SD), Chem. Phys. Letters 143, 413 (1988).

H.-J. Werner and P.J. Knowles, An Efficient Internally Contracted Multiconfiguration Reference Configuration Interaction Method, J. Chem. Phys. 89, 5803 (1988).

P. Pulay and T.P. Hamilton, UHF Natural Orbitals for Defining and Starting MC-SCF Calculations, J.Chem. Phys. 88, 4926 (1988).

OPTIMIZATION OF MINIMA AND SADDLE POINTS

Trygve Helgaker
Department of Chemistry
University of Oslo
P.O.B. 1033 Blindern
N-0315 Oslo 3
Norway

The optimization of minima and saddle points of a function in many variables is reviewed. Emphasis is on methods applicable to the calculation of electronic wave functions (ground and excited states) and the optimization of minima and transition states of molecular potential energy surfaces.

I. INTRODUCTION

We discuss in this paper the unconstrained optimization of stationary points of a smooth function f(x) in many variables. The emphasis is on methods useful for calculating molecular electronic energies and for determining molecular equilibrium and transition state structures. The discussion is general and practical aspects concerning computer implementations are not treated.

The field of minimization is well developed, see for example the monographs by Fletcher[1], by Gill, Murray and Wright[2], and by Dennis and Schnabel[3]. The optimization of saddle points such as excited electronic states and molecular transition states has received little attention outside the field of quantum chemistry and is less developed. In addition less information - experimental or intuitive - is usually available about saddle points, compounding this problem. In some respects the determination of saddle points resembles art more than technique. For previous reviews of methods for optimization of equilibrium geometries and transition states, see Schlegel[4] and Head and Zerner[5].

---

[1] R. Fletcher, *Practical Methods of Optimization* (Wiley, Chichester, 1980), Vol. 1

[2] P. E. Gill, W. Murray, and M. H. Wright, *Practical Optimization* (Academic, London, 1981)

[3] J. E. Dennis and R. B. Schnabel, *Numerical Methods for Unconstrained Optimization and Nonlinear Equations* (Prentice Hall, Englewood Cliffs, 1983)

[4] H. B. Schlegel, Adv. Chem. Phys. **69**, 249 (1987); H. B. Schlegel, in *New Theoretical Concepts for Understanding Organic Reactions*, edited by J. Bertrán and I. G. Csizmadia (Kluwer, Dordrecht, 1989)

II. STATIONARY POINTS

Before discussing methods for optimization of stationary points it is appropriate to state briefly the mathematical characterization of such points and lay out the basic strategies for their determination.

A. Characterization of Stationary Points

A stationary point x^* of a smooth function $f(x)$ in n variables may be characterized in terms of the derivatives at x^*. Sufficient conditions for a *minimum* are

$$\left.\begin{aligned} \nabla f(x^*) &= 0 \\ \mathrm{In}\nabla^2 f(x^*) &= [n, 0, 0] \end{aligned}\right\} \Rightarrow \text{minimum} \tag{2.1}$$

where the gradient and Hessian are given by

$$\nabla f = \begin{bmatrix} \frac{\partial f}{\partial x_1} \\ \vdots \\ \frac{\partial f}{\partial x_n} \end{bmatrix}, \quad \nabla^2 f = \begin{bmatrix} \frac{\partial^2 f}{\partial x_1 \partial x_1} & \cdots & \frac{\partial^2 f}{\partial x_1 \partial x_n} \\ \vdots & \ddots & \vdots \\ \frac{\partial^2 f}{\partial x_n \partial x_1} & \cdots & \frac{\partial^2 f}{\partial x_n \partial x_n} \end{bmatrix} \tag{2.2}$$

The inertia InM of a real symmetric matrix M in Eq. (2.1) is a triple

$$\mathrm{In}M = [\, \pi(M), \; \nu(M), \; \delta(M)] \tag{2.3}$$

where $\pi(M)$ is the number of positive eigenvalues of M, $\nu(M)$ the number of negative eigenvalues (sometimes referred to as the index of M), and $\delta(M)$ the number of zero eigenvalues. Equation (2.1) means that at a minimum the function has a vanishing gradient (zero slope) and a positive definite Hessian (positive curvature in all directions). However, Eq. (2.1) is not a necessary condition for a minimum. A point with inertia [n-k,0,k] may still be a minimum but to find out we must examine higher derivatives.

---

[5] J. D. Head and M. C. Zerner, Adv. Quantum Chem. **20**, 239 (1989)

Sufficient conditions for a k´th order *saddle point* are given by

$$\left.\begin{aligned} \nabla f(x^*) &= 0 \\ \mathrm{In}\nabla^2 f(x^*) &= [n-k, k, 0] \end{aligned}\right\} \Rightarrow \text{k´th order saddle point} \qquad (2.4)$$

For example, a first-order saddle point has one and only one direction of negative curvature. Clearly, to characterize a stationary point we must be able to calculate the gradient and the Hessian eigenvalues of f(x). The figures below illustrate a minimum and a saddle point in two dimensions.

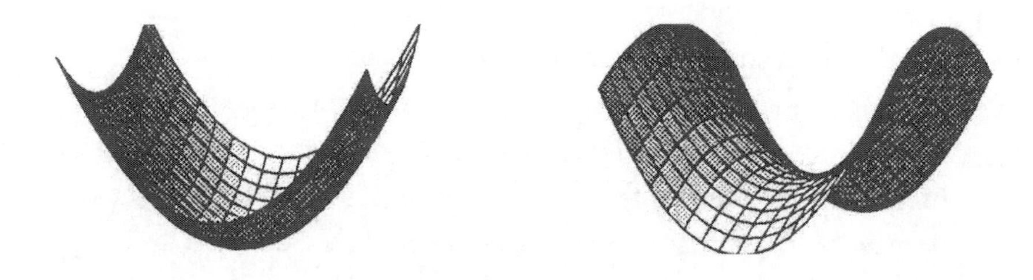

Both minima and saddle points are of interest. In the case of wave functions, the ground state is a minimum and the excited states are saddle points of the electronic energy function.[6] On potential surfaces minima and first-order saddle points correspond to equilibrium geometries and transition states. Higher-order saddle points on potential energy surfaces are of no interest.

B. Strategies for Optimization of Stationary Points

Optimizations may roughly be said to consist of two stages. In the first stage the purpose is to take us from some initial guess of the *optimizer* x* to a point in its neighborhood. In this *global* part of the search we may for example start with a Hessian with incorrect index and our goal is to locate an area with correct curvature. In the final *local* stage of the optimization our purpose is to determine the exact location of the optimizer

---

[6]However, not all minima and saddle points are satisfactory representations of electronic states. The decision as to whether a stationary point is a good approximation to an electronic state must be based on other criteria.

starting from a point in its immediate vicinity. A useful algorithm should handle both stages successfully. It should converge globally (i.e., from any starting point) and be fast in the local region.[7]

All practical methods start by constructing a *local model* of the function in the vicinity of the current estimate x_c of x^*. These models should accurately represent the function in some area around x_c, they should be easy to construct and yet flexible enough to guide us towards the stationary point. Such models are treated in Sec. III.[8]

In the *local* region we may expect the model to represent the function accurately in the neighborhood of x^*. Therefore, in this region the search is rather simple. We take a step to the optimizer of the local model, construct a new model and repeat until convergence. The different methods converge with a characteristic rate as discussed in Sec. IV.

In the *global* region the model should guide us in the right general direction towards x^*. This is relatively easy in minimizations since any step that reduces the function may be considered a step in the right direction. Global strategies for minimizations are treated in Sec. V.

In saddle point optimizations it is harder to judge the quality of a global step. Since a saddle point is a minimum in some directions and a maximum in others, the strategy is to identify these directions and take a step that increases and decreases the function accordingly. Such methods are discussed in Sec. VI.

III. LOCAL MODELS

We discuss in this section several models used in optimizations. Of these, the most successful are the quadratic model and its modifications, the restricted second-order model and the rational function model.

---

[7]It should be noted that there is no efficient way to ensure that the *global* minimum of a function is obtained in an optimization. In practice, we must be satisfied with methods which lead us to *local* minima.

[8]The parametrization of the function and the local models may significantly affect the performance of any optimization scheme. When choosing coordinates, we should reduce as much as possible coordinate couplings and higher-order dependencies.

A. The Linear Model

The simplest model is the *local linear or affine model*

$$m_A(x) = f(x_c) + \tilde{g}_c s \tag{3.1}$$

where

$$s = x - x_c \tag{3.2}$$

and g_c is the gradient of f(x) at the current point

$$g_c = \nabla f(x_c) \tag{3.3}$$

The linear model, which may also be constructed from an approximate gradient, is simple but not particularly useful since it is unbounded and has no stationary point. It contains no information about the curvature of the function. It is the basis for the *steepest descent method* in which a step opposite the gradient is determined by line search (*vide infra*).

The figure below illustrates the linear model for a function in one variable. The function has a minimum at $x^* = 8.5$ and we have plotted models expanded around 7 (dotted line), 7.4 (dashed line), and 8 (full line). Clearly, these models provide little information about the location of the minimum.

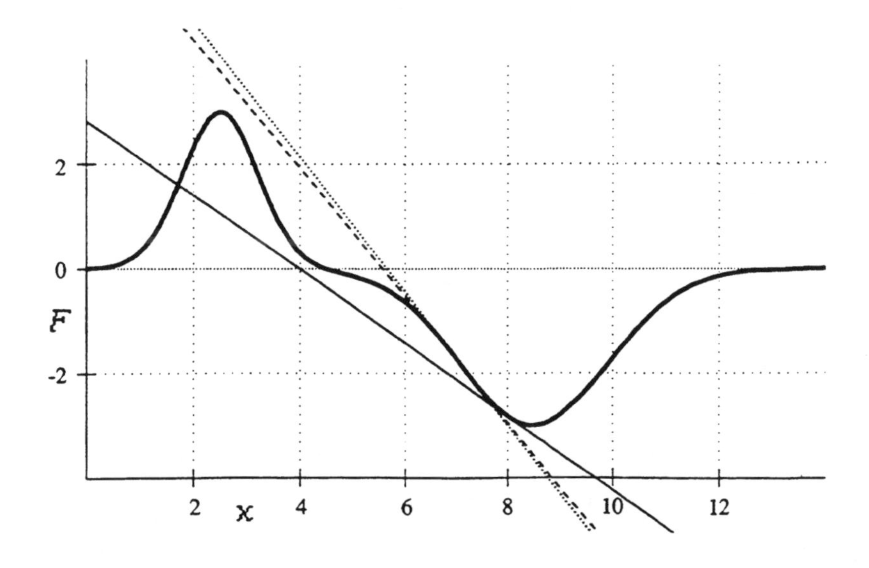

B. The Quadratic Model

The problem with the linear model is that it gives no information about the curvature of f(x). This is provided by the more useful *local quadratic model*

$$m_Q(s) = f(x_c) + \tilde{g}_c s + \tfrac{1}{2}\tilde{s} B_c s \tag{3.4}$$

The symmetric matrix B_c is either the exact Hessian at x_c

$$G_c = \nabla^2 f(x_c) \tag{3.5}$$

or some approximation to it. If the exact Hessian is used the quadratic

$$m_{SO}(s) = f(x_c) + \tilde{g}_c s + \tfrac{1}{2}\tilde{s} G_c s \tag{3.6}$$

is referred to as the *second-order (SO) model* since it is the second-order Taylor expansion of the function around x_c. The SO model is expensive to construct since it requires the gradient as well as the Hessian but it gives complete information about slope as well as curvature at x_c.

To determine the stationary points of the quadratic model we differentiate the model and set the result equal to zero. We obtain a linear set of equations

$$B_c s = -g_c \tag{3.7}$$

which has a unique solution

$$s = -B_c^{-1} g_c \tag{3.8}$$

provided B_c is nonsingular. The quadratic model therefore has one and only one stationary point. This is a minimum if B_c is positive definite. When B_c is the exact Hessian

$$s_N = -G_c^{-1} g_c \tag{3.9}$$

is called the *Newton step*. It forms the basis for *Newton's method* and its globally convergent modifications discussed later. When an approximate Hessian is used Eq. (3.8) is called the *quasi-Newton step*.

The figure below shows the SO models of the same function as for the linear model above. The model around 8 (full line) has a minimum close to the minimizer and the Newton step therefore provides a good estimate of x*. The model around 7.4 (dashed line) overshoots x*. The Newton step is not useful although the model has the correct convex shape. Finally, the model around 7 has the wrong shape and the Newton step takes us away from x*. Note that while there is little difference between the *linear* models at 7 and 7.4, the SO models differ strongly. We see that although the SO model gives more information about the function than does the linear model, its stationary point is not always a good estimate of x*.

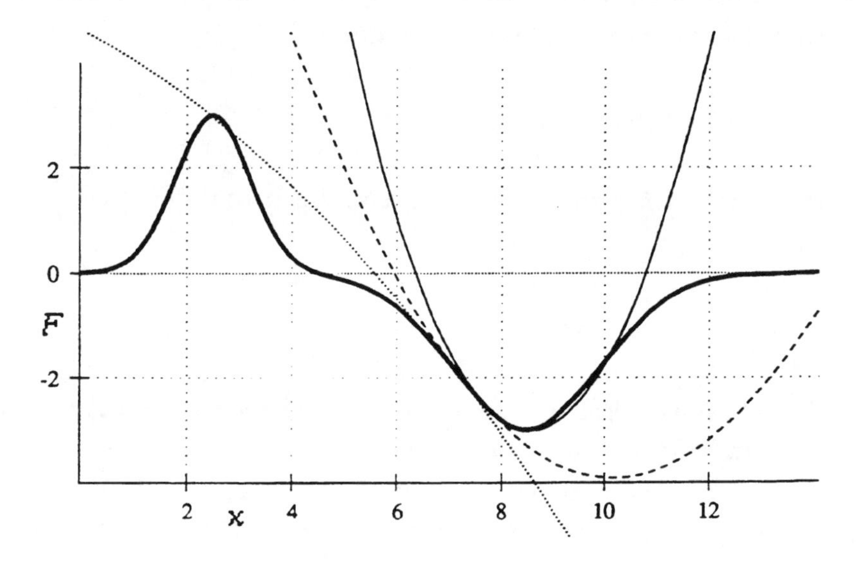

C. The Restricted Second-Order Model

The quadratic model is an improvement on the linear model since it gives information about the curvature of the function and contains a stationary point. However, the model is still unbounded and it is a good approximation to f(x) only in some region around x_c. The region where we can trust the model to represent f(x) adequately is called the *trust region*. Usually it is impossible to specify this region in detail and for convenience we assume that it has the shape of a hypersphere $|s| \leq h$ where h is the *trust radius*. This gives us the *restricted second-order (RSO) model*

$$m_{RSO}(s) = f(x_c) + \tilde{g}_c s + \tfrac{1}{2}\tilde{s} G_c s, \quad \tilde{s}s \le h^2 \tag{3.10}$$

using the exact Hessian.

The SO model has several stationary points. If the Newton step Eq. (3.9) is shorter than the trust radius $|s_N| < h$ then the RSO model has a stationary point in the interior. It also has at least two stationary points on the boundary $|s| = h$. To see this we introduce the Lagrangian

$$L(s,\mu) = m_{SO}(s) - \tfrac{1}{2}\mu(\tilde{s}s - h^2) \tag{3.11}$$

where μ is an undetermined multiplier. Differentiating this expression and setting the result equal to zero, we obtain

$$s(\mu) = -(G_c - \mu)^{-1} g_c \tag{3.12}$$

where the *level shift parameter* μ is chosen such that the step is to the boundary

$$\sqrt{\tilde{g}_c (G_c - \mu)^{-2} g_c} = h \tag{3.13}$$

To see the solutions to Eq. (3.13) more clearly we have plotted below its left- and right-hand sides as functions of μ for a function in three variables and with Hessian eigenvalues −1, 2, and 3. The step length function has poles at the eigenvalues as can be seen from Eq. (3.13) in the diagonal representation

$$\sum_i \frac{\phi_i^2}{(\lambda_i - \mu)^2} = h^2 \tag{3.14}$$

where λ_i are the eigenvalues and ϕ_i the components of the gradient along the eigenvectors. The eigenvalues determine the positions of the peaks and the gradient their width.

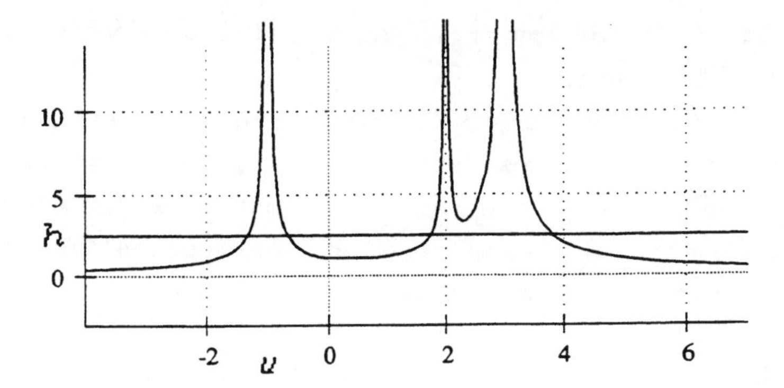

The solutions to Eq. (3.13) are found at the intersections of the two curves. Since the step goes to infinity at the eigenvalues and to zero at infinity there are at least two solutions: one with $\mu < \lambda_1$ (the smallest eigenvalue) and another with $\mu > \lambda_n$ (the largest eigenvalue). We may also have stationary points in each of the n–1 regions $\lambda_k < \mu < \lambda_{k+1}$. For large h we have two solutions in each region, for small h there may be no solution. If the model is constructed around a stationary point, the peaks are infinitely narrow and we have 2n stationary points on the boundary (along each eigenvector) in addition to the point in the interior.

It may be shown that if $\mu < \lambda_1$ then the solution to Eq. (3.13) is the global minimum on the boundary. In the diagonal representation the step may be written

$$\sigma_i = -\sum_i \frac{\phi_i}{\lambda_i - \mu} \tag{3.15}$$

where σ_i is the component of the step along the i´th eigenvector. We see that by selecting the solution $\mu < \lambda_1$ we take a step opposite the gradient in each mode.

The other solutions to Eq. (3.13) correspond to stationary points where the function is increased in some directions and reduced in others. For example, if we select a solution in the region $\lambda_2 < \mu < \lambda_3$ then the step is toward the gradient of the first two modes and opposite the gradient of all higher modes. The second-order change in the function may be written

$$\Delta m_{RSO} = \sum_i \frac{\phi_i^2 (\mu - \frac{1}{2}\lambda_i)}{(\lambda_i - \mu)^2} \tag{3.16}$$

and we see that the model decreases along all modes for which $\lambda_k > 2\mu$ and increases along all others.

The trust radius h reflects our confidence in the SO model. For highly anharmonic functions the trust region must be set small, for quadratic functions it is infinite. Clearly, during an optimization we must be prepared to modify h based on our experience with the function. We return to the problem of updating the trust radius later.

D. The Rational Function Model

The trust region was introduced since the SO model is a good approximation to the function only in some region around the expansion point. The resulting RSO model has stationary points on the boundary of the trust region and possibly a stationary point in the interior. These points are later used to construct globally convergent optimization algorithms.

There is another way to introduce restrictions on the step lengths in the global part of an optimization. The *rational function (RF) model* is given by[9]

$$m_{RF}(s) = f(x_c) + \frac{\tilde{g}_c s + \frac{1}{2}\tilde{s} G_c s}{1 + \tilde{s} S s} \tag{3.17}$$

which may be written in the form

$$m_{RF}(s) = f(x_c) + \frac{1}{2}\frac{\begin{bmatrix} \tilde{s} & 1 \end{bmatrix}\begin{bmatrix} G_c & g_c \\ \tilde{g}_c & 0 \end{bmatrix}\begin{bmatrix} s \\ 1 \end{bmatrix}}{\begin{bmatrix} \tilde{s} & 1 \end{bmatrix}\begin{bmatrix} S & 0 \\ 0 & 1 \end{bmatrix}\begin{bmatrix} s \\ 1 \end{bmatrix}} \tag{3.18}$$

where the metric S is symmetric matrix.[10] This model is bounded since large elements in the numerator are balanced by large elements in the denominator. Also, to second order the RF and SO models are identical since

---

[9]This method was used for optimization of electronic wave functions by A. Banerjee and F. Grein, Int. J. Quantum Chem. 10, 123 (1976) and D. R. Yarkony, Chem. Phys. Lett. 77, 634 (1981). It was applied to surface studies by A. Banerjee, N. Adams, J. Simons, and R. Shepard, J. Phys. Chem. 89, 52 (1985). It has, to the author's knowledge, not been discussed in textbooks on optimization.

[10]We here use the exact Hessian although the RF model may also be constructed from an approximate (updated) Hessian.

$$[1 + \tilde{s} S s]^{-1} = 1 - \tilde{s} S s + O(s^4) \tag{3.19}$$

Therefore, in the RF model we have added higher order terms to the SO model to make it bounded. The explicit form of the RF model depends on the matrix S which should reflect the anharmonicity of the function. Usually we do not have information to specify S in detail and we simply take it to be the unit matrix multiplied by a scalar S.[11]

To find the stationary points of the RF model we differentiate Eq. (3.18) and set the result equal to zero. We arrive at the eigenvalue equations

$$\begin{bmatrix} G_c & g_c \\ \tilde{g}_c & 0 \end{bmatrix} \begin{bmatrix} s \\ 1 \end{bmatrix} = \nu \begin{bmatrix} S & 0 \\ 0 & 1 \end{bmatrix} \begin{bmatrix} s \\ 1 \end{bmatrix} \tag{3.20}$$

Solving these equations we obtain n+1 eigenvectors and eigenvalues

$$\nu = \frac{[\tilde{s} \;\, 1] \begin{bmatrix} G_c & g_c \\ \tilde{g}_c & 0 \end{bmatrix} \begin{bmatrix} s \\ 1 \end{bmatrix}}{[\tilde{s} \;\, 1] \begin{bmatrix} S & 0 \\ 0 & 1 \end{bmatrix} \begin{bmatrix} s \\ 1 \end{bmatrix}} = 2\Delta m_{RF}(s) \tag{3.21}$$

corresponding to the n+1 stationary points of the RF model. From Eq. (3.21) we see that the minimum belongs to the lowest eigenvalue. In general, the k´th eigenvalue belongs to a saddle point of index k–1. Notice that the eigenvalues Eq. (3.21) give the change in the model rather than in the function when the step is taken. The stationary points of the RF model do not necessarily represent stationary points of f(x) but they are useful for constructing globally convergent optimization algorithms.

Since the coefficient matrix of the eigenvalue equations Eq. (3.20)

$$G_c^+ = \begin{bmatrix} G_c & g_c \\ \tilde{g}_c & 0 \end{bmatrix} \tag{3.22}$$

has dimension n+1 and contains the Hessian in the upper left corner it is called the *augmented Hessian*. The n+1 eigenvalues of the augmented

---

[11] For the same reason we assumed that the trust region of the RSO model is a simple hypersphere with an adjustable radius h.

Hessian λ_k^+ bracket the Hessian eigenvalues

$$\lambda_1^+ \leq \lambda_1 \leq \lambda_2^+ \leq \lambda_2 \leq \cdots \leq \lambda_n \leq \lambda_{n+1}^+ \tag{3.23}$$

The eigenvalues of Eq. (3.20) coincide with the eigenvalues of the augmented Hessian only when S equals unity.

To compare the RSO and RF models we expand Eq. (3.20) and obtain

$$s = -(G_c - \mu)^{-1} g_c \tag{3.24}$$

$$\tilde{g}_c s = \frac{\mu}{S} \tag{3.25}$$

where

$$\mu = \nu S \tag{3.26}$$

Inserting Eq. (3.24) in Eq. (3.25) we find

$$-\tilde{g}_c (G_c - \mu)^{-1} g_c = \frac{\mu}{S} \tag{3.27}$$

or in the diagonal representation

$$\sum_i \frac{\phi_i^2}{\mu - \lambda_i} = \frac{\mu}{S} \tag{3.28}$$

Plotting the left and right hand sides of this equation as functions of μ we obtain a figure such as the one below [using the same function as for Eq. (3.13)]. The left hand side goes to infinity at the Hessian eigenvalues. We have n+1 intersections, one for each stationary point of the rational model. Changing the metric S changes the slope of the straight line and therefore the intersections.

Once μ has been determined we calculate the step from the modified Newton equations Eq. (3.24). Therefore, the RF and RSO steps are calculated in the same way. The only difference is the prescription for determining the level shift. In the RSO approach μ reflects the trust radius h, in the RF model μ reflects the metric S. By varying h and S freely the same steps are obtained in the two models.

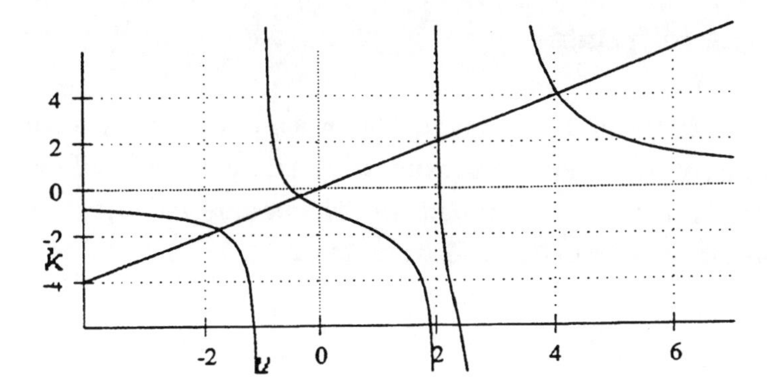

Close to a stationary point g_c vanishes. One of the eigenvalues of the augmented Hessian Eq. (3.22) then goes to zero and the rest approach those of G_c. The zero-eigenvalue step becomes the Newton step and the remaining n steps become infinite and parallel to the Hessian eigenvectors.

To summarize, in the RF approach we make the quadratic model bounded by adding higher-order terms. This introduces n+1 stationary points, which are obtained by diagonalizing the augmented Hessian Eq. (3.22). The figure below shows three RF models with S equal to unity, using the same function and expansion points as for the linear and quadratic models above. Each RF model has one maximum and one minimum in contrast to the SO models that have one stationary point only. The minima lie in the direction of the true minimizer.

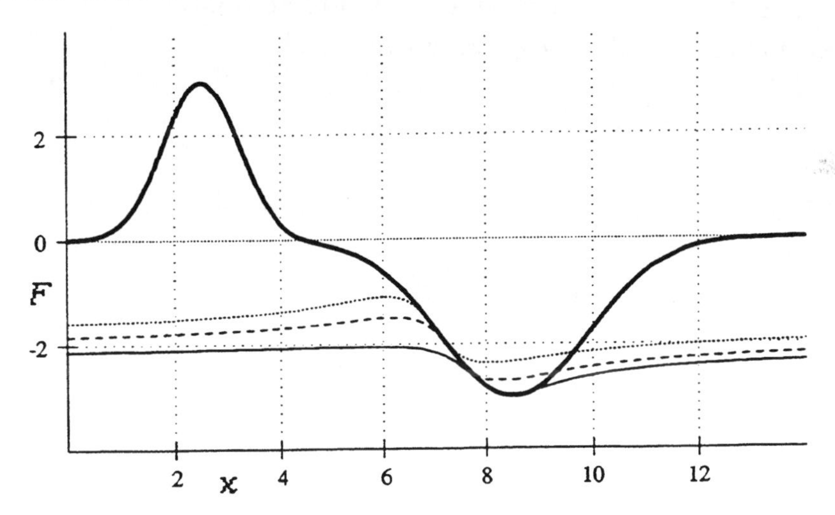

E. Hessian Updates

If the exact Hessian is unavailable or computationally expensive, we may use an approximation. Approximate Hessians are usually obtained by one of several *Hessian update methods*. The update techniques are designed to determine an approximate Hessian B_+ at

$$x_+ = x_c + s_c \tag{3.29}$$

in terms of the Hessian B_c at x_c, the gradient difference

$$y_c = g_+ - g_c \tag{3.30}$$

and the step vector s_c. Expanding g_c around x_+ gives

$$y_c = B_+ s_c + O(s_c^2) \tag{3.31}$$

which shows that the gradient difference y_c contains a component of the finite-difference approximation to the exact Hessian along the direction s_c. This finite-difference information together with structural characteristics of the exact Hessian is used to form the Hessian updates described below.

Based on the finite-difference formula Eq. (3.31) all Hessian updates are required to fulfill the *quasi-Newton condition*

$$y_c = B_+ s_c \tag{3.32}$$

and to possess the property of *hereditary symmetry*, i.e., B_+ is symmetric if B_c is symmetric. These requirements are fulfilled by the *Powell-symmetric-Broyden (PSB) update* given by

$$B_+ = B_c + \frac{(\tilde{s}_c s_c) T_c \tilde{s}_c + (\tilde{s}_c s_c) s_c \tilde{T}_c - (\tilde{T}_c s_c) s_c \tilde{s}_c}{(\tilde{s}_c s_c)^2} \tag{3.33}$$

where

$$T_c = y_c - B_c s_c \tag{3.34}$$

Notice that the construction of the updated Hessian involves simple matrix and vector multiplications of gradient and step vectors.

It is often desirable that the approximate Hessian is positive definite so that the quadratic model has a minimum. To ensure this we may use the *Broyden-Fletcher-Goldfarb-Shanno (BFGS) update* given by

$$B_+ = B_c + \frac{y_c \tilde{y}_c}{\tilde{y}_c s_c} - \frac{B_c s_c \tilde{s}_c B_c}{\tilde{s}_c B_c s_c} \tag{3.35}$$

which under certain weak conditions on the step vector has the property of *hereditary positive definiteness*, i.e., if B_c is positive definite, B_+ is positive definite. As we shall see, this is useful for minimizations even when the exact Hessian has directions of negative curvature. It is then more appropriate to speak of B_c as an effective rather than approximate Hessian.

There are other Hessian updates but for minimizations the BFGS update is the most successful. Hessian update techniques are usually combined with line search (*vide infra*) and the resulting minimization algorithms are called *quasi-Newton methods.* In saddle point optimizations we must allow the approximate Hessian to become indefinite and the PSB update is therefore more appropriate.

IV THE LOCAL REGION

If the quadratic model or one of its modifications are used, the local region presents no difficulties. All methods converge rapidly since they effectively reduce to Newton´s method in the local region. In this section we briefly discuss local convergence rates and stopping criteria.

A. Convergence Rates

In the local region, optimization methods may be characterized by their rate of convergence. Let x_k be a sequence of points converging to x^*

$$\lim_{k \to \infty} x_k = x^* \tag{4.1}$$

and let e_k be the error in x_k

$$e_k = x_k - x^* \tag{4.2}$$

Convergence is said to be linear if

$$\lim_{k \to \infty} \frac{|e_{k+1}|}{|e_k|} = a \qquad \leftarrow \text{ linear convergence} \tag{4.3}$$

for some (preferably small) number a, superlinear if

$$\lim_{k \to \infty} \frac{|e_{k+1}|}{|e_k|} = 0 \qquad \leftarrow \text{ superlinear convergence} \tag{4.4}$$

and quadratic if

$$\lim_{k \to \infty} \frac{|e_{k+1}|}{|e_k|^2} = a \qquad \leftarrow \text{ quadratic convergence} \tag{4.5}$$

Note that quadratic convergence implies superlinear convergence.

Quadratic convergence means that eventually the number of correct figures in x_c doubles at each step, clearly a desirable property. Close to x^* Newton´s method Eq. (3.9) shows quadratic convergence while quasi-Newton methods Eq. (3.8) show superlinear convergence. The RF step Eq. (3.20) converges quadratically when the exact Hessian is used. Steepest descent with exact line search converges linearly for minimization.

B. Stopping Criteria

Different stopping criteria may be used for optimizations. The most straightforward is to require the norm of the gradient to be smaller than some threshold

$$|g_c| \leq \varepsilon \tag{4.6}$$

but we may also test on the predicted second-order change in the function

$$\frac{1}{2}\left|\tilde{g}_c B_c^{-1} g_c\right| \leq \varepsilon \tag{4.7}$$

or on the size of the Newton step

$$\left|B_c^{-1} g_c\right| \leq \varepsilon \tag{4.8}$$

In addition, it is important to check that the structure of the Hessian is correct. For example, if a geometry minimization has been carried out with constraints on the symmetry of the molecule, the solution may turn out to be a saddle point when symmetry breaking distortions are considered.

V. STRATEGIES FOR MINIMIZATION

Global strategies for minimization are needed whenever the current estimate of the minimizer is so far from x^* that the local model is not a good approximation to $f(x)$ in the neighborhood of x^*. Three methods are considered in this section: the quadratic model with line search, trust region (restricted second-order) minimization and rational function (augmented Hessian) minimization.

A. Line Searches

In the global region the Newton or quasi-Newton step may not be satisfactory. It may, for example, increase rather than decrease the function to be minimized. Although the step must then be rejected we may still use it to provide a direction for a one-dimensional minimization of the function. We then carry out a search along the Newton step until an acceptable reduction in the function is obtained and the result of this *line search* becomes our next step.

All line searches start by defining a *descent direction*. Consider all vectors z which fulfill the condition

$$\tilde{z} g_c < 0 \tag{5.1}$$

Since

$$\left. \frac{df(x_c + tz)}{dt} \right|_{t=0} = \tilde{z} g_c < 0 \tag{5.2}$$

there must be a positive number τ such that

$$f(x_c + tz) < f(x_c) \quad \text{for } 0 \leq t \leq \tau \tag{5.3}$$

and z is therefore said to be a descent direction. The negative gradient obviously is a descent direction (often referred to as the *steepest descent* direction) as is the Newton or quasi-Newton step

$$s_N = - B_c^{-1} g_c \tag{5.4}$$

provided the Hessian is positive definite

$$\tilde{s}_N g_c = - \tilde{g}_c B_c^{-1} g_c < 0 \tag{5.5}$$

It is for this reason the positive definite BFGS update Eq. (3.35) is preferred over the PSB update Eq. (3.33) for minimizations. The positive definite Newton step is usually a better direction than steepest descent since it takes into account features of the function further away from x_c than does steepest descent.

Given the direction of search z_c at x_c we must find a satisfactory step along this direction. It would seem that the best is to minimize $f(x_c + tz_c)$ with respect to t and take the step

$$s_c = x_c + t^* z_c \tag{5.6}$$

where t^* is the minimizer. However, such *exact line searches* are expensive and not used in practice. Instead *inexact or partial line searches* are used to generate an *acceptable* point s_c along z_c. By acceptable we mean for example a point that fulfills the condition

$$f(x_c + tz_c) < f(x_c) + \tfrac{1}{2} t \tilde{g}_c z_c \tag{5.7}$$

for some $0 < t \leq 1$. The parameter t may be determined by interpolation. We first try the full Newton step. If this is not acceptable a smaller step is obtained by interpolation and tested. This *backtracking* process is repeated until an acceptable step is found.

Line searches are often used in connection with Hessian update formulas and provide a relatively stable and efficient method for minimizations. However, line searches are not always successful. For example, if the Hessian is indefinite there is no natural way to choose the descent direction. We may then have to revert to steepest descent although this step makes no use of the information provided by the Hessian. It may also be argued more generally that backtracking from the Newton step never makes full use of the available information since the Hessian is used only to generate the *direction* and not the *length* of the step. Alternative methods are provided by the RSO and RF models.

B. Trust Region Minimization

In the *trust region or restricted step method* we determine in each iteration the global minimizer of the RSO model Eq. (3.10). In the global region a step is taken to the boundary of the trust region

$$s(\mu) \;=\; -(G_c - \mu)^{-1} g_c \tag{5.8}$$

since this contains the minimizer of the model. In the local region the model has a minimizer inside the trust region and we take the Newton step

$$s(0) \;=\; -G_c^{-1} g_c \tag{5.9}$$

The method therefore reduces to Newton's method in the local region with its rapid rate of convergence.

When solving Eq. (5.8) the level shift parameter μ must be chosen such that $s(\mu)$ is the global minimizer on the boundary. From the discussion in Sec. III we know that μ must be smaller than the lowest Hessian eigenvalue. Also, μ must be negative since otherwise the step becomes longer than the Newton step. The exact value of μ may be found by bisection or interpolation.

The trust radius h is obtained by a feedback mechanism. In the first iteration some arbitrary but reasonable value of h is assumed. In the next iteration, h is modified based on a comparison between the predicted reduction in f(x) and the actual reduction. If the ratio between actual and predicted reductions

$$R_c \;=\; \frac{f_+ - f_c}{\tilde{g}_c s_c + \frac{1}{2}\tilde{s}_c G_c s_c} \tag{5.10}$$

is close to one h is increased. If the ratio is small the radius is reduced. If the ratio is negative the step is rejected, the trust region reduced and a new step is calculated from Eq. (5.8).

The trust region method is usually implemented with the exact Hessian. Updated Hessians may also be used but an approximate Hessian usually does not contain enough information about the function to make the trust region reliable in all directions. The trust region method provides us with the possibility to carry out an unbiased search in all directions at each step. An updated Hessian does not contain the information necessary for such a search.

It is interesting to note the difference between the RSO trust region method and Newton´s method with line search. In the trust region method we first choose the size of the step (the trust radius), and then determine the direction of the step (constrained minimization within the trust region). In Newton´s method with line search we first choose the direction of the step (the Newton direction) and next determine the size of the step (constrained minimization along the Newton direction). The trust region method is more robust (guaranteed convergence for smooth and bounded functions) and has no problems with indefinite Hessians. It is perhaps more conservative than line search since the size of the step is predetermined. However, line search requires additional energy calculations and is not equally well suited for handling indefinite Hessians. It is possible to combine features of trust region and line search minimizations by backtracking not along the Newton step but along the line generated by calculating trust region steps with different h (the Levenberg-Marquardt trajectory).

C. Rational Function Minimization

A global strategy for minimization may also be based on the RF model. In each iteration the eigenvalue equations

$$\begin{bmatrix} G_c & g_c \\ \tilde{g}_c & 0 \end{bmatrix} \begin{bmatrix} s \\ 1 \end{bmatrix} = \nu \begin{bmatrix} S & 0 \\ 0 & 1 \end{bmatrix} \begin{bmatrix} s \\ 1 \end{bmatrix} \tag{5.11}$$

are solved for the lowest eigenvalue ν and the corresponding eigenvector. This gives a step

$$s(\nu S) = -(G_c - \nu S)^{-1} g_c \tag{5.12}$$

to the minimizer of the rational function. This step is opposite the gradient in each mode if $\nu S < \lambda_1$. In the local region, ν goes to zero and the step approaches the Newton step. The parameter S may be used for step size control in the same way as h in trust region minimizations. However, S is usually set to one and the step is simply scaled down if it is unsatisfactory.

One advantage of the RF minimization over trust region RSO minimization is that we need only calculate the lowest eigenvalue and eigenvector of the augmented Hessian. In the trust region method we must first calculate the lowest eigenvalue of the Hessian and then solve a set of linear equations to obtain the step.

In conclusion, the trust region method is more intuitive than the RF model and provides a more natural step control. On the other hand, RF optimization avoids the solution of one set of linear equations, which is important when the number of variables is large.

VI. STRATEGIES FOR SADDLE POINT OPTIMIZATIONS

Optimizations of saddle points such as molecular transition states and excited electronic states are usually more difficult than minimizations. First, the methods to determine saddle points are less developed and less stable than methods for minimizations. Second, it is usually less clear where to start an optimization of a saddle point than a minimization.

In the immediate vicinity of a saddle point there is no problem. We proceed as for minimizations either by solving a set on linear equations to determine the Newton step Eq. (5.9), or by solving a set of eigenvalue equations to determine the near-zero eigenvalue solution Eq. (5.11). The difficulties are in the global part of the optimization.

The strategies for saddle point optimizations are different for electronic wave functions and for potential energy surfaces. First, in electronic structure calculations we are interested in saddle points of any order (although the first-order saddle points are the most important) whereas in surface studies we are interested in first-order saddle points only since these represent transition states. Second, the number of variables in electronic structure calculations is usually very large so that it is impossible to diagonalize the Hessian explicitly. In contrast, in surface studies the number of variables is usually quite small and we may easily transform to the diagonal representation. Because of these differences we first briefly discuss methods useful for excited states and then describe methods for transition states.

A. Methods for Excited Electronic States

As mentioned in Sec. II the general strategy in saddle point calculations is to take a step that increases the function in some directions while reducing it in others. In the RSO and RF approaches the step may be written

(6.1)

or in the diagonal representation

$$\sigma_i = -\sum_i \frac{\phi_i}{\lambda_i - \mu} \tag{6.2}$$

By choosing a level shift in the range $\lambda_k < \mu < \lambda_{k+1}$ we take a step which initially at least increases the function along the k lowest modes and reduces it along all higher modes. Therefore, if at each step we select a level shift in this range we may eventually expect to enter the local region of the k´th excited state.

Within the RSO framework we first determine the correct range for μ by calculating the k+1 lowest eigenvalues of the Hessian.[12] Next we select an appropriate level shift in this range and finally solve a linear set of equations to obtain the step. For example, to move towards the first excited state we calculate two eigenvalues and solve one set of linear equations. The level shift may be adjusted to hit the boundary with little extra effort. But as noted in Sec. III there are either two or no solutions in the desired range. Therefore, the level shift cannot always be chosen unambiguously.

In the RF model the (k+1)´th eigenvector gives a step in the right direction. To obtain the step we must therefore calculate the k+1 lowest eigenvalues of the augmented Hessian. For example, when optimizing the first excited state we calculate two eigenvalues but do not solve a set linear

[12]In electronic structure calculations it is usually not possible to calculate all eigenvalues of the Hessian. Instead, methods have been developed to calculate selected eigenvalues. To obtain the k´th mode we must first calculate all lower modes.

equations as in the RSO approach. The RF model is therefore the preferred one for wave functions.

For a given S the RF model avoids the ambiguity in selecting the level shift. Usually S is set to unity and the step is scaled down if it becomes too large. In principle, however, S may be used for step control[13] and by allowing for all possible values of S we obtain the same steps as in the RSO model.

B. Gradient Extremals

In surface studies we are interested only in first-order saddle points since these represent transition states but unlike electronic calculations there may be more than one first-order saddle point of interest. We must therefore develop methods that allow us to guide the search in the more promising directions to catch the different transition states.

Using the same method as for the first excited electronic state, we select a level shift in the region $\lambda_1 < \mu < \lambda_2$. This procedure may indeed lead to a transition state but in this way we always increase the function along the lowest mode. However, if we wish to increase it along a higher mode this can only be accomplished in a somewhat unsatisfactory manner by coordinate scaling. Nevertheless, this method has been used by several authors with considerable success.[14] The problem of several first-order saddle points does not arise in electronic structure calculations since there is only one first excited state.[15]

To develop a method to locate saddle points by selectively following one eigenvector we note that at a stationary point all n components of the gradient in the diagonal representation are zero:

---

[13]A modification of this scheme has been used for optimizing excited states of multiconfigurational self-consistent field wave functions, see H. J. Aa. Jensen and H. Ågren, Chem. Phys. **104**, 229 (1986).

[14]C. J. Cerjan, W. H. Miller, J Chem. Phys. **75**, 2800 (1981); J. Simons, P. Jørgensen, H. Taylor, and J. Ozment, J. Phys. Chem. **87**, 2745 (1983); D. T. Nguyen and D. A. Case, J. Phys. Chem. **89**, 4020 (1985); H. J. Aa. Jensen, P. Jørgensen, and T. Helgaker, J. Chem. Phys. **85**, 3917 (1986); J. Nichols, H. Taylor, P. Schmidt, and J. Simons, J. Chem. Phys. **92**, 340 (1990).

[15]Our *approximate* electronic wave function may have more than one saddle point. Nevertheless, if we select our initial guess carefully we should be relatively close to the saddle point that most closely represents the true excited state and the level-shifted Newton step should guide us reliably to this point.

$$\phi(x_{SP1}) = \begin{bmatrix} 0 \\ 0 \\ \vdots \\ 0 \end{bmatrix} \tag{6.3}$$

These n conditions define a point in n-dimensional space. We now move away from the stationary point in a controlled manner by relaxing only one of these conditions. For example, we may no longer require the second component of the gradient to be zero. We are then left with n–1 conditions, which define a line in n-dimensional space passing through the stationary point Eq. (6.3):

$$\phi[x_{GE}(t)] = \begin{bmatrix} 0 \\ \varphi(t) \\ \vdots \\ 0 \end{bmatrix} \tag{6.4}$$

We call this line a *gradient extremal (GE)*.[16] Unless the function increases indefinitely the gradient $\varphi(t)$ must eventually become zero or approach zero. It is therefore reasonable to expect that by following a GE we sooner or later hit a new stationary point

$$\phi(x_{SP2}) = \begin{bmatrix} 0 \\ 0 \\ \vdots \\ 0 \end{bmatrix} \tag{6.5}$$

If we start at a minimum this must be a saddle point. This observation is the basis for the GE algorithm: Transition states are determined by carrying out one-dimensional searches along GEs starting at a minimum. Since n GEs pass through each stationary point, there are 2n directions along which we may carry out such line searches.

The condition that only one component of the gradient differs from zero means that the gradient is an eigenvector of the Hessian at GEs:

$$G(x)\, g(x) = \mu(x)\, g(x) \tag{6.6}$$

[16]The term gradient extremal was introduced by D. K. Hoffmann, R. S. Nord, and K. Ruedenberg, Theor. Chim. Acta. **69**, 265 (1986). Gradient extremals have also been discussed by J. Pancir, Collect. Czech. Chem. Commun. **40**, 1112 (1975) and by M. V. Basilevsky and A. G. Shamov, Chem. Phys. **60**, 347 (1981).

We obtain the same equations by optimizing the squared norm of the gradient in the contour subspace where f(x) is equal to a constant k. Differentiating the Lagrangian

$$L(x,\mu) = \tilde{g}(x)\, g(x) - 2\mu[f(x) - k] \tag{6.7}$$

and setting the result equal to zero we arrive at Eq. (6.6). This means, for example, that the gradient extremal belonging to the lowest eigenvalue may be interpreted as a valley floor.

To develop a practical method for tracing gradient extremals we return to the restricted second-order model Eq. (3.10).[17] In this model the Hessian is constant

$$G_{SO}(x) = G_c \tag{6.8}$$

and the gradient is given by

$$g_{SO}(x) = g_c + G_c s \tag{6.9}$$

Inserting Eqs. (6.8) and (6.9) in the gradient extremal equation Eq. (6.6) we obtain

$$(\mu - G_c)(g_c + G_c\, s) = 0 \tag{6.10}$$

If μ is different from all Hessian eigenvalues we obtain the Newton step

$$g_c + G_c\, s = 0 \tag{6.11}$$

which takes us to the stationary point on the SO model. This is trivially a gradient extremal point. We therefore set μ equal to the k´th Hessian eigenvalue and obtain

$$(\mu_K - G_c)(g_c + G_c\, s) = 0 \tag{6.12}$$

which has the solutions

---

[17] P. Jørgensen, H. J. Aa. Jensen, and T. Helgaker, Theor. Chim. Acta **73**, 55 (1988).

$$x_k(t) = -P_k G_c^{-1} g_c + t v_k \tag{6.13}$$

Here v_k is the eigenvector

$$G_c v_k = \mu_k v_k \tag{6.14}$$

and P_k the projector

$$P_k = 1 - v_k \tilde{v}_k \tag{6.15}$$

Therefore the k´th GE is a straight line passing through the stationary point of the model in the direction of the k´th eigenvector.

Equation (6.13) forms the basis for a second-order saddle point algorithm. In each iteration we first identify the mode to be followed (*the reaction mode*), then take a projected Newton step to minimize the function along all other modes (*the transverse modes*), and finally take a step along the reaction mode until we reach the boundary of the trust region. If the projected Newton step takes us out of the trust region, we minimize the transverse mode on the boundary instead and do not take a step along the reaction mode. In the local region x^* lies within the trust region and we take the unprojected Newton step.

If the reaction mode becomes degenerate, this algorithm breaks down. We must then go to higher orders to take a step in the correct direction. Also, if there are many soft transverse modes the GE method may spend much time minimizing these.

C. Trust Region Image Minimization

In the GE method the minimization of the transverse modes comes first and the maximization of the reaction mode second. We now describe a method that weights minimization and maximization equally. We assume the existence of an *image* function with the following properties.[18] If the function to be optimized f(x) has the following gradient and eigenvalues at x

[18] The concept of image functions was introduced by C. M. Smith, Theor. Chim. Acta 74, 85 (1988).

$$\phi(\mathbf{x}) = \begin{bmatrix} \phi_1(\mathbf{x}) \\ \phi_2(\mathbf{x}) \\ \vdots \\ \phi_n(\mathbf{x}) \end{bmatrix} \qquad \lambda(\mathbf{x}) = \begin{bmatrix} \lambda_1(\mathbf{x}) \\ \lambda_2(\mathbf{x}) \\ \vdots \\ \lambda_n(\mathbf{x}) \end{bmatrix} \tag{6.16}$$

then the gradient and eigenvalues of the image function $\bar{f}(\mathbf{x})$ are

$$\bar{\phi}(\mathbf{x}) = \begin{bmatrix} -\phi_1(\mathbf{x}) \\ \phi_2(\mathbf{x}) \\ \vdots \\ \phi_n(\mathbf{x}) \end{bmatrix} \qquad \bar{\lambda}(\mathbf{x}) = \begin{bmatrix} -\lambda_1(\mathbf{x}) \\ \lambda_2(\mathbf{x}) \\ \vdots \\ \lambda_n(\mathbf{x}) \end{bmatrix} \tag{6.17}$$

Hence, in the diagonal representation the gradient and Hessian are identical except for opposite sign in the lowest mode.[19] Therefore, a first-order saddle point of the function coincides with a minimum of the image and we may determine the transition state by minimizing the image function.

To minimize the image function we use a second-order method since in each iteration the Hessian is needed anyway to identify the mode to be inverted (*the image mode*). Line search methods cannot be used since it is impossible to calculate the image function itself when carrying out the line search. However, the trust region RSO minimization requires only gradient and Hessian information and may therefore be used. In the diagonal representation the step Eq. (5.8) becomes

$$s(\mu) = -\frac{\phi_1}{(\lambda_1 + \mu)} v_1 - \sum_{i \neq 1} \frac{\phi_i}{(\lambda_i - \mu)} v_i \tag{6.18}$$

The only difference between the steps along the image mode and the transverse modes is the sign of the level shift. The level shift $\mu < -\lambda_1$ is determined such that the step is to the boundary of the trust region. Equation (6.18) forms the basis for the trust region image minimization (TRIM) method for calculating saddle points.[20]

[19] In the following we assume that the function and its image always differ in the first element (the lowest eigenvalue) although we may construct image functions by changing sign of any mode.

[20] T. Helgaker, Chem. Phys. Lett. **182** , 503 (1991).

To compare the image and GE methods, notice that in the diagonal representation the quadratic model may be written

$$m_{SO}(s) = f(x_c) + \sum_i m_i(\sigma_i) \tag{6.19}$$

where

$$m_i(\sigma_i) = \phi_i \sigma_i + \frac{1}{2}\lambda_i \sigma_i^2 \tag{6.20}$$

In RSO minimizations we minimize Eq. (6.19) within the trust region. If instead we wish to maximize the lowest mode and minimize the others we may use the model

$$\overline{m}_{SO}(s) = f(x_c) - m_1(\sigma_1) + \sum_{i \neq 1} m_i(\sigma_i) \tag{6.21}$$

since by reducing $-m_1(\sigma_1)$ we increase $m_1(\sigma_1)$. Equation (6.21) is the SO model of the image function.

We can now see the difference between the GE and TRIM methods. In the GE algorithm we first minimize the transverse modes and then maximize the reaction mode. In the TRIM method we minimize and maximize simultaneously by introducing an auxiliary function Eq. (6.21). The underlying idea of the GE method are lines connecting stationary points. The idea behind the TRIM method is an auxiliary function (the image function) whose minima coincide with the saddle points of the original function.

An image function does not exist for all functions. It exists by definition for all quadratic functions. It also exists trivially for functions in one variable since an image is obtained simply by changing the sign of the function. In any case, the image concept is useful for formulating an algorithm.

D. Rational Function Mode Following

In the GE algorithm we select one eigenvector as the reaction mode and follow this towards the transition state. A similar *mode following* technique has also been developed within the RF framework.[21] However, the RF

---

[21] J. Baker, J. Comput. Chem. 7, 385 (1986); A. Banerjee, N. Adams, J. Simons, and R. Shepard, J. Phys. Chem. 89, 52 (1985).

model Eq. (3.17) is not sufficiently flexible for mode following. Instead we use the model

$$m_{RF}(x) = f(x_c) + \frac{m_1(\sigma_1)}{1+\sigma_1^2} + \frac{\sum_{i\neq 1} m_i(\sigma_i)}{1+\sum_{i\neq k} \sigma_i^2} \tag{6.22}$$

where the first mode is the reaction mode. We have separated the model in two parts representing the reaction mode and the transverse modes. Each term is divided by a quadratic to make it bounded. The two metrics in Eq. (6.22) are set to unity. Note that two heuristic parameters appear in Eq. (6.22) rather than one as in the methods previously discussed.

The model Eq. (6.22) has 2n stationary points. To see this we differentiate the model and obtain two independent sets of equations

$$\begin{bmatrix} \lambda_1 & \phi_1 \\ \phi_1 & 0 \end{bmatrix} \begin{bmatrix} \sigma_1 \\ 1 \end{bmatrix} = v_R \begin{bmatrix} \sigma_1 \\ 1 \end{bmatrix} \tag{6.23}$$

$$\begin{bmatrix} \lambda_2 & & 0 & \phi_2 \\ & \ddots & & \vdots \\ 0 & & \lambda_n & \phi_n \\ \phi_2 & \cdots & \phi_n & 0 \end{bmatrix} \begin{bmatrix} \sigma_2 \\ \vdots \\ \sigma_n \\ 1 \end{bmatrix} = v_T \begin{bmatrix} \sigma_2 \\ \vdots \\ \sigma_n \\ 1 \end{bmatrix} \tag{6.24}$$

Equation (6.23) has two solutions for the reaction mode and Eq. (6.24) has n solutions for the transverse modes. To determine a transition state we choose the maximum of Eq. (6.23) and the minimum of Eq. (6.24). If the step becomes too big it is scaled down.

The three methods discussed here all work by the same principle. The Hessian is diagonalized and the reaction mode is identified. A step is then taken such the function is increased along the reaction mode and decreased along the transverse modes. The methods differ in the way this maximization and minimization is carried out.

VII. CONCLUSIONS

We have discussed the general features of methods for optimizing minima and saddle points. It should be remembered that the implementation of these methods involves adapting the general strategies to a specific problem. It is therefore difficult to give hard and fast rules for optimizations. Also, the relative performance of the various methods is difficult to measure - it depends on the implementation of the methods and problem at hand. Therefore, no such comparisons have been given here.

Acknowledgments

I wish to thank Hans Jørgen Aa. Jensen, Poul Jørgensen, and Bernhard Schlegel for discussions.

Accurate Calculations

and

Calibration

Peter R. Taylor[1]
ELORET Institute
Palo Alto, CA 94303
USA

December 21, 1991

[1]Mailing address: NASA Ames Research Center, Moffett Field, CA 94035-1000, USA

Chapter 1

Approximations and Error Analysis

1.1 Introduction

It is generally true that quantum chemists would like to perform all calculations to the highest possible accuracy. However, this is often computationally feasible only for small systems, because of the rapid increase in computer time and perhaps storage requirements of the calculations with increasing size of the molecule. In many cases it is not even necessary to perform such calculations for the purposes of prediction or interpretation, since the accuracy to which results need be known is often inversely proportional to the size of the system. That is, for diatomic molecules it is generally necessary to compute spectroscopic constants or intensities very accurately, because these quantities can often be determined to high accuracy by experiment. For systems of a dozen or more atoms it is seldom necessary to have such accurate results, because the experimental accuracy is lower and the problems quantum chemists are called on to answer will consequently often involve lower accuracy.

The most important thing to keep in mind is that quantum chemistry is much more concerned with energy differences than with total energies. Of course, one way to obtain accurate energy differences is to have very accurate total energies, but this is neither the only way nor the most efficient way. A more realistic approach is to try to describe different systems to the same level of accuracy, so that energy differences are reliable even if the total energies are not. As a reminder of what this entails, we can recall that the binding energy of the N_2 molecule is 0.3% of the total energy, and for F_2 is 0.03%. Describing two systems (in this case separated atoms and molecule) to similar accuracy so that useful binding energy estimates can be obtained is clearly a very demanding task.

Different properties will require different standards as to what constitutes an accurate result. For example, a common standard for binding energies in chemically bonded systems is an accuracy of 1 kcal/mol, often referred to as *chemical accuracy*. For prediction of results for experimentalists, or distinguishing between different experimental results, this is a reasonable, if exacting, standard. On the other hand, a binding energy for a van der Waals molecule accurate to 1 kcal/mol would very likely be completely useless, because the uncertainty would be comparable to or larger than the quantity itself. Conformational energy differences are often easier to calculate than binding energies, which is just as well since for large, relatively floppy molecules an accuracy of perhaps 0.1 kcal/mol or better may be required. However, for predictive purposes the accuracy of the geometry of such a molecule can usually be much less than is the case for small systems, where we may need results accurate to a few thousandths of an Ångström.

It is difficult to lay down firm standards of what is an acceptable uncertainty in a quantum chemical result, since this can vary considerably from case to case. It is part of the quantum chemist's job to decide how accurately a given result must be obtained for his/her purposes, as we shall discuss in this course. However, the accuracy that can be achieved in principle is limited by several fundamental approximations that are made in deriving conventional quantum chemical methodology, and we begin by considering these approximations.

1.2 Fundamental Approximations

It is instructive to consider first the approximations made in arriving at the usual form of the molecular Hamiltonian used in electronic structure calculations:

$$H = -1/2 \sum_{i} \nabla_i^2 + \sum_{i>j} r_{ij}^{-1} - \sum_{i,A} Z_A r_{iA}^{-1} + \sum_{A>B} Z_A Z_B R_{AB}^{-1}, \tag{1.1}$$

where lower-case indices refer to (moving) electrons and upper-case indices to (fixed) nuclei. First, this equation corresponds to point particles interacting nonrelativistically via purely Coulombic forces. In reality, nuclei are known to have finite size, and the electronic motion is known to be relativistic. There are thus two immediate approximations made in deriving Eq. 1.1. The effect of relativity on the total energy scales roughly as Z^4/c^2 for an atom with nuclear charge Z, where c is the velocity of light (approximately 137 a.u.). Thus we would expect relativity to become important by about sodium ($Z = 11$). However, the effect of relativity on energy differences or properties is harder to gauge. The effect of nuclear size is also difficult to estimate without explicit calculation.

Second, Eq. 1.1 corresponds to electrons moving in the field of fixed nuclei — the clamped-nucleus Born-Oppenheimer approximation. This approximation derives from a perturbation theory analysis in which the perturbation parameter is $M^{-\frac{1}{4}}$, where M is the nuclear mass in atomic units (electron mass unity). Thus even for the worst case of a hydrogen nucleus this quantity is about 0.15, and for heavier atoms it is smaller. The Born-Oppenheimer approximation is thus generally a very good one, although the effects of its breakdown can be crucial in some spectroscopic or dynamical phenomena involving potential surface near-crossings, for example.

At this stage, therefore, we can state with some confidence that if our concern is with lighter elements and with single electronic states isolated from other states, the assumptions of fixed point-charge nuclei and nonrelativistic electronic motion are very reasonable ones. Both are susceptible of verification by explicit calculation, although this is not easy, as we shall discuss further in Chapter 6.

1.3 Practical Approximations

We may thus hope to obtain reliable estimates of molecular structure and properties by determining the eigenfunctions Ψ and eigenvalues E of the Schrödinger equation

$$H\Psi = E\Psi \tag{1.2}$$

where the Hamiltonian is given by Eq. 1.1. Note that it is at this point that we begin to think, conventionally, of making approximations, but the fact that approximations have been made already should be kept in mind. The reason for making approximations in solving Eq. 1.2 is that it defies analytical solution. One approach is to construct a *trial wave function* $\tilde{\Psi}(\alpha)$ that depends on a set of variable parameters $\{\alpha\}$. If we then make the functional

$$\tilde{E}(\alpha) = \left\langle\tilde{\Psi}(\alpha)|H|\tilde{\Psi}(\alpha)\right\rangle / \left\langle\tilde{\Psi}(\alpha)|\tilde{\Psi}(\alpha)\right\rangle \tag{1.3}$$

stationary with respect to variations of $\{\alpha\}$, the "energy" $\tilde{E}(\alpha)$ converges from above on the true energy from Eq. 1.2, and the wave function converges (in the mean) on the true wave function, as the parameter set $\{\alpha\}$ is expanded to completeness.

The simplest form of this *variational approach* is for $\tilde{\Psi}(\alpha)$ to depend linearly on the parameter set $\{\alpha\}$. If we write this linear expansion as

$$\tilde{\Psi} = \sum_k \Phi_k \tilde{c}_k, \tag{1.4}$$

substitute it into Eq. 1.3, and make the energy stationary with respect to variations of the linear coefficients $\tilde{c}$, we obtain a *secular determinant*

$$\left|H_{kl} - \tilde{E}S_{kl}\right| = 0 \tag{1.5}$$

for the energy and a set of equations

$$\sum_l \left(H_{kl} - \tilde{E}S_{kl}\right) \tilde{c}_l = 0 \tag{1.6}$$

for the coefficients. Here

$$H_{kl} = \langle \Phi_k | H | \Phi_l \rangle \tag{1.7}$$

and

$$S_{kl} = \langle \Phi_k | \Phi_l \rangle \tag{1.8}$$

are respectively Hamiltonian and overlap matrix elements between the functions $\{\Phi_k\}$. These are fixed functions that depend on the coordinates of all N electrons in the molecule, they are referred to as the *N-particle basis*. They can conveniently be taken as orthonormal,

$$S_{kl} = \delta_{kl}, \tag{1.9}$$

in which case the variational equations correspond to the eigenvalue problem

$$\mathbf{H}\tilde{\mathbf{c}} = E\tilde{\mathbf{c}}. \tag{1.10}$$

If the N-particle basis were a complete set of N-electron functions, the use of the variational approach would introduce no error, because the true wave function could be expanded exactly in such a basis. However, such a basis would be of infinite dimension, creating practical difficulties. In practice, therefore, we must work with incomplete N-particle basis sets. This is one of our major practical approximations. In addition, we have not addressed the question of how to construct the N-particle basis. There are no doubt many physically motivated possibilities, including functions that explicitly involve the interelectronic coordinates. However, any useful choice of function must allow for practical evaluation of the N-electron integrals of Eq. 1.7 (and Eq. 1.8 if the functions are nonorthogonal). This rules out many of the physically motivated choices that are known, as well as many other possibilities. Almost universally, the N-particle basis functions are taken as linear combinations of products of *one-electron functions* — orbitals. Such linear combinations are usually antisymmetrized to account for the permutational symmetry of the wave function, and may be spin- and symmetry-adapted, as discussed elsewhere:

$$\Phi_k = \mathcal{O} \prod_i^N \phi_{ki}(x_i), \tag{1.11}$$

where $\mathcal{O}$ is some group-theoretical operator and the functions $\{\phi\}$ are termed molecular orbitals (MOs). The MOs are generally constructed as orthonormal linear combinations of a one-particle basis ("atomic orbitals"):

$$\phi_{ki} = \sum_\mu \chi_\mu C_{\mu,ki}. \tag{1.12}$$

We will return to the determination of the coefficient matrix C later.

Thus the one-particle basis determines the MOs, which in turn determine the N-particle basis. If the one-particle basis were complete, it would at least in principle be possible to form a complete N-particle basis, and hence to obtain an exact wave function variationally. This wave function is sometimes referred to as the *complete CI* wave function. However, a complete one-particle basis would be of infinite dimension, so the one-particle basis must be truncated in practical applications. In that case, the N-particle basis will necessarily be incomplete, but if all possible N-particle basis functions are included we have a *full CI* wave function. Unfortunately, the factorial dependence of the N-particle basis size on the one-particle basis size makes most full CI calculations impractically large. We must therefore commonly use truncated N-particle spaces that are constructed from truncated one-particle spaces. These two truncations, N-particle and one-particle, are the most important sources of uncertainty in quantum chemical calculations, and it is with these approximations that we shall be mostly concerned in this course. We conclude this section by pointing out that while the analysis so far has involved a configuration-interaction approach to solving Eq. 1.2, the same N-particle and one-particle space truncation problems arise in non-variational methods, as will be discussed in detail in subsequent chapters.

1.4 Error Analysis

At this point it is useful to discuss how we can best quantify any errors in our calculations. Let us suppose we have a computed quantity $\{P_i; i = 1, n\}$ for each of n molecules or properties, and a set of reference results $\{P_i^0; i = 1, n\}$. These may be obtained from reliable experiments or highly accurate calculations: the source is not important for our reasoning. A popular approach to estimating uncertainty in results is by using the root-mean-square (RMS) error

$$\frac{1}{n}\left[\sum_i (P_i - P_i^0)^2\right]^{\frac{1}{2}}, \tag{1.13}$$

This is the basis for least-squares methods of functional approximation, for example, and is certainly useful as a some measure of scatter in data points. However, as we shall see, the errors we must guard against in quantum chemistry tend to be *systematic errors*: the failure of some approximation we have made. In this sense, it is equally important to know the worst-case error in our quantities. An appropriate analogy is with a numerical analyst writing a routine to evaluate some function. For a user, the important issue is "how large is the *maximum* error when I use this routine?". We should therefore consider also the maximum error

$$\max_i |P_i - P_i^0|. \tag{1.14}$$

The RMS error of Eq. 1.13 is related to the 2-norm of the vector of differences $P_i - P_i^0$, while that of Eq. 1.14 is related to the ∞-norm [1]. We are not asserting that the latter should replace the former as an estimate of uncertainties in calculations, but that both error estimates should be considered if the uncertainties in predicted values for which there are no standards are to be realistic. We should note that a more elaborate approach to comparing computed results to a set of standard values has been devised by Maroulis and co-workers using principles of information theory [2]. Their work has been little used in practice, but it provides a convenient way of comparing many values.

Of course, while the mathematical edifice built up in the last paragraphs is very useful, its applicability is predicated on the availability of "reference values" to calibrate our approximate results. This is perhaps the most desirable way to estimate our uncertainties, and we shall consider it at length in the next chapter. However, there will undoubtedly be cases where there are no reference values available. In such circumstances the quantum chemist must rely on his/her knowledge of what is known generally about the effects of certain approximations, and try to deduce their effects in the case of interest. We shall therefore try to draw some general conclusions about our various approximations.

Chapter 2

Computational Methods

2.1 Electron Correlation

The simplest truncation of the N-particle space is to use only one N-electron basis function — a single configuration. If the energy is optimized with respect to the MO coefficients of Eq. 1.12 we then have the Hartree-Fock self-consistent field (SCF) method. This special case is sufficiently important that the difference

$$E_{\text{exact}} - E_{\text{HF}} \tag{2.1}$$

is defined as the *correlation energy* [3]. Here E_{HF} is the Hartree-Fock energy, that is, the SCF result in a complete one-particle basis set. The SCF approach is often useful in itself for qualitative or semiquantitative studies, but is rarely adequate for quantitative accuracy. We are then faced with computing the correlation energy, or solving the correlation problem, as it is sometimes stated.

If we view electron correlation as an inadequacy of the single configuration Hartree-Fock approximation, we can identify two different effects. The first is the influence of other configurations that are low-lying in energy and that mix strongly with the Hartree-Fock configuration. These give rise to *nondynamical* correlation, which can usually be dealt with by multiconfigurational SCF techniques. Nondynamical correlation is often small in closed-shell molecules near their equilibrium geometry, but it increases enormously in importance as a molecule is distorted and bonds are formed or broken. It can also be important in open-shell systems like excited states or transition metals. *Dynamical* correlation arises from the r_{ij}^{-1} term in the Hamiltonian operator. This term is singular as $r_{ij} \rightarrow 0$, and mathematical studies of the properties of exact wave functions show that they must contain cusps in r_{ij} to cancel this singularity. Hence treating dynamical correlation requires describing this cusp behaviour. As we shall see, the slow convergence of the expansion of a two-electron cusp in products of one-electron functions is responsible for much of the difficulty in describing dynamical correlation.

Another correlation effect that is sometimes noted arises from the fact that the SCF MOs are optimized in an uncorrelated calculation and might therefore relax if correlation was included. For our purposes this "effect of correlation on orbitals" can be regarded (and treated) as a part of nondynamical correlation. We will now enumerate methods for treating both nondynamical and dynamical correlation: they will be surveyed only briefly as all are treated in great detail elsewhere. It should be understood that there is no sharp dividing line between nondynamical and dynamical correlation, and methods for treating one will undoubtedly account in some part for the other. It is usually most efficient to use different techniques for each, however.

2.2 Nondynamical Correlation

There are two popular ways of accounting for nondynamical correlation. The first is to explicitly include several configurations in the SCF procedure: multiconfigurational SCF (MCSCF) methods [4]. In early applications of MCSCF schemes the configurations to be included were selected by the user; this was relatively simple for cases of breaking a single bond in a simple system, but the rules for choosing configurations became ever more elaborate for larger or more complicated molecules. The most common approaches currently are probably the complete active space SCF (CASSCF) scheme [5], where the user selects a set of chemically important *active* MOs and the MCSCF configurations are defined as a full CI within that space, or simplified generalized valence bond (GVB) methods [6], in which pairwise excitations are made from bonding to antibonding MOs. The CASSCF scheme can lead to very long configuration expansions; the simple GVB schemes yield fewer configurations but are less satisfactory for multiple bonds and open-shell systems.

The second approach to treating nondynamical correlation has an air of the ostrich about it: ignore the spin symmetry of the wave function and use unrestricted Hartree-Fock (UHF) theory as the single configuration description [7]. Since the UHF wave function comprises one spin-orbital for each electron, a molecular UHF wave function should dissociate to atomic UHF wave functions, for example. This is certainly not the case for spin-restricted Hartree-Fock (RHF) molecules and atoms in general. And there is an attractive simplicity about UHF — no active orbitals to identify, and so forth. However, where nondynamical correlation would be important in an RHF-based treatment, the UHF method will suffer from severe spin-contamination, while where nondynamical correlation is not important the RHF solution may be lower in energy than any broken-symmetry UHF solution, so potential curves and surfaces may have steps or kinks where the spin symmetry is broken in the UHF treatment.

2.3 Dynamical Correlation: Configuration Interaction

Configuration interaction is probably the oldest treatment for recovering dynamical correlation [8]. We simply generate a list of configurations and use them in a linear variational calculation of the type discussed in Chapter 1. We obtain an upper bound to the exact energy from this procedure. A common way of constructing the configurations is to classify them relative to the Hartree-Fock configuration Ψ_0:

$$\Psi_{\mathrm{CI}} = c_0\Psi_0 + \sum_i \sum_a \Psi_i^a c_i^a + \sum_{i>j} \sum_{a>b} \Psi_{ij}^{ab} c_{ij}^{ab} \cdots \quad . \tag{2.2}$$

Here $i, j \ldots$ index MOs occupied in Ψ_0, while $a, b \ldots$ index MOs unoccupied in Ψ_0, and Ψ_{ij}^{ab} denotes a configuration obtained by exciting two electrons in MOs i and j to MOs a and b. (In reality multiple spin and symmetry couplings may be possible unless we use single determinants as our configurations, but we can ignore this here.) If all levels of excitation up to N-fold are used for an N-electron system we have a full CI wave function. If this is done for a complete (infinite) one-particle basis we have a complete CI expansion of the exact wave function. In practice the expansion in Eq. 2.2 must be truncated. This is commonly done after the double excitation level (CI with single and double excitations, CISD), as higher than double excitations cannot interact with Ψ_0 through the Hamiltonian.

One attractive feature of the CI method is that it can be fairly straightforwardly extended to handle the case where nondynamical correlation is important enough to require a multiconfigurational Ψ_0. This gives the multireference CI (MRCI) approach. Although the number of configurations in the expansion can become very large, MRCI calculations can in principle be applied at any geometry and can be used for excited electronic states. This is not true of the CISD method.

2.4 Dynamical Correlation: Perturbation Theory

Unlike the CI case, where for a long time there has been general agreement as to what treatment to use (although not necessarily as to how it should be implemented computationally), until recently there were several different approaches to the use of perturbation theory in treating dynamical correlation. For the last few years, however, one approach, Møller-Plesset (MP) perturbation theory, has been dominant [7]. It employs the MP partitioning of the Hamiltonian

$$H = F + V \tag{2.3}$$

where F is the Fock operator for the SCF treatment used (RHF for closed-shell molecules and UHF for open-shell cases). The energy is then expanded as a perturbation series in V. The nth-order treatment is denoted MPn. The simplest level,

MP2 (note that with the partitioning of Eq. 2.3 there is no first-order correction to the energy), is probably the cheapest available treatment of dynamical correlation.

The convergence behaviour of the MP expansion is obviously crucial to the success of the method, and numerous investigations have been carried out [9, 10]. Neither MP2 nor MP3 are entirely satisfactory. These implicitly require the first-order wave function, which involves only double excitations. MP4, if carried out completely, involves single, double, triple and quadruple excitations. It is more expensive than CISD, say, but often produces better results. MP5 and higher orders are likely to be impractically expensive.

A comparison between perturbation theoretic and truncated CI methods is difficult, because the latter include some terms effectively to infinite order, but obviously omit some terms in lower orders. A pragmatist who prefers to do perturbation theory "until it converges" is likely to prefer infinite-order schemes, while a purist may prefer to obtain a result that is exact through a finite order in perturbation theory.

One area in which CI methods appear to be thoroughly superior to perturbation theory is in the treatment of multireference problems: problems with substantial nondynamical correlation effects. Even UHF-based single reference perturbation theory methods may not cope with some such situations, and multireference perturbation theory, despite many efforts over the years, still appears to be far from developing a general flexible approach that is competitive with MRCI. Transition-metal chemistry, in particular, is a graveyard for UHF-based MP methods.

2.5 Dynamical Correlation: Coupled-cluster Methods

Coupled-cluster methods [11] are infinite-order nonlinear treatments that can be related to CI methods as follows. We rewrite the CI expansion Eq. 2.2 formally as

$$\Psi_{\mathrm{CI}} = (1 + C_1 + C_2 + \ldots)\Psi_0, \tag{2.4}$$

where we have used *intermediate normalization* in which $\langle\Psi_0|\Psi_{\mathrm{CI}}\rangle = 1$. Eq. 2.4 has all the single excitations grouped into the C_1 term, which includes both the CI coefficients and some sort of excitation operators (which we shall not elaborate on here) that generate the singly-excited configurations.

Consider now employing not the linear formulation of Eq. 2.4, but the following *exponential* approach

$$\Psi_{\mathrm{CC}} = \exp(T)\Psi_0, \tag{2.5}$$

which we can expand as

$$\Psi_{\mathrm{CC}} = (1 + T_1 + T_2 + T_1^2 + T_1T_2 + T_2^2 + T_3 \ldots)\Psi_0. \tag{2.6}$$

We note first that if we employ all levels of excitation, up to N-fold, the wave function is equivalent to a full CI wave function, although the formulation is considerably more complicated because of the nonlinear terms. However, it is more practical to truncate at some finite order. Suppose we retain only the operators T_1 and T_2. Clearly, the resulting Ψ_{CC} includes not only single and double excitations, but higher excitations of *all orders*, because of the higher powers of T_1 and T_2. The coefficients of these higher-order excitations, however, are given exclusively as products of the coefficients of the single and double excitations. Such higher-order excitations are termed *disconnected clusters*; their coefficients (usually referred to as cluster amplitudes) are products of the amplitudes of *connected clusters*.

The cluster amplitudes can be determined formally by substituting Ψ_{CC} into the Schrödinger equation and then projecting onto the space of SCF, singly-excited, doubly-excited configurations, etc. The same level of excitation is required for projection as for the highest T_i operators included in the wave function. The resulting problem can be cast as a set of nonlinear simultaneous equations for the cluster amplitudes. Details are given by Bartlett [11], for example.

A fixed excitation level of coupled-cluster treatment (like CCSD, coupled-cluster with single and double excitations) is not substantially more expensive than the same excitation level of CI calculation, but the results are generally superior because of the inclusion of the effects of disconnected clusters. Coupled-cluster methods can also be related to perturbation theory. It is possible to obtain MP2 and MP3 energies as a by-product of a CCSD calculation, for example. Full MP4 includes a contribution from (connected) triple excitations that is not included in CCSD: this has led several authors to explore the possibility of adding the MP4 connected triples energy to the CCSD energy to obtain a treatment accurate through fourth order of perturbation theory. When the MP4 energy expression is used with CCSD doubles amplitudes rather than the MP amplitudes the resulting treatment is denoted CCSD+T [12, 13]. An extension [14] that includes part of the fifth-order term has been much used recently, this method is denoted CCSD(T). Finally, we should note that Pople and co-workers have arrived at a set of equations closely related to CCSD, termed quadratic CI, QCISD [15]. The QCISD equations can be obtained by dropping several (usually small) terms from the CCSD equations. Under most circumstances the two methods yield similar results; with the perturbational inclusion of corrected triples the QCISD(T) and CCSD(T) methods agree even more closely [16]. In most of what follows any reference to coupled-cluster methods can be assumed to apply to QCI as well.

2.6 Comparison of Formal Properties

It is appropriate at this point to compare some formal properties of the three general approaches to dynamical correlation that we have introduced: configuration interaction, perturbation theory, and the coupled-cluster approach. First, we note that taken far enough (all degrees of excitation in CI and CC, infinite order of perturbation theory) all three approaches will give the same answer. Indeed, in a complete one-particle basis all three will then give the exact answer. We are concerned in this section with the properties of truncated CI and CC methods and finite-order perturbation theory.

The CI energy is obtained from a variational calculation and is thus an upper bound to the exact energy. Since neither the perturbation theory energy nor the coupled-cluster energy is variational, neither is an upper bound. This advantage of CI has been much debated — since energy differences are generally of more interest in chemistry than total energies the advantage does not seem to be great. More important perhaps than having an upper bound to the exact energy is to have a treatment that is bounded at all, and here there is a clear disadvantage to perturbation theory (coupled-cluster methods are not necessarily bounded either, although problems seem to arise much less than with perturbation theoretic approaches). It used to be felt that there was a clear advantage to approaches like CI, that made a functional stationary, in developing analytical energy derivative methods. This now appears to be less of an issue.

A much more important issue than variational bounds (although an issue that many workers have been reluctant to recognize) concerns the number of electrons in the correlation treatment. The primary reason for using the exponential approach in Eq. 2.5 is that this guarantees *correct scaling* of the correlation energy with the number of particles (see, for example, Bartlett [11]). For instance, the correlation energy of an arbitrary number of mutually noninteracting two-electron systems depends (correctly) linearly on the number of two-electron systems when treated by CCSD. Correct scaling is more general, however: it means that a CC treatment of any number of noninteracting subsystems of any size will give an answer that is the sum of CC energies of the subsystems. This is true of perturbation theory too, but it is *not* generally true of truncated CI. This suggests caution in the application of CI methods to systems with more than a few electrons — the scaling errors grow rapidly. For the case of noninteracting two-electron systems the CI result scales as the square root of the number of systems, for instance.

The property of correct scaling with particle number is termed "size-extensivity" [17]. CC and perturbation theory are size-extensive, while CI is not. A related, but not identical, property is termed "size-consistency" [18]: this requires that the energy of two noninteracting systems be the sum of the individual system energies, as in

$$E_{AB}(r_{AB} \to \infty) = E_A + E_B \tag{2.7}$$

for fragments A and B. Note that Eq. 2.7 implies a requirement that the treatment dissociate AB correctly: a UHF treatment will usually satisfy Eq. 2.7 while an RHF treatment may not. (Note that both RHF and UHF are size-extensive.) Hence a perturbation theory treatment based on RHF may not be size-consistent for a given process, although the same treatment based on UHF may be. Size-consistency is perhaps a more useful requirement in practice than size-extensivity, but an easy way to obtain size-consistency, at least formally, is of course to apply a size-extensive correlation treatment to a size-consistent reference function. Where this is not possible it is essential to compute the separated fragment energy from a "supermolecule" treatment that comprises both fragments far apart, but lack of correct scaling may still create problems.

One way of achieving size-consistency for a dissociation process is to use an MCSCF wave function as the reference. Unfortunately, as noted above, there are as yet no general multireference perturbation theory or multireference coupled-cluster treatments that can be applied to such an MCSCF reference function. For rather few electrons, as we shall see, the MRCI approach performs acceptably.

2.7 Approximate Methods for Correct Scaling

The issue of correct scaling with particle number has led to the development of numerous approximate schemes for accounting for size-extensivity effects. We list some of them here in increasing order of formal sophistication.

The simplest schemes for correcting the behaviour of the CI correlation energy with particle number are of a type first suggested by Langhoff and Davidson [19] and usually termed a *Davidson correction*. For the CISD case we have

$$\Delta E = E_{\mathrm{corr}}(1 - c_0^2), \tag{2.8}$$

where c_0 is the coefficient of the SCF configuration in the CISD wave function. The correction ΔE is added to the correlation energy — it is conventional to denote this by a suffix +Q, as in CISD+Q. A justification for the use of Eq. 2.8 can be developed using perturbation theory [20]. Note that the Davidson correction incorrectly predicts a non-zero size-extensivity correction even for a two-electron system. It can also be shown that the neglect of higher-order terms used in deriving Eq. 2.8 will lead to incorrect behaviour for large numbers of electrons. Thus the method is best avoided for very few or for many electrons.

A number of workers have developed similar corrections to Eq. 2.8 but modifying the behaviour in both limits [21, 22]. These modified Davidson corrections have found little favour, although extensive series of comparisons suggest they are superior to the original [23].

The Davidson correction is often applied to multireference CI results as well as the CISD case. Two formulas can be envisaged for the multireference case:

$$\Delta E = E_{\mathrm{corr}}(1 - \sum_R c_R^2), \tag{2.9}$$

where c_R is the coefficient of reference configuration R in the MRCI wave function, and

$$\Delta E = E_{\mathrm{corr}}(1 - \sum_R c_R^0 c_R), \tag{2.10}$$

where c_R is the coefficient of reference configuration R in the MRCI wave function and c_R^0 is the coefficient in the *reference* wave function. In both cases E_{corr} is the difference between the MRCI energy and the energy obtained with only the reference configurations. The formula of Eq. 2.10 can be justified to some extent by formal comparisons with coupled-cluster methods [24], while Eq. 2.9 was arrived at simply by analogy with Eq. 2.8 [25]. It is therefore a perennial problem for the purists that Eq. 2.9 works better in practice!

Perhaps the oldest approach to correcting for incorrect scaling in CISD is the use of "coupled electron-pair approximations" (CEPA) [26, 27], which can usually be related to coupled-cluster methods. They involve modifying the CI secular equation to give a set of *linear* equations for the unknown coefficients, although these may depend implicitly on the energy, necessitating an iterative procedure. The simplest is the so-called linearized coupled-cluster method. It can be derived from the CCSD equations simply by dropping all nonlinear terms. We should note that the use of the exponential formulation in the first place to obtain the CCSD equations guarantees size-extensivity, no matter how many terms we choose to omit [11]. Progressive inclusion of more terms leads to a variety of CEPA methods, although in the earliest derivations these were generally obtained by adding terms to the CISD equations so that they displayed correct scaling for special cases (like mutually noninteracting two-electron systems).

Very similar in spirit to CEPA, but formulated as a functional to be made stationary, is the coupled-pair functional (CPF) approach of Ahlrichs and co-workers [28]. CPF can be viewed as modifying the CISD energy functional to obtain size-extensivity for the special case of noninteracting two-electron systems. One disadvantage of some of the CEPA methods is that, unlike CISD or CCSD, the results are not invariant to a unitary transformation that mixes occupied orbitals with one another. CPF is designed to display such invariance, at least for the model case of transforming between completely localized and completely delocalized orbitals for noninteracting two-electron systems. CPF has been extended by Chong and Langhoff in their modified CPF (MCPF) approach [29]. In MCPF some terms in the CPF functional are adjusted to reduce the tendency of CPF to overestimate size-extensivity effects in the presence of nondynamical correlation. This is very useful for systems with substantial near-degeneracy effects, such as transition metal compounds, but an unfortunate side-effect is that the results can be strongly dependent on the form of the occupied MOs, as even the approximate invariance of CPF to unitary transformations on the occupied MOs is lost.

Gdanitz and Ahlrichs devised a simpler variant of CPF, the averaged coupled-pair functional (ACPF) approach [30]. This produces results very similar to CPF for well-behaved closed-shell cases and is completely invariant to a unitary transformation on the occupied MOs. Its big advantage is that it can be cast in a multireference form. Multireference ACPF is probably the most sophisticated treatment of the correlation problem currently available that can be applied fairly widely, although it can encounter difficulties with the selection of reference spaces, as discussed elsewhere.

2.8 Diagnostics of N-particle Space Problems

Having enumerated a generous selection of correlation treatments, an important issue must be how to judge when these methods are appropriate. The calibration of these methods by comparison and evaluation is considered in the next chapter: we are concerned here with ways in which we can *deduce* the inappropriateness of a given level of treatment just from the calculation itself. Perhaps the simplest such diagnostic is to examine the coefficients of the various configurations in the wave function. In a single-reference treatment, if the reference CSF coefficient becomes too small, or if certain excited CSF coefficients become too large, we can expect the single-reference treatment to fail, as some sort of near-degeneracy effect or problem with nondynamical correlation is clearly indicated. Some examination of CI coefficients, CC amplitudes, or the norms of wave functions in perturbation theory should always be performed. However, there are cases in which this is not adequate, because of defects in the generation of the MOs, as we now discuss.

We can readily envisage that a single-configuration SCF calculation on a particular system may produce MOs which reflect certain biases characteristic of the SCF method. For example, using ground-state SCF orbitals may result in some low-lying excited states appearing at much higher excitation energies than they should. They may then mix into the correlated wave function less strongly, and thus the wave function will appear to be much more dominated by the single reference CSF than it really is. This situation is commonplace in transition-metal-containing molecules, since the separations between the various atomic states and their contribution to the wave function are seldom correct at the SCF level. A very simple case is CuH, whose closed-shell ground state is strongly dominated by the configuration derived from the atomic $d^{10}s$ asymptote when SCF MOs are used. However, this picture is false: the d^9s^2 asymptote should also contribute strongly to the wave function, but the form of the SCF orbitals is biased against this state and it contributes much less than it should.

It might be hoped that this orbital bias would hardly influence the results of correlation treatments, at least if single excitations (which primarily account for orbital relaxation) are included. Unfortunately, this depends on the correlation treatment. Considerable experience that has accumulated suggests that size-extensive (or approximately size-extensive) infinite-order methods are much less affected by orbital bias than other treatments. For example, a single-reference CISD wave function (based on SCF MOs) for CuH appears to be dominated by the SCF determinant and shows no other CSFs with large weights in the expansion. Going from CISD to CPF the picture changes considerably — the weight of the SCF determinant decreases and several important excited CSFs appear. Thus the CPF treatment gives a much better impression of the multiconfigurational effects in the CuH ground state than does CISD, and so does CCSD.

The need to overcome such orbital bias and account for multiconfigurational effects is all part of the need to treat nondynamical correlation adequately, of course. It would be useful to have a simple way of identifying potential difficulties with nondynamical correlation when performing a single-reference calculation. One approach to this was formulated by Lee and Taylor [31]: since the orbital bias is best reflected by the single excitations, they suggested taking the norm of the single excitation cluster amplitudes in the CCSD wave function (in intermediate normalization), scaled to be independent of the number of electrons correlated, as a diagnostic of the importance of nondynamical correlation. Specifically,

$$\mathcal{T}_1 = \|\mathbf{t}_1\|/\sqrt{N}, \tag{2.11}$$

where $\mathbf{t}_1$ is the norm of the single excitation cluster amplitudes and N is the number of electrons correlated. Lee and Taylor showed by comparisons for a large range of systems that values of the $\mathcal{T}_1$ *diagnostic* larger than 0.02 indicated nondynamical effects large enough that the results of single-reference treatments (limited to single and double excitations) should be regarded with caution. We note that the value of $\mathcal{T}_1$ has the same invariance properties with respect to separate unitary transformations on the occupied and virtual orbitals as the CCSD wave function itself, so it is invariant to whether canonical or localized orbitals are used.

We can list a few sample values of $\mathcal{T}_1$. For good single-reference closed-shell systems $\mathcal{T}_1$ is commonly less than about 0.01 (0.0065 for Ne and 0.0104 for HF). For a notoriously difficult molecule like FOOF $\mathcal{T}_1$ is 0.031, while for CuH $\mathcal{T}_1$ is 0.036. The value of $\mathcal{T}_1$ decreases slightly as the one-particle basis set is improved, although this usually affects only the fourth decimal place. We have also investigated the corresponding diagnostic for the QCISD treatment [16], obtained by substituting the QCI single excitation amplitudes into Eq. 2.11 and denoted $\mathcal{Q}_1$. In general, $\mathcal{Q}_1$ is similar in value to $\mathcal{T}_1$ for good single-reference cases, but since the terms omitted from QCISD but present in CCSD confer better stability on the latter, $\mathcal{Q}_1$ increases more rapidly than $\mathcal{T}_1$ as nondynamical correlation becomes more important. Nevertheless, both diagnostics provide useful information about the importance of nondynamical correlation, and certainly any case in which the diagnostic values are larger than 0.02 is grounds for concern about the adequacy of a QCISD or CCSD treatment.

We have already pointed out that one measure of the importance of near-degeneracies, as a part of nondynamical correlation, is the weight of excited CSFs in the correlated wave function. CPF (or QCISD or CCSD) is much better for this purpose than CISD, because the latter is less sensitive to orbital bias and hence is more affected by incorrect separation between reference and excited CSFs. The ACPF method displays an extreme sensitivity to inadequacies in the reference CSF space, and demonstrates this by amplifying the coefficients of the excited CSFs missing from the reference space enormously [32]. In cases of very important CSFs missing from the reference, the ACPF calculation can converge to a root of the wrong configurational structure, with almost no weight on the reference CSFs, but considerable weight on the excited CSFs. The ACPF method is thus very good at indicating dissatisfaction with the reference space, the only problem is that including the extra CSFs in the reference space may lead to impossibly large calculations. This issue of ACPF reference spaces will be treated in more detail elsewhere, as will other approaches to coping with it.

2.9 Disclaimer

The exclusion of a given correlation treatment from this survey should not necessarily be taken as a condemnation. Constraints of time and space preclude the inclusion of all treatments in current use: I have simply attempted to cover as many as possible while concentrating on those with which I am most familiar. I stand by my assertion that there is no generally useful multireference perturbation theory or coupled-cluster treatment available yet (although there have been many brave attempts). I have excluded exotica such as methods that can be applied only to cases of two or three electrons. Finally, it should be understood that for our present purposes a difference of implementation does not constitute a difference of methodology. Thus the MRD-CI approach of Buenker, Peyerimhoff and co-workers [33], or the CIPSI method [34], for example, are regarded here as multireference CI methods.

Chapter 3

Calibration of Methods

3.1 Comparison with Reference Values

The most obvious way to calibrate a particular method is to compare it with some known "reference" values. The choice of these values requires some attention, however. Experimental results might seem to be an obvious choice, but on closer examination we can see a number of objections to such a comparison.

First, there is the obvious objection that there may be no experimental values for the properties and systems of interest. Second (and almost as obvious) is the possibility that the experimental values are wrong. Third, the "experimental" values may in fact be derived from experimental measurements by a number of steps that involve assumptions or other theoretical calculations. All of these objections are important, but in one sense they are orthogonal to the real issue: *what if our calculations contain multiple sources of error that can cancel with one another*? We know already that any truncated one-particle space and truncated N-particle space treatment has two sources of error, these two truncations. And there is no reason to suppose that the error from these two sources cannot cancel, indeed, from the early days of large-scale correlated atomic wave functions there is good evidence that they *do* cancel [35]. Hence even if there are absolutely reliable experimental values for the properties and molecules we want to consider, using them to calibrate theoretical methods may be useless unless we can establish whether we have a cancellation of errors or not.

If comparison with experiment is not appropriate, what should be used for reference values? Clearly, the desirable thing would be to eliminate as many sources of error as possible. For instance, if we wish to establish the reliability of an SCF treatment in a given one-particle basis, we could use numerical Hartree-Fock results for diatomic molecules (that is, essentially complete basis results) as benchmarks. Any difference between the finite basis and numerical results would presumably be due to inadequacies in the former, as otherwise the same approximations are made in both treatments. It is crucial to understand that this approach gives much more information than would comparison with experiment. Since our SCF treatment includes no correlation effects, it would be impossible to say from comparison with experiment what the effect of one-particle basis truncation was. We might even observe perfect agreement with experiment if the one-particle basis truncation effect exactly cancelled with the neglected correlation effects.

This calibration of one-particle basis truncation illustrates the value of benchmark calculations that are "exact" within certain approximations, like numerical Hartree-Fock. Of course, it is generally more of interest to calibrate correlated calculations. One source of benchmarks for correlation treatments is obtained by employing a truncated one-particle space but then using the full N-particle space that can be generated from it: a full CI calculation. This corresponds to the exact result within the given one-particle space, in effect, to the solution of the Schrödinger equation projected onto some subspace of the full Hilbert space. Unfortunately, full CI expansions grow rapidly with the number of MOs included and the number of electrons correlated. Nevertheless, significant advances in full CI methodology in the last seven or eight years have made it possible to perform calculations with over a billion Slater determinants. This allows benchmark calculations correlating up to ten or more electrons.

Before presenting some illustrative examples of full CI benchmarks, we should note one point about the one-particle spaces used. In general, while it might be possible to perform full CI calculations with small one-particle basis sets but correlating many electrons, such calculations are of little value. Basis sets smaller than double zeta plus polarization (DZP) recover so little of the correlation energy (and particularly little of the angular correlation effects that require higher angular momentum functions for their description) that the results of full CI calculations in such basis sets should be viewed with considerable caution.

3.2 Comparison with Full CI Calculations

A useful example that illustrates many features of full CI calibrations is the energy separation between the ground ($^3B_1$) and first excited ($^1A_1$) states of the methylene radical, CH_2 [36]. In Table 3.1 we list results obtained in a DZP basis with a variety of computational methods. Only the six valence electrons are correlated in these calculations.

We may note a number of features of these results. First, nondynamical correlation is known to be much more important in the singlet state than in the triplet. Hence it is not surprising that the multireference methods perform better here than the single reference methods. Nevertheless, the errors in the single reference meth-

Table 3.1: $CH_2\,{}^1A_1 - {}^3B_1$ Separation: Full CI Comparisons (kcal/mol)

Method	Separation	Error
SCF/CISD	14.63	2.66
SCF/CISD+Q	12.35	0.38
SCF/CPF	12.42	0.45
MCSCF/MRCI[a]	12.20	0.23
MCSCF/MRCI+Q	12.03	0.06
CASSCF/MRCI[b]	11.97	0.00
CASSCF/MRCI+Q	11.79	−0.18
Full CI	11.97	—

[a] Two configuration singlet, RHF triplet.
[b] Full valence CASSCF.

ods that are approximately size-extensive (CISD+Q and CPF) are relatively small. For this case, in which only six electrons are correlated, the multireference Davidson correction seems to overestimate the result. An ACPF calculation with the same configuration space would also overshoot slightly. In cases in which CASSCF reference spaces can be used to generate the MRCI wave function, other comparisons show that the Davidson correction is not to be preferred until eight or more electrons are correlated. On the other hand, if a CASSCF reference space is too large and the reference configurations are selected according to some criterion of importance, it may be desirable to include the Davidson correction even when fewer than eight electrons are correlated. The results of Table 3.1 obtained with a two-reference singlet and single reference triplet illustrate this point.

We turn now to a comparison of various methods with full CI for the water molecule at several different geometries [10, 37]. This allows us to examine how well different methods describe the geometry dependence of the energy, and hence how they will perform in calculations of potential energy surfaces, for example. Total energies obtained with eight electrons correlated are compared with full CI in Table 3.2. We can see from these results that all methods except CISD and MP2 perform very well at equilibrium. As the bonds are stretched the performance of most of the single-reference methods begins to deteriorate, although the infinite-order triples-corrected methods CCSD(T) and QCISD(T) still agree well with full CI even when the bonds are stretched to twice their equilibrium value. The best overall agreement with full CI is obtained from the MRCI+Q method. However, all these comparisons are based on the total energy. It is more instructive to compare the geometry dependence of

Table 3.2: H_2O Full CI Comparisons (E_h).

Method	r_e	1.5^*r_e[a]	2^*r_e[b]
CISD	−76.243772	−76.040984	−75.876606
CISD+Q	−76.254549	−76.067003	−75.942257
CPF	−76.252504	−76.064365	−75.956222
MP2	−76.243659	−76.048095	−75.898603
MP4	−76.255705	−76.065641	−75.937409
QCISD	−76.252745	−76.062040	−75.930888
QCISD(T)	−76.256007	−76.069591	−75.953530
CCSD	−76.252502	−76.061247	−75.930862
CCSD(T)	−76.255907	−76.069407	−75.956901
MRCI	−76.254108	−76.069363	−75.950517
MRCI+Q	−76.257805	−76.072943	−75.953731
Full CI	−76.256624	−76.071405	−75.952269

[a] Both bonds stretched to 1.5 times equilibrium.
[b] Both bonds stretched to twice equilibrium.

the energies directly. Table 3.3 displays for each method the difference between the energy at a distorted geometry and at equilibrium.

The agreement between the full CI and MRCI+Q variation of energy with geometry is essentially perfect. Even the uncorrected MRCI values differ from the full CI results by less than 1 mE_h. The single-reference CCSD(T) and QCISD(T) methods perform well, although not quite as well as the multireference treatments. For the smaller geometry change the CPF, CISD+Q, CCSD and QCISD results are good, but on stretching the bonds to twice the equilibrium value the energy change is underestimated by CPF and overestimated by the remainder. The MP2 results are quite unsatisfactory. We may note also that results (not shown here) for the MPn and QCI/CC treatments based on a UHF reference wave function are in considerably poorer agreement with the full CI results than the RHF-based treatments shown in these tables. This is disappointing in view of the expectation that UHF will cope with bond-breaking better than RHF — the problem appears to be the strong spin-contamination in the UHF wave function at the stretched bond lengths used. Comparisons indicate that the same problem is observed for other molecules at similarly distorted geometries.

Table 3.3: H_2O Energy Difference Comparisons (E_h).

Method	$1.5^*r_e - r_e$	$2^*r_e - r_e$
CISD	0.202788	0.367166
CISD+Q	0.187546	0.312292
CPF	0.188139	0.296282
MP2	0.195564	0.345056
MP4	0.190064	0.318296
QCISD	0.190705	0.321857
QCISD(T)	0.186416	0.302477
CCSD	0.191255	0.321640
CCSD(T)	0.186500	0.299006
MRCI	0.184745	0.303951
MRCI+Q	0.184862	0.304074
Full CI	0.185219	0.304355

3.3 Conclusions from Full CI Comparisons

A systematic series of benchmark calculations on a variety of molecules and properties was undertaken by the NASA Ames group, and a comprehensive review is available [38]. In addition, these results have been used for comparison purposes by other authors, as new methods are developed. We can summarize the findings as follows. First, MRCI methods (with a Davidson correction as required) are almost universally in excellent agreement with the full CI results. These MRCI calculations are usually based on valence shell CASSCF calculations, although they may involve selection of reference configurations. The major exception to this success story is in the calculation of electron affinities of electronegative species like O and F. Here configurations that involve excitations out of the valence shell must be included in the reference space for the MRCI+Q and full CI results to agree. This situation can also arise in molecules containing such atoms, especially when ionic configurations are important.

Second, where nondynamical correlation is unimportant (that is, mainly closed-shell molecules near their equilibrium geometries), size-extensive (or nearly size-extensive) treatments like CPF, CCSD, and CISD+Q perform well. However, as bonds are stretched, or in species where nondynamical correlation is important because of near-degeneracies or open-shell effects, the performance of these methods deteriorates fairly rapidly. As might be expected, the CCSD method generally deteriorates slowest. In general, low-order perturbation theory methods do not agree well with full CI; for well-behaved systems MP4(SDTQ) (that is, full fourth order) gives good results but as bonds are stretched even the UHF-based formulation becomes unacceptable. The best single-reference-based method is undoubtedly CCSD(T) (or QCISD(T)), which produces results in excellent agreement with full CI even in cases with substantial nondynamical correlation effects. We shall have more to say about the performance of CCSD(T) later.

Third, we can draw a vital conclusion about correlated calculations from these benchmarks. They have established that MRCI (or MRCI+Q) calculations on systems with up to ten electrons correlated, for example, involve little or no error from truncation of the N-particle basis, at least for calculations in DZP (or somehat larger) basis sets. If we assume that this conclusion is independent of any coupling between the one-particle and N-particle space, an assumption that is supported by the (very limited) available evidence, we can conclude that in *any* one-particle basis there will be little or no truncation error in MRCI calculations. It then follows that we should expect that MRCI in a complete basis would agree with complete CI, that is, with the exact result. Hence if our MRCI calculations do not suffer from errors as a result of N-particle space truncation we infer that the main source of errors (if any) must be the one-particle basis. This had been suggested on numerous previous occasions — our best correlation treatments handle the correlation problem very well and the errors in our best results actually reflect inadequacies in the one-particle basis. We shall turn our attention to one-particle basis sets in the next chapter.

3.4 Other Calibration Methods

While the use of full CI calculations as a calibration of different correlation treatments is obviously a very desirable way to proceed, they may not always be feasible. In particular, as the number of electrons increases the size of the full CI calculation grows so rapidly that it becomes impossible. In such a case any calibration has to be based much more on assumption. First, we can assume that methods such as MRCI+Q or MRACPF will generally perform as well as they have been shown to do in the full CI comparisons. In that case, if it is feasible to perform such calculations on the system of interest, we can do that and assume that there will be little or no error in the correlation treatment. This may not be possible either in an acceptable one-particle basis. We may be able to perform a multireference calculation in a small basis and then use it to calibrate another treatment that can be applied in a larger basis — this is certainly preferable to performing no calibration study at all. The use of methods like CISDTQ for this purpose has been suggested [39]: this may be useful near equilibrium geometries, but at distorted geometries, especially in multiply-bonded systems, increasing nondynamical correlation effects will make such methods unreliable.

Another approach to calibration is to perform full CI calculations on a system that is simpler than the molecule of interest, but related to it. For example, the conclusions of a full CI calibration of methods for estimating the CH bond energy in CH_4 could be applied to the CH bond energies of C_4H_{10}, for which a full CI calibration is likely to be impossible for some time. This reasoning can be extended along the lines of the previous paragraph: comparing not with full CI for a related system but with elaborate multireference treatments, say. The important observation here is *not* that it becomes necessary to compromise our calibrations. It is that it is seldom the case that we cannot find a calibration calculation or chain of reasoning that allows us to determine what is an acceptable correlation treatment. We shall illustrate this in detail in Chapter 5. And after some experience has been acquired, if no explicit calibration method can be devised that experience can be used to make value judgments as to how best to approach a given system, although in most cases this is likely to be the least reliable approach to "calibrating" a correlation treatment.

Chapter 4

Basis Sets

4.1 The One-particle Basis: Convergence Issues

The choice of one-particle space is perhaps the most important decision in setting up any calculation, since ultimately this choice determines the reliability of the calculation. Nothing can overcome limitations in the one-particle basis. Some formal limitations, such as the lack of a nuclear cusp or improper long-range behaviour of Gaussian functions, have turned out to be of relatively little importance in practice. Indeed, the universality of Gaussian functions in computational quantum chemistry illustrates this well — Slater-type functions are hardly used today, even for diatomics, and we shall not discuss them here at all. Much more important are limitations arising from the convergence of our results with the *size* of the basis set.

One of the most important steps in the development of Gaussian basis sets was the realization that, while many primitive (i.e., individual) Gaussian functions were required to provide an acceptable representation of an atomic orbital, the relative weights of many of these primitives were almost unchanged when the atoms were combined into molecules. The relative weights of these primitives could therefore be fixed from some prior calculation (on the atom, say) and only the overall scale factor for this *contracted Gaussian function* need be determined in the molecular calculation. However, even with contraction to reduce the effective dimension of the basis set, many basis functions were required. In retrospect this seems inevitable, although it has been a source of much disappointment over the years. We can consider separately the convergence of results excluding and including dynamical correlation with the one-particle space.

At the SCF or MCSCF level, the basis set requirements are fairly simple. We can imagine that the occupied molecular orbitals are given as a simple linear combination of atomic orbitals: this corresponds to a minimal basis set. The results so obtained are fairly crude, but by admitting extra functions to represent the atomic orbitals more flexibly (split-valence, double zeta, etc) we can obtain a much better description. However, some effects require going beyond the occupied atomic orbitals:

the lower symmetry in molecules allows higher angular momentum atomic functions to contribute. Since their contribution allows the charge density in bonds to polarize they are usually termed *polarization functions*. A similar effect can be achieved by allowing the atomic orbitals to "float" off the nuclei, but this is complicated to automate and has not been much used [40]. Basis sets such as double zeta plus polarization (in which a single d set is added on first-row atoms and a single p set on hydrogen) generally give SCF (or MCSCF when nondynamical correlation is included) results for energies and energy-related properties like spectroscopic constants that are in good agreement with those of much larger basis sets. More flexible sets like triple zeta valence plus two sets of polarization functions (perhaps with f functions on the heavier atoms as well) produce results close to the *Hartree-Fock limit* (the complete one-particle space single configuration result). Properties such as dipole moments or polarizabilities are discussed in more detail in Chapter 5. We should also note here that these statements apply mainly to lighter atoms, where there is commonly not much change in the atomic orbitals on formation of a molecule. In the transition metals, for example, there may be relatively drastic relaxation effects in the occupied d orbitals when a molecule is formed. Larger valence basis sets are then required even for qualitative accuracy: a minimal basis representation of the d orbitals is quite useless for transition metals in molecules [41].

In contrast to the fairly rapid convergence of SCF or MCSCF results with one-particle basis size, the convergence of dynamical correlation results is very much slower. If we think in terms of a configuration expansion, correlation effects are described by replacing occupied orbitals with orbitals of the same spatial extent but with additional nodes. In an atom, for example, we need correlating orbitals with additional angular or radial nodes, giving rise to *angular correlation* and *radial correlation*, respectively. For an atom with occupied $2p$ orbitals we must augment the basis with d, f, and higher functions (each l value appearing several times in order to achieve "radial saturation") to describe angular correlation; radial correlation would require multiple p functions with $3p, 4p, \ldots$ nodal structure, but similar to $2p$ in their radial extent.

Perturbation theory studies of atoms have established that atomic correlation converges as $(l+\frac{1}{2})^{-4}$ with angular quantum number l, and convergence with respect to the radial quantum number is also slow. This is, of course, not unexpected in view of the difficulty in describing a two-electron cusp in the wave function by an expansion in products of one-electron functions. It is interesting to note that the radial convergence does not improve with the use of exponential radial terms, that is, from using Slater-type functions. In fact, such functions may be no more advantageous than Gaussians in terms of their ability to represent correlating orbitals. There is, of course, no reason to suppose that this slow convergence of dynamical correlation will not also occur in molecules, and indeed it does. In fact, since a basis set for use in dynamical correlation calculations must at least begin to describe both radial and angular correlation effects, we cannot hope to obtain sensible results with smaller than double zeta plus polarization (or split-valence plus polarization) sets. We can regard these in essence as "minimal basis sets for correlated calculations". This will result in large basis sets for "minimal" correlated treatments of larger molecules, as well as for accurate calculations on smaller systems, but compromise here exacts a heavy price. It is therefore the description of dynamical correlation that places the greatest demands on our basis sets. Fortunately, basis sets that are able to describe dynamical correlation well generally also provide very good SCF results, so we can concentrate our efforts on dynamical correlation with confidence that this will not degrade the SCF or MCSCF level description.

4.2 Traditional One-particle Basis Sets

As discussed elsewhere, traditional Gaussian basis sets are almost always segmented contracted sets: each primitive Gaussian contributes to only one contracted function. These sets usually belong to one of two families. The first comprises contractions of the type systematized by Dunning [42], based on the primitive sets of Huzinaga [43], or sometimes the smaller primitive sets of Roos and Siegbahn [44] or the larger sets of van Duijneveldt [45]. The second comprises "Pople-type" basis sets, with contraction schemes and primitive sets developed by Pople and co-workers [7].

Dunning-type contractions are characterized by considerable flexibility in the valence part of the primitive space. Typically, the outermost primitive functions are not contracted at all, contraction being reserved for the inner parts of the valence orbitals and the core orbitals. The commonest contracted set of this type is probably the [4*s* 2*p*] contraction of the (9*s* 5*p*) set. Unfortunately, there are at least two such "double zeta" contraction schemes in use, as well as an erroneous one. Some care may be required to reproduce results asserted to be obtained with "a Huzinaga-Dunning [4*s* 2*p*] basis". Because of the relatively flexible contraction scheme these basis sets usually perform well, especially when large primitive sets such as van Duijneveldt's (13*s* 8*p*) sets are used. However, it should be noted that such primitive sets are difficult to contract this way without significant loss of accuracy at the atomic SCF level, unless very large contracted sets are used.

Pople-type basis sets are generally characterized by heavier contraction (that is, fewer contracted functions for a given primitive set) and consequently somewhat lower accuracy. The earliest set was a minimal basis contraction (STO-3G) of a

(6s 3p) primitive set, followed by split-valence contractions, 4-31G and 6-31G, of (8s 4p) and (10s 4p) primitive sets, respectively. More recent Pople-type contracted sets attempt to increase the flexibility of at least the valence-shell description by expanding the basis, although sets like 6-311G have been criticized as implying a triple zeta description of the valence shell that is not realized in practice [46]. An important feature of all these sets is that the valence-shell s and p primitives have the same exponents. This substantially improves the performance of some integral programs, at the cost of imposing a constraint on the basis.

Neither of these basis set families is satisfactory for accurate calculations without the addition of polarization functions. Various ad hoc rules have been developed over the years for polarization exponents. Since SCF polarization is less sensitive to exponent choice than correlation, it is reasonable to let the latter determine the exponents. By fitting the results of correlated calculations on closed-shell hydrides, Ahlrichs and Taylor [47] arrived at the following formulas for d exponents (α_d)

$$\alpha_d = 0.02Z^2 \quad \mathrm{B-Ne} \tag{4.1}$$

and

$$\alpha_d = 0.077Z - 0.69 \quad \mathrm{Al-Ar}, \tag{4.2}$$

together with a suggested α_p of 1.0 for hydrogen. Investigations of higher angular momentum functions suggest multiplying the d exponents by 1.2 to approximate a correlation-optimized f exponent (and similarly by 1.2^2 for the g exponent, etc.). For hydrogen an exponent of 1.0 seems appropriate for all l values. Of course, when adding an f function one normally splits the d function into at least two. For sets of the type TZ2p and TZ2pf, that is, [5s 3p 2d] or [5s 3p 2d 1f], a ratio of 3 between the two exponents after the split is adequate.

As implicitly noted at the end of the previous paragraph, Huzinaga-Dunning and similar basis sets are commonly represented by a notation like [5s 3p 2d 1f], although acronyms like TZ2pf are also used. The addition of polarization functions to Pople-type sets is operationally the same as discussed above, but gives rise to an amazingly complicated notation of asterisks and suffixes like (D) and (3d). It is sometimes impossible (at least in the author's experience) to reconstruct some of these basis sets without actually having the GAUSSIAN*xx* program on hand.

One vexed question concerning polarization sets is the number of functions in a given shell, as discussed elsewhere. The early quantum chemistry codes employed Cartesian Gaussian functions, so that a "d" set actually comprises the five spherical harmonic d functions and a 3s function generally termed a *contaminant*. The reason for this emotive terminology is that with multiple polarization sets the contaminants, which are not orthogonal to the original s functions (or p functions for the $4p$ contaminants for a Cartesian f set, and so on), can create severe linear dependence problems. These may be geometry-dependent, and hence can appear partway through a series of calculations. There are repeated suggestions to reoptimize sp basis sets with the contaminants present, but this is not very practical. Where the possibility exists to use spherical harmonics and to eliminate the contaminants, it should be exploited. If the loss of the contaminants changes the results substantially, the original sp set used would appear to be deficient; such deficiencies should then be investigated and corrected.

4.3 Recent Developments in One-particle Basis Sets

There has been increased effort in the last five years to design better one-particle basis sets. One motivation for these efforts was described in Sec. 3.3: full CI benchmark calculations have clearly established that the main weakness even in our largest calculations is the basis set. A particular cause for concern is the use of atomic SCF calculations to optimize orbital exponents and define contraction coefficients. If the ultimate goal is to perform correlated calculations, it would seem preferable to include correlation in the construction of the basis set. Another cause for concern is the haphazard approach often taken to improving the basis set used in a calculation. Going from a Huzinaga-Dunning $[4s\ 2p\ 1d]$ to a $[5s\ 3p\ 2d\ 1f]$ based on a larger primitive sp set and, commonly, on different ad hoc rules for choosing the polarization exponents, is generally (and not unreasonably) regarded as substantially improving the one-particle basis. Yet it is extremely difficult to quantify just what this improvement is, or whether any trend (or even extrapolation) can be based on these results. The smaller set is not a subset of the larger set, so just how they are related is not clear.

One approach to designing basis sets for correlated calculations is to use large (ideally, saturated) primitive Gaussian sets and then to contract them using correlated atomic calculations. This is the *atomic natural orbital* (ANO) approach [48, 49] — the contracted basis functions to be used in molecular calculations are the natural orbitals obtained from, say, a CISD calculation on the atom using the primitive Gaussian basis. (The choice of correlation treatment is not restricted: for an atom with substantial nondynamical correlation effects MRCI would be preferable, while for an atom with many electrons a size-extensive treatment would be advisable.) Note that in this procedure the polarization functions — that is, higher angular momentum functions — are included at the outset. As discussed elsewhere, the natural orbital occupation numbers are a measure of the importance of each natural orbital, and it is convenient to use the occupation numbers as a truncation criterion. In this way, retaining ANOs with occupation numbers larger than 10^{-4} leads to sets of the size [$5s\,4p\,3d\,2f\,1g$] for first-row atoms, while a criterion of 10^{-3} leads to [$4s\,3p\,2d\,1f$] sets. It should be emphasized that although these sets have the delightful aesthetics of an exercise in numerology, the sizes are obtained directly from the structure of the NO occupation numbers!

In addition to treating correlation explicitly from the outset, ANO basis sets also answer the second criticism levelled above. Provided the same primitive set is used, a smaller ANO set (that is, one obtained from a larger occupation number threshold) is a genuine subset of a larger set. A sequence of ANO sets may therefore be expected to converge towards the result that would be obtained without contraction. As the primitive basis approaches completion, we should approach the complete one-particle space result with our sequence of ANO sets. A number of studies of convergence of the *contraction error* — the difference between the uncontracted basis result and a contracted basis result — in ANO basis sets have appeared [48, 49, 50, 51], as well as many application calculations [38] and some interesting variations in design methodology [52].

A second recent development in basis set design is Dunning's work on optimization at the correlated level in atoms [53]. This work resembles the traditional approach, but instead of being carried out at the SCF level to determine exponents and contraction coefficients, it is carried out using some correlation treatment. This approach has some features in common with the ANO contractions described above, and was to some extent catalyzed by the success of ANOs. Optimized primitive Gaussians are added to a "core" basis obtained from SCF calculations in order to describe the correlation effects, and the resulting combined set is used for molecular calculations. These "correlation-consistent" (*cc*) basis sets are generally similar in size to ANO sets, but since fewer primitive Gaussians are used the integral evaluation is faster.

Optimization of the basis set for atoms inevitably introduces some bias towards the atomic description, although this effect is seldom noticeable in basis sets of the traditional types described in Sec. 4.2. Bias towards the atoms can appear, however, if very small ANO sets (or, presumably, *cc* sets) are used. An ANO set contracted to split-valence plus polarization, [$3s\ 2p\ 1d$], is probably too contracted to provide a good description of molecular binding. In particular, the d NO may be quite different in shape from what is required to describe polarization and correlation in the molecule. At least two d orbitals should be included to properly allow for these effects. Hence ANO or *cc* basis sets smaller than, say, [$4s\ 3p\ 2d$] should probably not be used for molecular calculations.

Central to both ANO and *cc* basis sets is the method of *general contrac-*

tions [54], in which a given primitive can contribute to any contracted functions. (Even though the *cc* polarization sets are uncontracted, general contraction is recommended for the *sp* sets.) Such contraction can be simulated by a segmented contraction program, but only at considerable (often impractical) expense. General contraction is still something of a rarity in quantum chemistry programs, but it has many advantages in addition to the use of ANO or *cc* contractions, as discussed elsewhere.

4.4 Molecular Properties

The larger basis sets described in Sec. 4.2 or the more modern sets of Sec. 4.3 should yield good energies, and (assuming the N-particle treatment is appropriate) can therefore be expected to yield good values for spectroscopic constants like bond distances, vibrational frequencies, etc. However, other properties such as dipole moments, polarizabilities and so on will not necessarily be described well. In order to assess the basis set requirements for a given property, it is desirable to think of the form of the one-electron operator involved. For example, the quadrupole moment is an operator of the form $\mathbf{r}\mathbf{r}^T$. Thus this operator has a "radial" behaviour like r^2, which introduces a weighting or sampling bias towards the outer regions of the charge density into expectation values like

$$\left\langle \Psi | \mathbf{r}\mathbf{r}^T | \Psi \right\rangle . \tag{4.3}$$

Thus we can expect the more diffuse basis functions to affect the quadrupole moment. Further, since the quadrupole moment operator behaves like a tensor operator with $l = 2$, we can expect angular momentum functions with higher l values than the occupied orbitals to be important. Together, these observations suggest that diffuse higher angular momentum functions will be important to obtaining good quadrupole moments.

Such analyses can be carried out for any operator (see Sec. 5.3 for a detailed discussion of polarizabilities, for example), but some general conclusions can be stated here. First, the diffuse higher angular momentum functions that may be needed for multipole moments have exponents that are considerably smaller (by a factor of three, say) than the typical single polarization function exponents discussed in Sec. 4.2. Hence the latter functions will not provide the polarization needed for properties. Even splitting the single functions into two will usually not accomplish this, although for dipole moments the results may be acceptable. More important for the higher multipole moments may be adding diffuse functions of the same symmetry as the occupied orbitals to the basis. These properties depend critically on the charge distribution at much larger distances than affects the energy, so improving the description of the wave function in this region is crucial. The easiest approach is to add diffuse functions with exponents obtained as a *even-tempered* sequence, that is, with exponents of the form $\alpha\beta^k$. The value of α here can be taken as the outermost exponent of the original primitive set, with β chosen as 1/3 for smaller primitive sets like (9*s* 5*p*) and as 2/5 for larger sets like (13*s* 8*p*).

Second, careful consideration must sometimes be given not only to the form of the operator but also the way different wave function terms contribute to the property value. For example, the electric field gradient (EFG) at a given nucleus A has the form $\mathbf{r}_A\mathbf{r}_A^T/r_A^5$, suggesting a strong weighting (r^{-3}) towards the innermost regions of the charge distribution. However, as pointed out by Moccia and Zandomeneghi [55], the charge density in the core regions is close to spherical, and so the contribution to the EFG averages essentially to zero. Despite the form of the operator, therefore, the main contribution to the EFG comes from the inner valence region. Hence very high exponent high angular momentum functions (the operator again has an $l = 2$ dependence) functions are *not* required here since core polarization does not contribute substantially. However, higher angular momentum functions are required to describe the valence contribution. Fortunately, in many cases the functions added to describe angular correlation can also perform this task.

So far we have not discussed the one-particle space requirements for describing excited states. When these are of valence character, the basis set requirements are similar to those of ground states. However, if Rydberg character is present in the excited state, diffuse functions must be added to describe the spatially extended Rydberg orbitals. Note that if the character of an excited state or states is not known, it is imperative to provide enough flexibility in the basis to describe any Rydberg character. If the state is in fact a valence state, adding diffuse orbitals will not cause any problems (other than perhaps the size of the resulting calculation, or linear dependencies if the basis is already very large). On the other hand, omitting the diffuse functions will result in excluding any possibility of Rydberg states, and may give a completely false picture. This is an especially important point to keep in mind when studying geometry variation of excited states, as Rydberg/valence mixing is often strongly geometry dependent. If the basis set is not flexible enough, there will be a gross variation in reliability from one part of the excited state potential energy surface to another.

Rydberg character leads naturally into another reason for adding diffuse functions to the basis: the description of negative ion character in the wave function. The charge density in negative ions is usually considerably more extended, spatially, than for neutral species. Hence the accurate description of electron affinities and negative ions requires augmentation of the basis with diffuse functions. If the molecular wave function is strongly ionic at some geometries (or for some excited states) the same augmentation is desirable for the negative ion centre. It might be thought that some adjustment of the basis would also be required for positive ions or positive ion character, but it is generally true that for moderate to large basis sets the positive ion is described about as well as the neutral species, at least for lighter elements. For the transition metals this may not always be the case.

4.5 The Choice of One-particle Basis

The reader will have observed that this chapter has contained no illustrative comparisons of basis sets of the type that allowed us to classify N-particle space treatments as acceptable or unacceptable. This is partly because there is a much wider variety of one-particle basis sets in use than correlation treatments, so such a comparison would be a much larger and more complicated undertaking (see, for example, the H_2CO study by Davidson and Feller [56]). It is also partly due to the fact that the choice of one-particle basis set — the most crucial choice to be made in setting up a quantum chemical calculation — is the step that the average quantum chemist is most cavalier about and most prejudiced about. No comparison, other than an impractically exhaustive one covering a large range of properties, is likely to convince any such person of the error of their ways; I am not willing to attempt such a daunting task. Anyone who believes that one can obtain the right answer for the right reasons with a DZP basis set can (and probably should) skip the rest of this chapter.

In choosing a basis set the paramount but conflicting issues are accuracy and computational cost. These are obviously inversely related, and there is little more to be said about it. However, computational cost alone should not determine what basis set is used: this is especially true for black-box programs with a library of canned basis sets, where selecting the largest basis compatible with one's budget may produce sets unsuitable for the task in hand. Instead, the requirements of the N-particle space treatment and any properties to be computed should be considered. If only the SCF method is to be employed, larger basis sets can be used and any necessary diffuse functions for particular properties can be included. If the treatment includes dynamical correlation, a DZP set is the minimum acceptable starting point, and if property values are to be calculated it will be necessary to extend such a set. In this section I shall review a few of my own favourites among traditional basis sets, and then make some general statements about the newer basis sets described in Sec. 4.3.

For a DZ level basis for the first row, the $(7s\ 3p)$ sets of Roos and Siegbahn [44] or van Duijneveldt [45] can be contracted to $[4s\ 2p]$. A DZP basis can be constructed by adding polarization exponents as described in Sec. 4.2, but it is probably preferable to use the Huzinaga-Dunning set [42]: $(9s\ 5p)$ contracted to $[4s\ 2p]$

and augmented with a polarization function. For hydrogen a [2*s*] contraction of a (3*s*) or (5*s*) primitive set can be combined with these. For more accurate studies the (11*s* 7*p*) sets of Huzinaga [43] or van Duijneveldt [45], or the latter's (13*s* 8*p*) set can be used, contracted to [6*s* 4*p*] for the smaller primitive set and to [8*s* 5*p*] for the larger. Such large primitive sets are quite difficult to contract without significant loss of energy when a segmented contraction is used. Multiple polarization functions such as (2*d* 1*f*) are appropriate for use with these contracted sets. I have always been fond of a [5*s* 3*p*] contraction of the (9*s* 5*p*) primitive set as an economical halfway house between DZP and larger sets, but most of my colleagues find no favour with this, preferring Huzinaga's (10*s* 6*p*) primitive set [43] if a [5*s* 3*p*] contraction is to be used.

These sets are for the first-row atoms. For the second row the choices are rather limited, especially when economy is a motivation. The (10*s* 6*p*) primitive sets of Roos and Siegbahn [44] can be contracted to [6*s* 4*p*], that is, double zeta, or the larger (12*s* 9*p*) primitive sets of Huzinaga [43] can be contracted to [6*s* 5*p*], following McLean and Chandler [57]. These larger basis sets are essentially valence triple zeta. Note that in order to obtain satisfactory results, even for the atom, with the contracted basis, all of these schemes require duplication of one or more primitives. We are thus moving away from a strictly segmented contraction here. Polarization functions can be added according again to the rules given in Sec. 4.2. For heavier elements on the right-hand side of the Periodic Table the choice of basis sets is even more limited. I have used Dunning's sets for the third and fourth rows [58], but these are very difficult to contract without substantial loss in energy and accuracy, unless a general contraction scheme (or substantial duplication of primitives) is used.

For the first transition-metal row, a number of primitive sets have been suggested, but the most commonly used are those of Wachters [59]. These fairly large sets, (14*s* 9*p* 5*d*), are not, in themselves directly suitable for use in molecules, as discussed extensively by Wachters and by others since [60]. One of the most important modifications that must be made is to augment the *d* primitives. These are optimized in most cases for atomic ground states of the form $d^n s^2$, and the *d* basis is not flexible (diffuse) enough to provide a reasonable description of the *d* orbital in states of the form $d^{n+1}s$ or d^{n+2}. Since these states can make a substantial contribution to the binding in transition-metal systems, at least one more *d* primitive must be added. One or more *p* primitives must also be added to describe the 4*p* orbital, which may also contribute to the bonding. After augmentation contracted basis sets of the form [8*s* 6*p* 4*d*] or [8*s* 6*p* 3*d*] result, although with general contraction a [6*s* 5*p* 3*d*] set at least as accurate could be used. A single *f* polarization primitive is not as effective for the transition metals as for the lighter elements, but a two- or three-component Gaussian fit to a Slater-type *f* function, or some other contracted *f* function works very well.

For heavier elements primitive basis sets are available, but their contraction is a real art, at least if segmented contractions are used. The use of general contractions is far preferable and is much easier to carry out.

As can be seen by glancing over this display of bias, a number of basis sets have been omitted, with a particularly glaring example being the Pople-type basis sets discussed in Sec. 4.2. Since I have never used an integral program that could exploit the economies of using the same exponents for different angular momentum functions, I have never been concerned with putative computation time benefits from this source. On the other hand, even the largest basis sets of this type appear to be inferior to the larger sets mentioned here, and, more importantly, the Pople-type sets are much less satisfactory for the second row than for the first, a deficiency that can be expected to grow with atomic number, and one that reflects the increasing importance of the constraint on exponents.

For more accurate calculations, I use ANO contractions, or sometimes the *cc* sets of Dunning [53]. For the first row the ANOs are based on the ($13s$ $8p$) sets of van Duijneveldt [45], or occasionally the ($18s$ $13p$) sets of Partridge [61]. These are augmented with polarization sets of the form ($6d$ $4f$ $2g$), although when basis set saturation is in doubt we have sometimes used ($6d$ $5f$ $4g$). The polarization exponents are even-tempered sequences with the Ahlrichs and Taylor recommended exponents [47] as geometric mean and a ratio of 2/5 between successive exponents. The ($6d$ $5f$ $4g$) polarization set has sometimes been augmented further with a ($3h$ $2i$) set, but this is too expensive for other than benchmark-level calculations. The contracted sets are of the form [$4s$ $3p$ $2d$ $1f$] or [$5s$ $4p$ $3d$ $2f$ $1g$], as discussed in Sec. 4.3. For the second row, the ($18s$ $13p$) sets of Partridge [62], augmented as for the first row, give excellent results: using general contractions the core orbitals can be contracted to minimal AOs, so that the final contracted sets are of the form [$5s$ $4p$ $2d$ $1f$], etc. Calculations on heavier elements are therefore not much more expensive than with lighter elements, at least after the integral evaluation stage. We have used the new optimized primitive sets of Partridge [61] or Partridge and Fægri [63] in conjunction with ANOs for elements up to I or Xe in correlated calculations (up to Pb in relativistic Hartree-Fock studies) and have found them to perform very well. Contraction of all but the valence orbitals to a minimal AO description is a very convenient consequence of the use of general contractions.

The correlation-consistent basis sets of Dunning are only beginning to appear, but they show great promise as a way to achieve accuracy similar to that of ANOs, at less computational expense in the integral evaluation (they are generally the same size, in terms of contracted functions, as the ANO sets). It is interesting to contrast the optimistic conclusions of Dunning [53] with the much less favourable conclusions reached by Ahlrichs and co-workers [64] about correlation-optimized polarization exponents in molecular investigations. I should also mention the ANO sets generated by the Lund group [52]: these are based on somewhat smaller primitive polarization spaces than those used by Almlöf and Taylor [48], and an ingenious system of averaging over the neutral atom, various cationic and anionic states, and states perturbed by an electric field is used in obtaining the ANOs. These newer ANO sets give better results for properties than the original ones, although this may involve using larger contracted sets. Calculation of properties using ANO basis sets is a complicated issue, as we now discuss.

Since the optimum orbitals for describing correlation are similar in size to the occupied orbitals, but feature additional radial and angular nodes, it is not surprising that adding more atomic natural orbitals to a basis does little to improve the flexibility of that basis in the outermost regions of the charge density. Properties that depend on the valence region or inner-valence region, like EFGs, are described very well by ANO sets. Properties, such as multipole moments, that are sensitive to the charge density at longer distances, are not so well described. Our approach [50, 51, 65] has been to uncontract the outermost primitive function from the ANO contractions to achieve this flexibility, thus increasing the size of the contracted set. The Lund group's averaged ANOs [52] are more flexible in the outer regions as a result of the averaging procedure, but since they may be somewhat larger to achieve the same accuracy in the energy the final contracted set is the same size. In principle, Dunning's *cc* sets [53], which have the outermost primitives uncontracted anyway, should be superior to either, but for sets of the same final size we have not always found this to be true. We note that this problem of describing some properties is unique to ANOs among current (large) basis sets, as in order to achieve any sort of reasonable contraction error most large segmented sets involve leaving virtually all of the more diffuse exponents uncontracted anyway.

We conclude this rather lengthy survey with a simple example: the application of different basis sets to the singlet/triplet separation in the CH_2 radical [66], which was used as a comparison of N-particle space treatments in Sec. 3.2.

It is interesting to note here that at the time of these calculations the best value for the separation was virtually the same as the $[4s\ 3p\ 2d\ 1f/3s\ 2p\ 1d]$ result, but was obtained with a basis roughly twice the size. In order to emphasize the importance of computing all terms, I have included theoretical estimates of the zero-point vibration and core-valence correlation contribution to the splitting. It can be seen that accounting for only zero-point contributions would give a theoretical T_0 value of 8.97 kcal/mol, in almost exact agreement with experiment. However, core-valence

Table 4.1: $CH_2\,{}^1A_1 - {}^3B_1$ Separation: ANO Basis Results (kcal/mol)

Basis	Separation
[3*s* 2*p* 1*d*/2*s* 1*p*]	11.33
[4*s* 3*p* 2*d* 1*f*/3*s* 2*p* 1*d*]	9.66
[5*s* 4*p* 3*d* 2*f* 1*g*/4*s* 3*p* 2*d*]	9.13
Expt (T_0)	9.02 (±0.01)
Zero-point	−0.16
Core-valence	0.35

correlation increases the theoretical value to 9.32 kcal/mol, somewhat larger th the experimental value. Much of the remaining discrepancy is probably still due one-particle space incompleteness. Much more detailed analyses of such problems a given in the next chapter.

Chapter 5

Case Studies

5.1 Dissociation Energy of N_2

A principal reason for studying the binding energy of N_2 is that the correlation contribution is very large, about half of the total. N_2 thus provides an excellent test of methods for obtaining accurate binding energies. We consider calibration of the correlation treatment first. Unfortunately, N_2 is too large to allow reliable full CI benchmarks with all ten valence electrons correlated at present, but full CI results in a DZP basis obtained by Bauschlicher and Langhoff [67] with six electrons correlated are given in Table 5.1. The results of Table 5.1 show that none of the economical

Table 5.1: Spectroscopic Constants of N_2: Full CI Comparisons

Method	r_e (Å)	ω_e (cm^{-1})	D_e (kcal/mol)
CISD	1.112	2436	191.4
CISD+Q	1.119	2373	198.6
CISDT	1.115	2411	195.1
CISDTQ	1.122	2343	201.4
CISDQ	1.120	2361	198.0
CPF	1.118	2382	196.6
MCPF	1.119	2370	197.3
MRCI[a]	1.123	2334	201.6
MRCI+Q	1.123	2333	202.1
Full CI	1.123	2333	201.8

[a] six-electron six active orbital CASSCF reference.

single-reference methods investigated reproduces the full CI results especially well; CISDTQ gives good agreement with the full CI but this method is impractically expensive in large basis sets. Neither CPF nor MCPF work as well as they did for systems like CH_2, suggesting that these methods will perform better for single bonds than for multiple bonds. Once again, the multireference methods are in excellent agreement with the full CI, although D_e is somewhat too large at the MRCI+Q level. Let us now look in detail at basis set effects on D_e at the multireference level.

Table 5.2 shows the N_2 dissociation energy obtained with a sequence of ANO basis sets at the MRCI level [68]. We should note that all ten valence electrons are correlated in these calculations, which are based on a six-electron six active orbital CASSCF reference.

Table 5.2: Dissociation Energy of N_2: Basis Set Comparisons

Basis set	D_e (kcal/mol)
(13*s* 8*p* 6*d* 4*f* 2*g* 1*h*) set	
[4*s* 3*p* 2*d* 1*f*]	216.0
[5*s* 4*p* 3*d* 2*f*]	219.6
[6*s* 5*p* 4*d* 3*f*]	220.8
[6*s* 5*p* 4*d* 3*f*] + diffuse *sp*	221.0
[5*s* 4*p* 3*d* 2*f* 1*g*]	222.1
[6*s* 5*p* 4*d* 3*f* 2*g*]	223.6
[6*s* 5*p* 4*d* 3*f* 2*g* 1*h*]	224.3
(18*s* 13*p* 6*d* 5*f* 4*g* 3*h* 2*i*) set	
[5*s* 4*p* 3*d* 2*f* 1*g*]	222.3
[5*s* 4*p* 3*d* 2*f* 1*g* 1*h*]	223.4
[5*s* 4*p* 3*d* 2*f* 1*g* 1*h* 1*i*]	224.0
[6*s* 5*p* 4*d* 3*f* 2*g*]	223.7
[6*s* 5*p* 4*d* 3*f* 2*g* 1*h*]	224.5
[6*s* 5*p* 4*d* 3*f* 2*g* 1*h* 1*i*]	225.0

Two different sets of primitive Gaussians were used to construct the ANO basis sets here, so estimates of the effect of expanding both the radial and angular functions can be made. We can see by comparing the [5*s* 4*p* 3*d* 2*f* 1*g*] results in the two primitive sets that the smaller primitive set is already close to radial saturation, at last as far as the D_e value is concerned. In fact, this also appears to be true for the [6*s* 5*p* 4*d* 3*f* 2*g* 1*h*] sets, so it seems that only one primitive *h* function is needed. The effect of adding the *h* ANO is to increase D_e by almost 1 kcal/mol, so it appears that *h* functions are required to achieve "chemical accuracy" for the binding energy in N_2. Adding an *i* ANO increases D_e by only 0.5 kcal/mol, and this number is probably something of an overestimate as we do not add any lower angular momentum ANOs at the same time. Finally, the original primitive s and p sets seem to cover an adequate range of exponents, as adding a diffuse sp set hardly affects D_e. We should note that this question must be explicitly investigated: a comparison between contracted sets based on ($13s$ $8p$) and ($18s$ $13p$) primitive sets is not adequate in itself to answer the question as the outermost exponents of the two sets are not substantially different.

Before using our largest basis sets to obtain an estimate of D_e, there are several other issues we should consider [68]. To begin with, we have only a six-electron full CI calibration of our N-particle space treatment. In that calibration the MRCI+Q result overshot the full CI D_e value. However, since the lack of size-extensivity in MRCI would be expected to become apparent with eight or more electrons correlated, it seems very unlikely that MRCI+Q will overshoot when ten electrons are correlated. In support of this contention, we note that the MRACPF and MRCI+Q D_e values are almost identical when ten electrons are correlated, at least when the multireference calculation is based on a six-electron six active orbital CASSCF treatment [68]. Curiously enough, if the CASSCF treatment is expanded to a full valence space (ten electrons in eight active orbitals) and this space is used in multireference calculations, the Davidson correction (and MRACPF) *decreases* D_e relative to the uncorrected MRCI value. This seems counterintuitive: we would expect a correction for size-extensivity to affect the molecule more than the separated atoms. The explanation appears to lie in the treatment of the separated atoms — with the larger CASSCF space as a reference the MRCI calculation includes the dominant atomic excitations not just on one atom but on both atoms simultaneously. Hence the MRCI treatment at dissociation is close to size-consistent already, and inclusion of the +Q correction results in an artificial lowering of the separated atom energy through double counting of some correlation effects [68].

In addition, none of these calculations involve correlation of more than ten electrons, so no correlation effects from the core electrons are included at all. Explicit inclusion of the core electrons at the CPF level was found to increase D_e by about 0.7 kcal/mol in calculations by Ahlrichs and co-workers [69], while in calculations by Almlöf and co-workers [68] the same increase was obtained by a completely different technique (inclusion of only core-valence correlation effects, as described in Sec. 6.2). Hence it appeared safe to assume that core correlation would increase D_e by less than 1 kcal/mol. However, recent calculations by Werner and Knowles [70] give a larger effect of about 1.5 kcal/mol, so this question is not yet settled.

In Table 5.3 we list MRCI+Q spectroscopic constants for N_2 obtained with one of our largest basis sets. Of the remaining errors compared to experiment, we can estimate that about 0.001 Å of the error in the bond length arises from relativistic effects [71]. These would also increase the frequency slightly. Explicit calculations

Table 5.3: Spectroscopic Constants of N_2

Method	r_e (Å)	ω_e (cm^{-1})	D_e (kcal/mol)
Computed[a]	1.100	2353	225.2
Experiment	1.098	2359	228.4

[a] [$6s\ 5p\ 4d\ 3f\ 2g\ 1h$] basis MRCI+Q values.

using first-order perturbation theory show that relativistic effects on D_e are negligible. The error in D_e at first sight appears to be disappointingly large. However, this is not the largest basis set used in comparisons above, and we know from those results that an i ANO, for example, would increase D_e by perhaps 0.4 kcal/mol. We note also that core correlation will add another 0.7 kcal/mol (at least), although when adding these effects to the D_e value of Table 5.3 we should also apply a correction of about −0.3 kcal/mol for basis set superposition error [68]. Including all these terms gives a D_e value of 226.0 kcal/mol. In addition, we can estimate from the results of Table 5.2 that additional *spdfghi* ANOs might increase D_e by 0.5 kcal/mol, while higher angular functions *in toto* might add another 0.3 kcal/mol. Our best estimate is therefore 226.8 kcal/mol: we can estimate the uncertainty in this result due to the various corrections we have applied to be about 1 kcal/mol, although it is very unlikely that our estimated value could be too large. Since the experimental result, which is certainly well established and very accurate, is 228.4 kcal/mol, our estimate is in error by 1.8 kcal/mol, which is well outside the uncertainty from our corrections. The remaining discrepancy could be due in large part to an underestimate of core correlation effects, in view of the results of Werner and Knowles [70]. These authors obtain a final computed D_e value within 1 kcal/mol of experiment, although we should note that they do not correct for basis set superposition error, arguing that it is neither necessary nor desirable to do so.

Finally, we should mention some recent coupled-cluster results on N_2. Using large ANO basis sets (up to [$6s\ 5p\ 4d\ 3f\ 2g$]) Lee and Rice [72] have obtained excellent ground-state spectroscopic constants using the CCSD(T) method. The computed bond length is similar to the MRCI+Q result of Table 5.3, while the harmonic frequency is even closer to experiment, within 2 cm^{-1}. We can speculate that higher angular functions and small contributions like relativistic effects will shorten the bond and increase the frequency, so that the complete basis set CCSD(T) bond length would be a whisker too short and the frequency slightly too large, but probably not by more than the small difference between experiment and CCSD(T) in Lee and Rice's computed values. Thus the CCSD(T) method yields spectroscopic constants in remarkably good agreement with multireference values and hence (in a large basis) in good agreement with experiment. This impressive performance of the CCSD(T) method is also seen for the binding energy: Scuseria has used his new open-shell version of the method to compute this quantity, obtaining almost perfect agreement with MRCI+Q results in the same basis [73]. Although a single-reference-based method like CCSD(T) could not be applied to the entire potential curve of N_2 (at least not using an RHF reference function), the performance of the method on one of the most difficult first-row molecule binding energies is extremely encouraging. This is especially so in view of the mediocre performance of methods like CPF in predicting the D_e value of N_2 [69].

5.2 Properties of Small Be Clusters

It would be elegant, but disingenuous, to present this study in a logical order: the true sequence of events was much less logical and it is more honest to present that here. The original motivation was to examine the trimer and tetramer of beryllium (and magnesium) with the CCSD method, and to compare the results with those from MRCI. The work was started in 1988, well before we had the capability of including any effects of connected triple excitations in the coupled-cluster treatment.

The Be atom has one of the largest near-degeneracy correlation effects of any atom in the Periodic Table, because of the very important $2s^2 \rightarrow 2p^2$ double excitation. As a consequence, nondynamical correlation is a dominant feature in the electronic structure of small molecules containing Be, and thus it does not seem likely that single-reference-based correlation treatments would be very successful for systems like Be_3 and Be_4. However, while it seems reasonable to proceed immediately to multireference treatments, it is worth considering just what that entails. A full valence CASSCF treatment of a cluster Be_n would involve $2n$ electrons in $4n$ active orbitals (an s and p shell on each Be). For $n = 3$ six electrons in twelve orbitals already generates about 4000 CSFs for the totally symmetric singlet, while for Be_4 the CASSCF wave function comprises over 220 000 CSFs. Such an MCSCF calculation is feasible (just), but the question arises as to whether it will be possible to follow such a CASSCF calculation with any sort of MRCI treatment. It is certainly clear that MRCI will be an extremely expensive computational method to apply to small Be clusters. We may note here that previous calculations had established that Be_3 has an equilateral triangular structure while Be_4 is tetrahedral — both molecules have singlet ground states.

Table 5.4 shows results for the bond length and binding energy of Be_3 obtained with a variety of computational methods [74]. Two basis sets were used: ANO sets

Table 5.4: Bond Lengths and Dissociation Energies for Be_3.

Method	Basis	r_e (a_0)	D_e (kcal/mol)
SCF	[4*s* 2*p* 1*d*]	(4.472)[a]	−3.2
CISD+Q	[4*s* 2*p* 1*d*]	(4.472)	8.9
CCSD	[4*s* 2*p* 1*d*]	4.472	4.2
CPF	[4*s* 2*p* 1*d*]	4.488	8.6
CASSCF	[4*s* 2*p* 1*d*]	(4.372)	−2.3
MRCI(0.025)	[4*s* 2*p* 1*d*]	4.373	14.2
MRCI(0.01)	[4*s* 2*p* 1*d*]	4.372	14.1
MRCI(CAS)	[4*s* 2*p* 1*d*]	4.372	14.1
SCF	[5*s* 3*p* 2*d* 1*f*]	(4.240)	−1.4
CCSD	[5*s* 3*p* 2*d* 1*f*]	4.240	11.3
CPF	[5*s* 3*p* 2*d* 1*f*]	4.202	14.9
CASSCF	[5*s* 3*p* 2*d* 1*f*]	(4.199)	0.2
Valence(0.025)	[5*s* 3*p* 2*d* 1*f*]	(4.199)	0.5
MRCI(0.025)	[5*s* 3*p* 2*d* 1*f*]	4.199	22.4

[a] Bond lengths in parentheses were not optimized.

of the form [4*s* 2*p* 1*d*] and [5*s* 3*p* 2*d* 1*f*]. These are contractions of a (12*s* 7*p* 4*d* 3*f*) primitive set using NOs averaged from CISD calculations on the ground $1s^2 2s^2$ (^1S) and excited $1s^2 2s 2p$ (^1P) states of Be. This approach takes account of the near-degeneracy but also ensures that the valence *p* orbital has a radial extent that allows it to participate properly in binding. ANOs obtained from the $1s^2 2s^2$ (^1S) state have *p* orbitals that are too tight for effective *sp* hybridization.

A variety of N-particle space treatments were used, including single-reference CISD+Q and CPF, as well as several levels of multireference CI. These are distinguished by different selection thresholds for inclusion of reference configurations, as given in parentheses. Thus MRCI(0.025) indicates that the reference space comprised those occupations that gave rise to at least one configuration (that is, spin-coupling) with a coefficient greater than 0.025 in absolute magnitude in the CASSCF wave function. MRCI(CAS) denotes a calculation in which all CASSCF occupations appear in the reference list. The selection procedure was performed once near r_e and once at long distance and the resulting reference lists were merged. The effect of the multireference Davidson correction was very small for the properties considered here, and the results are not reported.

Table 5.4 demonstrates that Be_3 is not bound at all at the SCF level. However, the inclusion of only nondynamical correlation in the CASSCF calculation does not give any binding either. On the other hand, methods that are primarily suitable for treating dynamical correlation, like CISD, CCSD, or CPF, are in poor agreement with MRCI for the binding energy, although the bond lengths are fair. It therefore appears that while the binding in Be_3 derives entirely from dynamical correlation, this dynamical correlation is only described well if the calculation also treats the nondynamical correlation accurately. This is consistent with our rather pessimistic reasoning above. We may note that there is a very substantial basis set effect on the computed binding energy at all levels of treatment.

It is essential to use selection of reference configurations in the MRCI procedure with the larger basis set, as the MRCI(CAS) calculation would involve several million CSFs. The result labelled Valence(0.025) is obtained with only the "valence" configurations — that is, it is the MRCI(0.025) wave function with all excitations to the virtual orbitals omitted. This wave function will be considerably smaller than the CASSCF wave function, but if we are to describe the nondynamical correlation acceptably it should produce similar results. This is one reason for not using a selection threshold of (0.05): the Valence(0.05) binding energy was found to be quite different from the CASSCF value [74].

One reasonable question is why the CCSD method performs so poorly, and why it appears inferior to less complete treatments like CISD+Q or CPF. This is actually a rather subtle consequence of the exact size-extensivity of the CCSD method. The CCSD method is exact for a two-electron system, hence it is equivalent to a full CI for Be atom. Since the method is also size-extensive, it is exact for any number of noninteracting Be atoms. However, CISD+Q is not exact for the atom and is not exactly size-extensive, so it provides a poorer description of the separated atom limit than CCSD. It thus yields a larger binding energy because errors are present at both equilibrium and dissociated geometries, whereas in the CCSD case they are present only at equilibrium. The CPF method is constructed to be exact for certain noninteracting aggregates of two-electron systems and should behave more like CCSD, although it is not obvious that this happens here.

We may also note here that the MCPF method initially gave a binding energy of only about 8 kcal/mol in the larger basis. This result is in error because MCPF is not invariant to a unitary transformation on the occupied orbitals. Our calculations were performed originally in C_{2v} symmetry, and at long distances the two Be $2s$-derived orbitals that transform according to the a_1 irreducible representation can mix essentially arbitrarily. (At distances where the Be atoms interact with one another, D_{3h} symmetry orbitals are obtained, but at long distances there is no interaction to resolve the higher symmetry.) This results in an erroneous MCPF result for the dissociated limit. Since CCSD and CISD are invariant to a mixing of the occupied orbitals no such problem arises there, and CPF, while not exactly invariant, should be close to it. However, this arbitrariness at long distance also creates difficulties for the MRCI calculations. No problem arises for the MRCI(CAS) case, but when references are to be selected at two different geometries and the lists merged, it is essential that the MOs at the two geometries be related in some well-defined way. If the MOs at infinite separation are essentially arbitrary, the CSFs at infinity will not correspond in any one-to-one way to those near r_e, and the merged reference lists will not be consistent. One way to avoid this problem (and the difficulty with MCPF) is to impose D_{3h} symmetry and equivalence restrictions on the MOs [75]. This does not affect the SCF and CASSCF energies at r_e or at long distance, but the form of the MOs will be different. Since there is only one occupied orbital of any given D_{3h} symmetry species in the SCF wave function at long distance, there is no arbitrariness in the form of the MOs. The same situation arises for the CASSCF calculations.

We turn now to the binding energy results obtained for Be_4 [74], which are displayed in Table 5.5. The bond length was held fixed at 3.9 a_0, which is the CCSD optimimum value in a large segmented basis set. There is a qualitative difference

Table 5.5: Dissociation Energies for Be_4.

Method	Basis	D_e (kcal/mol)
SCF	[4*s* 2*p* 1*d*]	32.9
CCSD	[4*s* 2*p* 1*d*]	42.9
CASSCF	[4*s* 2*p* 1*d*]	34.9
Valence(0.05)	[4*s* 2*p* 1*d*]	34.8
MRCI(0.05)	[4*s* 2*p* 1*d*]	56.1
Valence(0.025)	[4*s* 2*p* 1*d*]	34.1
MRCI(0.025)	[4*s* 2*p* 1*d*]	56.0
SCF	[5*s* 3*p* 2*d* 1*f*]	40.0
CCSD	[5*s* 3*p* 2*d* 1*f*]	63.5
CASSCF	[5*s* 3*p* 2*d* 1*f*]	45.0
MRCI(0.05)	[5*s* 3*p* 2*d* 1*f*]	77.3

between Be_4 and Be_3 in that the former is substantially bound at the SCF level.

However, the correlation contribution to binding is still enormous. Like Be_3, the CASSCF method gives little more binding than SCF, showing again that nondynamical correlation makes little direct contribution to the binding, but comparison of the CCSD and MRCI results shows that it is essential to describe the nondynamical correlation properly to obtain a reliable estimate of the dynamical contribution to binding in small Be clusters. An MRCI reference selection threshold of 0.05 performs considerably better for Be_4 than it did for Be_3; since there are well over 200 000 CSFs in the CASSCF wave function it is impossible at present to perform an MRCI(CAS) calculation, and even an MRCI(0.01) calculation is not feasible. It should be noted that T_d symmetry and equivalence restrictions were imposed on the MOs in the SCF and CASSCF calculations, to avoid problems with the MOs at large separations.

Based on the results of Tables 5.4 and 5.5, we can make an estimate of what the true binding energies of these clusters are [74]. Our best directly computed results must be underestimates, given the substantial increases seen when the basis set is enlarged, although it is very probable that further extension of the basis set would not affect the results nearly as much (explicit comparisons with large segmented basis sets support this contention). For Be_3 Lee and co-workers estimated a D_e value of 24 kcal/mol — this value was asserted to be uncertain by about 2 kcal/mol, thus incorporating the computed value into the uncertainty range. However, is seems unlikely that the true value can be less than 24. For Be_4 a D_e value of 83 kcal/mol was suggested, uncertain by perhaps 3 kcal/mol, although again the estimate is likely to be low rather than high. As a final note on this phase of the Be cluster investigation, it is interesting to compare the binding energies to the bulk values. The binding energy per atom in bulk Be is 78 kcal/mol, while in Be_3 and Be_4 it is 8.0 and 20.8 kcal/mol, respectively. (Convergence is slow: a computed binding energy per atom for Be_5 gives 22.0 kcal/mol.) On the other hand, comparing binding energy per "bond", that is, per pairwise interaction, we obtain 8.0 kcal/mol for Be_3 and 13.8 for Be_4, the latter comparing favourably with the bulk value of 13. Hence convergence here appears to be better.

The conclusions at this point are that the single-reference-based correlation treatments applied to Be_3 and Be_4 are not in satisfactory agreement with MRCI, especially for the binding energy. This is not unexpected: the T_1 diagnostic for both clusters is about 0.035 (we recall from Sec. 2.8 that values larger than about 0.02 indicate substantial nondynamical correlation effects). For comparison, the diagnostic for Be atom is 0.021, so that the nondynamical effects are even more important in the clusters than in the atom. At this stage it would appear that the only acceptable approach is MRCI, and this is too expensive to use for larger clusters, or even for more calculations (say, for a force field) on the trimer and tetramer. However, after this phase of our investigations was complete Lee had extended the coupled-cluster program to calculate CCSD(T) energies. The results of applying the method to Be_3 and Be_4 were something of a surprise [16, 76], as can be seen in Table 5.6. Not only

Table 5.6: Bond Lengths and Dissociation Energies for Be_3 and Be_4.

	Method	Basis	r_e (a_0)	D_e (kcal/mol)
Be_3	CCSD	[5*s* 3*p* 2*d* 1*f*]	4.239	11.3
	CCSD(T)	[5*s* 3*p* 2*d* 1*f*]	4.217	20.4
	MRCI(0.025)	[5*s* 3*p* 2*d* 1*f*]	4.200	22.5
Be_4	CCSD	[4*s* 2*p* 1*d*]	4.041	44.2
	CCSD(T)	[4*s* 2*p* 1*d*]	4.060	58.5
	MRCI(0.025)	[4*s* 2*p* 1*d*]	4.054	59.2
	CCSD	[5*s* 3*p* 2*d* 1*f*]	3.900	63.5
	CCSD(T)	[5*s* 3*p* 2*d* 1*f*]	3.921	79.5

are the CCSD(T) bond lengths in better agreement with MRCI than were the CCSD values, the CCSD(T) binding energies are an enormous improvement over the CCSD values. For Be_4 the difference in D_e values between CCSD(T) and MRCI is actually less than 1 kcal/mol. It would appear from the results of Table 5.6 that the CCSD(T) method is close to a perfect solution for our Be clusters: it agrees exceptionally well with MRCI but is very much cheaper. Indeed, it would be possible to treat much larger clusters with the method, possibly as large as a dozen atoms.

While the field of quantum chemistry is substantially populated with pessimists, I have never considered myself one of them. Indeed, I have tended to the viewpoint that cheerful optimism, punctuated occasionally (or even frequently!) by disappointment when something fails to work, is considerably more fun than gloomy pessimism coupled with being right some (or even most!) of the time. Nevertheless, gift horses are so infrequent in our field of research that the performance of CCSD(T) here seemed too good to be true. We therefore embarked on a series of calibration calculations on Be_3 to try to understand whether CCSD(T) was getting the right answer for the right reason.

Since we would assume that for a six-electron system like Be_3 an MRCI calculation should reproduce the result of a full CI calculation, we initially reasoned as follows. If CCSD(T) is in excellent agreement with MRCI (and hence presumably with the exact N-particle space treatment) there are two possibilities. First, it may be that the only thing missing from the CCSD treatment was the connected triples contribution. That is, the fourth-order (and partial fifth-order) contribution of connected triple excitations is extremely important, but higher-order terms in the triples and all terms in connected quadruples and higher terms are not. This does not seem altogether likely, but it is one possible scenario. The other is that there is a cancellation between the higher-order effects of triples and the higher excitations, so that CCSD(T) is fortuitously good for these systems. One obvious way to test this is to perform full CCSDT calculations: a coupled-cluster treatment in which all excitations though connected triples are included explicitly. Thus the effect of triple excitations is included to all orders. If CCSD(T) and CCSDT agree closely, it can be concluded that the effect of higher-order terms in the triples is negligible, and that presumably the contribution of higher excitations must therefore be negligible too.

We performed a series of calculations comparing CCSD(T) and several other methods with both full CI and CCSDT treatments [77]. It is not necessary to list all the results here, we shall merely summarize the conclusions. Comparisons of the CCSD(T) and CCSDT results indicate that the (T) correction accounts for over 80% of the connected triples contribution. Since the CCSDT binding energy for Be_3 is about halfway between the CCSD(T) and MRCI (or full CI) values, it seems that about half the error in the CCSD(T) binding energy comes from the higher-order effects of connected triples and half from higher excitations. Both of the latter effects increase the binding energy, so there is no cancellation of errors in the CCSD(T) result. For this case, at least, the best calibrations and tests we can perform support the hypothesis that CCSD(T) provides results for the Be clusters that approach the quality of MRCI, but at considerably less computational cost.

Once the reliability of CCSD(T) had been established, we could proceed with confidence to use it to predict vibrational frequencies for Be_3 and Be_4. In order to obtain the best possible prediction to aid experimentalists, a full quartic force field was generated for each molecule [76], using finite differences of computed energies, and fundamental frequencies were obtained via second-order perturbation theory. In Table 5.7 we list the CCSD(T) fundamental frequncies and, for comparison, the CCSD, CCSD(T) and MRCI harmonic frequencies.

The CCSD(T) and MRCI harmonic frequencies are in excellent agreement with one another. We should note that the reference configurations for the MRCI are the same as those used for the binding energy, that is, they are selected at r_e and at long distance. Full D_{3h} or T_d symmetry was imposed on those CASSCF wave functions in order to avoid arbitrary rotations of the MOs at long distance. Thus for Be_4, for example, no geometries with symmetry lower than tetrahedral were used in selecting reference configurations, and it is possible that some of the small differences between CCSD(T) and MRCI may result from the omission of some reference configurations

Table 5.7: Vibrational Frequencies (cm^{-1}) for Be_3 and Be_4[a].

Be_3	Method	Basis	a'_1 mode	e' mode	
	CCSD	[5s 3p 2d 1f]	433	407	
	MRCI(0.025)	[5s 3p 2d 1f]	490	427	
	CCSD(T)	[5s 3p 2d 1f]	480	417	
	CCSD(T)[b]	[5s 3p 2d 1f]	459	400	
Be_4	Method	Basis	a_1 mode	e mode	t_2 mode
	CCSD	[4s 2p 1d]	597	445	534
	MRCI(0.025)	[4s 2p 1d]	602	451	529
	CCSD(T)	[4s 2p 1d]	602	436	527
	CCSD	[5s 3p 2d 1f]	667	480	581
	CCSD(T)	[5s 3p 2d 1f]	662	469	571
	CCSD(T)[b]	[5s 3p 2d 1f]	639	455	682

[a] Harmonic frequency unless otherwise indicated.
[b] Fundamental frequency.

that become more important at distorted geometries. The basis set effects on going from [4*s* 2*p* 1*d*] to [5*s* 3*p* 2*d* 1*f*] are large, but it is probable that residual basis set effects on the frequencies are small. The fundamental frequencies should certainly be accurate to within 20 cm^{-1}, and are very likely accurate to within 10.

The infrared intensity (within the double harmonic approximation) of the e' mode in Be_3 is very small and this mode is unlikely to be observable in a direct experiment, although an indirect observation via photodetachment from Be_3^- might be possible. The t_2 mode in Be_4 has a substantial intensity and may well be directly observable. The positive anharmonic correction for this mode appears to be a genuine effect: it is clear that frequency predictions for experimentalists based on harmonic frequencies would be completely useless for this case. More recently, we have applied the same computational methods to Mg_3, Mg_4, Ca_3, and Ca_4 [78, 79]. The same excellent agreement between CCSD(T) and MRCI structures, binding energies, and harmonic frequencies is seen for all these molecules. The magnesium clusters are very weakly bound and the frequencies are small. The calcium clusters are more strongly bound, but the mass effect leads again to small frequencies. It is unlikely that any of these species will be directly observable by infrared measurements.

While this beryllium cluster work could have been written up here in the logical sequence 1) calibration of CCSD(T), 2) application of CCSD(T) to the clusters, such a portrayal would be false to history. I have tried to present what was done in the sequence in which we did it, in order to give some sense of how many accurate investigations are conducted, and how the first approach or approaches taken may not be the best ones.

5.3 Polarizabilities and Hyperpolarizabilities of the Rare-gas Atoms

If a closed-shell atom is placed in an electric field of strength F, it undergoes a change in energy of

$$\Delta E = -\frac{1}{2}\alpha F^2 - \frac{1}{24}\gamma F^4 \ldots, \tag{5.1}$$

where α is the dipole polarizability, γ is the second hyperpolarizability, etc. The calculation of the polarizability and hyperpolarizability is a very demanding task, because the major part of the response to an applied field comes from adjustments in the outer part of the charge density. These effects are sensitive to electron correlation, require diffuse functions in the one-particle basis for an adequate description, and require diffuse higher angular momentum functions as well, as we can reason as follows. We can express the polarizability as a second-order perturbation energy sum over states

$$\alpha = \sum_{K \neq 0} \langle 0|\mu|K\rangle\langle K|\mu|0\rangle/(E_0 - E_K), \tag{5.2}$$

where $|0\rangle$ is the ground state with energy E_0 and the sum runs over all excited states. μ is the dipole operator. If $|0\rangle$ involves occupied orbitals with angular momentum l, the states that will couple through μ will have angular momentum $l+1$. Since it is mainly the outer part of the charge density that is affected by the field, we will need diffuse basis functions with angular momentum $l+1$. Similar reasoning for the hyperpolarizability, a fourth-order term, shows that diffuse functions with angular momentum $l+2$ will be required. Note that all this reasoning applies to the SCF case as well as to correlated treatments.

Hyperpolarizabilities of molecules are an active field of study, because they determine a variety of important nonlinear optical effects. However, the deduction of hyperpolarizabilities from these nonlinear optical experiments is a very complicated business. The experiments are carried out at finite frequencies, so some extrapolation is required to obtain an estimate of the static (zero frequency) value. Also, there is a contribution to the hyperpolarizability from molecular vibrations, so even after extrapolation to zero frequency the "experimental" result may not correspond to what is desired. It is therefore very useful to study atoms, both theoretically and experimentally, since this vibrational contribution is necessarily absent. Calculations of accurate static values can be compared with the experimental extrapolations to zero frequency to check the reliability of this process.

In order to provide accurate static values for this purpose, we have recently undertaken a series of calculations of the polarizability and hyperpolarizability of the rare gases. An extensive basis set investigation was performed for Ne, and we shall consider these results in detail. We shall also discuss aspects of the correlation treatment and computational methodology. We begin by considering methods for the calculation of the polarizability and hyperpolarizability.

The polarizability is the second derivative of the energy with respect to an applied electric field, and the hyperpolarizability is the fourth derivative. If analytic energy derivatives of high enough order are available for the N-particle space treatment of interest, the desired results can be obtained with a single calculation. This is usually not possible. If lower-order derivatives are available it may be possible to obtain the desired results by a mixture of analytic and numerical differentiation. For our purposes, no suitable energy derivatives were available and the differentiation had to be performed numerically. We can contemplate doing this by two methods, finite differences or fitting. In the former method we simply expand the energy response, Eq. 5.1, in terms of finite field values and determine the constants α and γ by finite difference expressions. Some care is required to choose the field strengths so as to minimize contamination by higher-order terms but to retain as much numerical precision as possible, and the computed energy results must be converged very tightly for use in the finite difference expressions. In the fitting method, a set of perturbed energies is computed for a range of finite field strengths, and the results are least-squares fitted to a polynomial in field strength. By using a range of strengths, and including higher-order terms in the fit, stable values for the desired constants can be obtained. In practice, we have found little difference between the two approaches, and we are confident that the results to be quoted are precise to at least 0.5%, that is, the errors in the fitting or finite differences are no greater than this.

The question of what field strengths to use in calculations of this sort will be discussed elsewhere, so we will touch on it only briefly here. From the point of view of achieving maximum precision, we would like the energy change Eq. 5.1 to be as large as possible. However, the contamination of the results by contributions from higher polarizabilities increases rapidly (because of their higher power dependence) with increasing field strength, and to reduce this effect we would like the field strengths to be as small as possible. Several authors have discussed appropriate compromises between these opposed needs, and a particularly good exposition can be found in the paper by Cernusak *et al* [80]. In general, two field strengths giving energy lowerings of about 10 and 100 μE_{h} respectively, or somewhat larger, are suitable. For the Ne atom this corresponds to fields of 0.003 and 0.01 a.u. Obviously, estimation of these field strengths requires some estimates of the polarizability and hyperpolarizability — it is seldom that there are no estimates at all, although they may be quite unreliable. In such cases it may be preferable to carry out a series of SCF calculations first in order to establish (roughly) the appropriate field strengths. I should mention that using large fields can lead to bizarre problems, for example, a field gradient (quadrupolar field) strength appropriate to computing the quadrupole polarizability of Ne will, when applied to Be, lead to a ground-state configuration of $1s^2 2p^2$! The quadrupole polarizability of Be is much larger than that of Ne (as are the dipole polarizabilities), and the effect of an applied field gradient is therefore enormously greater. This is a good example of the need for exploratory calculations on the field strength.

While previous studies have established the importance of correlation effects in determining the polarizability and hyperpolarizability, we can obtain a great deal of information about basis set effects in Ne by considering SCF values only to begin with. All of the basis sets to be discussed are derived originally from a (13*s* 8*p* 6*d* 4*f*) primitive set contracted to [4*s* 3*p* 2*d* 1*f*] using ANOs. The outermost *spdf* primitives were then released from the contraction, giving a basis denoted [4+1*s* 3+1*p* 2+1*d* 1+1*f*] and labelled basis A in the tables. Further uncontraction of the outermost *sp* primitives yielded a [4+1+1*s* 3+1+1*p* 2+1*d* 1+1*f*] set labelled B and uncontraction of the *d* primitives a [4+1+1*s* 3+1+1*p* 2+1+1*d* 1+1*f*] set labelled C. SCF polarizabilities obtained with these sets augmented with additional diffuse functions (obtained by even-tempered extrapolation from the outermost exponents of the original basis with a successive ratio of 2/5) are listed in Table 5.8 [81].

Table 5.8: Ne SCF Polarizability and Hyperpolarizability (a.u.).

Basis	α	γ
A	2.15	17.5
A+(1*s* 1*p*)	2.34	49.2
A+(2*s* 2*p* 1*d*)	2.34	56.3
A+(2*s* 2*p* 1*d* 1*f*)	2.34	66.9
A+(2*s* 2*p* 1*d* 2*f*)	2.34	70.2
A+(3*s* 3*p* 2*d* 3*f*)	2.34	71.9
A+(3*s* 3*p* 2*d* 3*f* 2*g*)	2.34	72.2
A+(3*s* 3*p* 2*d* 3*f* 2*g* 1*h*)	2.34	72.2
B+(3*s* 3*p* 2*d* 3*f*)	2.34	72.2
C+(3*s* 3*p* 2*d* 3*f*)	2.38	71.2

A number of conclusions can be drawn from Table 5.8. The convergence of α is very rapid once an additional diffuse *sp* set has been added, the only change then observed in α is a slight increase when an extra *d* primitive is released from the ANO contraction in going from basis B to basis C. The convergence of γ with basis set is much slower, and a relatively large set of diffuse functions, ($2s\ 2p\ 1d\ 2f$), must be added. (The main reason it is unnecessary to add two diffuse *d* functions is that the outermost *d* exponent of the original $6d$ primitive set is diffuse enough already.) *g* and higher angular functions would not contribute at all to the SCF hyperpolarizability if analytic differentiation were used. The small changes observed when *g* functions are added therefore reflect contributions from higher-order contaminants in our differentiation procedure. Finally, the polarizability and hyperpolarizability in our largest sets are very close to the Hartree-Fock limit values.

Correlation contributions to the Ne polarizabilities were estimated using the CCSD approach with the same basis sets as used above [81]. (We note in passing that these Ne CCSD hyperpolarizability results [81] are all somewhat too large, as is the CCSD(T) value. Details are given in an erratum that will be published in early 1992.) The results are given in Table 5.9.

Table 5.9: Ne CCSD Polarizability and Hyperpolarizability (a.u.).

Basis	α	γ
A	2.35	22.4
A+($1s\ 1p$)	2.60	68.7
A+($2s\ 2p\ 1d$)	2.60	83.8
A+($2s\ 2p\ 1d\ 1f$)	2.60	97.2
A+($2s\ 2p\ 1d\ 2f$)	2.60	103.1
A+($3s\ 3p\ 2d\ 3f$)	2.61	107.3
A+($3s\ 3p\ 2d\ 3f\ 2g$)	2.61	108.1
A+($3s\ 3p\ 2d\ 3f\ 2g\ 1h$)	2.61	108.2
B+($3s\ 3p\ 2d\ 3f$)	2.61	109.9
C+($3s\ 3p\ 2d\ 3f$)	2.64	108.7

The convergence of α is again rapid once a diffuse *sp* set has been added. The convergence of γ is slow, although it is almost the same as was observed in the SCF case. Functions of higher angular momentum than *f* contribute little to γ, less than 1%, even though the correlation contribution to γ is very large. In fact, comparison of Tables 5.8 and 5.9 shows that the basis set requirements for converging both polarizabilities, at least in terms of adding diffuse functions, are the same at the SCF and CCSD levels. This is a rather encouraging conclusion, it means that we can perform our studies on augmenting basis sets for hyperpolarizabilities at the SCF level, which is much cheaper, and then use the resulting sets in correlated calculations.

While the CCSD method is certainly reliable enough for basis set investigations, it is probably not capable of achieving an accuracy of much better than 10% for γ, which is not enough for our purposes. On the other hand, the CCSD(T) method agrees almost perfectly with full CI energy benchmarks for Ne, and in an extended basis with up to multiple i functions the CCSD(T) correlation energy appears to be within 2% of the estimated valence-shell correlation energy [81]. We have therefore used the CCSD(T) method to compute polarizabilities for all the rare gases [81, 82]. The basis sets are all obtained by the same general approach as for Ne basis C: full details are given by Rice *et al.* [82]. Results are given in Tables 5.10 and 5.11.

Table 5.10: Rare gas CCSD(T) Polarizability (a.u.).

	computed	estimated[a]	experiment
He	1.38	1.38	
Ne	2.69	2.67±0.03	2.669
Ar	11.21	11.2±0.1	11.08
Kr	17.16	17.0±0.2	16.79
Xe	27.99	27.4±0.5	27.16

[a] See text.

Table 5.11: Rare gas CCSD(T) Hyperpolarizability (a.u.).

	computed	estimated[a]	experiment
He	43.6	43.6	(43.1)[b]
Ne	118.3	119±4	119±2
Ar	1231	1220±30	1167±6
Kr	2830	2810±90	2600
Xe	7180	7030±200	6888

[a] See text.
[b] Theoretical value (Bishop and Pipin).

In general the agreement between our best estimated values and agreement is excellent for the polarizabilities and very good for the hyperpolarizabilities. A word is in order as to how these estimates were obtained, and where the uncertainties come from. In general, the best directly computed value we have is from CCSD(T) calculations in a large basis set, correlating only the valence electrons. These results are felt to be essentially at the basis set limit, and the correlation treatment is assumed to be close to a full CI. The effects of correlating the core electrons have been explored for all systems [82], correlating up to 26 electrons (the M and N shells in Kr or the N and O shells in Xe). Basis sets augmented with up to f functions chosen to describe core correlation were used in conjunction with the CCSD and/or MP2 methods. The difference between the valence-correlated and core and valence-correlated results were added to the best computed numbers to correct for core correlation. Similarly, for Kr and Xe we computed relativistic corrections to the polarizabilities at the SCF level, and added these to the best computed results [82]. As can be seen from Tables 5.10 and 5.11, these various corrections are relatively small. The uncertainty estimates arise largely from the uncertainties in these corrections. For example, our relativistic correction to the Kr polarizability is −0.08 a.u., obtained from first-order perturbation theory. Using a more complete Dirac-Fock treatment Kolb and co-workers [83] obtained only +0.01 a.u. We have therefore assumed that our relativistic corrections could be in error by as much as 100%.

Although the computed and experimentally estimated [84] (extrapolated to zero frequency) hyperpolarizability for Ne are in perfect agreement, the results for Ar do not agree within their respective uncertainties. Actually, this is due to an experimental uncertainty estimate that is almost certainly too optimistic. The uncertainty quoted for the experimental value is derived from the experimental statistics, that is, it is a measure of random error in the measurement. It contains essentially no contribution from any possible source of systematic error. In fact, Shelton believes that a more realistic uncertainty would be ±20 or perhaps ±30 [85]. If that were the case there would be no disagreement between theory and experiment. This is an excellent illustration of the dangers of relying on a given experimental estimate or uncertainty. It is always necessary to ascertain exactly what the experimentalist means by his/her error bars.

5.4 Geometrical Structure of C_3^+

Quantum chemical calculations had established that two structures of the C_3^+ ion, a linear $^2\Sigma_u^+$ state and a cyclic (almost equilateral triangular) $^2B_2$ state were very similar in energy. Experiments were first interpreted in terms of the cyclic structure, but it was pointed out that the experimental analysis could not exclude a linear structure with a very large amplitude bending motion, so several *ab initio* studies were undertaken to try to settle this question [86, 87, 88].

Single-reference CI calculations were performed by Grev and co-workers [86]. They attempted to calibrate their calculations by performing CISD+Q in their full basis (which was only of DZP quality), CISDTQ in a truncated basis obtained by eliminating ten of the virtual MOs, and "full CI" in a severely truncated basis obtained by eliminating 33 virtual MOs. They corrected their full-basis CISD+Q result by the difference between CISDTQ and CISD+Q in the truncated basis and the difference between CISDTQ and full CI in the severely truncated basis. Unfortunately, the full CI is simply a full valence CI — eleven electrons in twelve orbitals — and hence gives very little information about dynamical correlation. Nevertheless, they had at least made an attempt to perform some calibration. On the basis of these calculations, Grev and co-workers asserted that the linear geometry was some 7 kcal/mol above the cyclic structure; they assigned an uncertainty of ±4 kcal/mol to this value, which in view of their small basis might seem a little optimistic.

Even with this uncertainty, their result was in sharp contrast with QCISD(T) calculations performed independently by Raghavachari [87] and by Martin and co-workers [88]. These studies considered basis set extension including f functions. Raghavachari obtained a separation of 3 kcal/mol (the positive sign indicates the cyclic structure is lower), but argued (partly on the dubious grounds that the bending potential from linear to cyclic displayed a maximum near 100°) that QCISD(T) was unreliable for this case and artefactually favoured the linear structure. He estimated the separation would be larger, in agreement with the results of Grev *et al* [86]. Martin and co-workers simply used their calculations to make a best estimate of 2 kcal/mol for the separation, pointing out that with any reasonable estimate of the uncertainty this meant it was not possible to decide whether the molecule was cyclic or linear [88]. In view of the very substantial multireference character of the wave function, as exemplified by the reference configuration weights published by Grev and co-workers [86], it seemed highly desirable to examine this problem with more elaborate N-particle space treatments and thereby to provide a better calibration of the single-reference studies. We therefore reexamined this problem with MRCI methods [89].

One interesting development that took place during the MRCI studies was the announcement by Grev and co-workers that their CISDTQ results were incorrect [90]. Using correct CISDTQ values to revise their adjustments to the CISD+Q results they obtained an estimate of 4±4 kcal/mol for the separation. This was now consistent with the assertion of Martin and co-workers that it was not possible to decide whether the minimum energy structure was cyclic or linear.

The first set of MRCI calculations [89] was performed using the same DZP basis set as Grev and co-workers: results are presented in Table 5.12, together with several single-reference results.

Table 5.12: C_3^+ DZP Basis Cyclic/Linear Energy Separation (kcal/mol).

Method	separation	% ref($^2\Sigma_u^+$)[a]	% ref($^2B_2$)[a]
CASSCF	1.46		
Val(0.05e)[b]	2.93	94.2	93.4
Val+Q(0.05e)	3.73		
Val(0.025)	1.19	96.4	95.8
Val+Q(0.025)	1.79		
Val(0.01)	1.32	98.6	98.4
Val+Q(0.01)	1.38		
MRCI(0.05e)	3.20	90.6	90.1
MRCI+Q(0.05e)	4.37		
ACPF(0.05e)	4.74	88.3	87.2
MRCI(0.025)	1.50	92.2	91.8
MRCI+Q(0.025)	2.74		
ACPF(0.025)	2.82	90.7	90.0
MRCI(0.01)	1.60	93.7	93.7
MRCI(0.01)	1.68		
CISD	15.69		
CISD+Q	9.41		
QCISD	9.38		
QCISD(T)	0.87		
MP4(SDTQ)	–24.63		

[a] Weight of reference configurations.
[b] See text for meaning of 'e' suffix.

The multireference results of Table 5.12 were all based on full valence CASSCF calculations with eleven electrons in twelve active orbitals. This produces a large configuration expansion (about 85 000 CSFs) so it is not possible to perform MRCI(CAS) calculations. Reference configuration lists were selected at the cyclic and linear geometries (taken from MP2 optimized structures) and then merged. The core electrons were not correlated in any of the calculations. One complication in the CASSCF calculations should be pointed out. Since the cyclic state arises as the $^2B_2$ component, in C_{2v} symmetry, of a $^2E'$ state in the D_{3h} symmetry (equilateral triangular) structure, it would be desirable to obtain MOs with D_{3h} symmetry and equivalence restrictions in a calculation on the equilateral structure. As discussed elsewhere, the $^2B_2$ state calculation will not display these properties. However, by performing a state-averaged CASSCF calculation, averaging over the $^2B_2$ and $^2A_1$ states in C_{2v} symmetry, full D_{3h} symmetry and equivalence restricted MOs will be obtained. Unfortunately, while the $^2B_2$ state correlates with $^2\Sigma_u^+$ in linear symmetry, the $^2A_1$ state correlates with one component of a $^2\Pi$ state. Hence to obtain MOs that display $D_{\infty h}$ symmetry and equivalence restrictions we must include the other component of the Π state (another $^2B_2$ state in C_{2v}) in the state averaging. Hence we must average over three roots in the CASSCF calculation, which thus becomes rather expensive.

We note first that the valence CI results agree well with the CASSCF results only for thresholds of 0.025 or smaller. We may therefore assume that the nondynamical correlation effects are only described well in calculations using these smaller thresholds. This is also suggested by the weight of the reference CSFs in the valence CI wave functions. Dynamical correlation was included for all three selection thresholds. In the ACPF calculations, the use of the reference list originally obtained at a selection threshold of 0.05 led to a suprisingly large separation and a very poor weight for the reference CSFs in the ACPF wave function for the $^2B_2$ state. This was traced to five valence CSFs that displayed very large coefficients in the ACPF wave function. Once these additional CSFs were included as references (calculations denoted by 0.05e) the ACPF results were in better agreement with the MRCI+Q results and the weight of the reference CSFs in the wave function was much more satisfactory. This sensitivity to selection of reference configurations is characteristic of ACPF and occasionally creates considerable difficulty, as we mentioned in Sec. 2.8. Note that there are no problems with the reference lists selected at 0.025 or 0.01.

Overall, the results of Table 5.12 suggest that the MRCI+Q or ACPF results with reference selection at a threshold of 0.05 are not satisfactory. Convergence of the results with selection threshold is not complete even by a threshold of 0.025, in fact, which is unusual. The MRCI+Q and ACPF results are in very good agreement with one another at smaller thresholds. Surprisingly, at a threshold of 0.01 they are also very similar to the CASSCF result.

Results obtained with a larger [$5s$ $3p$ $2d$ $1f$] segmented contracted basis are given in Table 5.13 [89]. The effect of expanding the basis set is to increase the separation for all methods. Since the MRCI(0.025) calculation comprises about six million CSFs it is near the limit of what is currently feasible. The MRCI(0.01) calculation would involve over twenty million CSFs, which is impractically large. Our best directly computed number is probably the MRCI+Q(0.025) result of a little over 5 kcal/mol. We can improve on this estimate by using the small basis results, however.

Table 5.13: C_3^+ [5*s* 3*p* 2*d* 1*f*] Basis Cyclic/Linear Energy Separation (kcal/mol).

Method	separation
CASSCF	3.18
Val(0.025)	2.94
Val+Q(0.025)	3.54
MRCI(0.025)	3.92
MRCI+Q(0.025)	5.18
QCISD	12.62
QCISD(T)	3.42

We note that tightening the selection threshold from 0.025 to 0.01 reduced the separation by about 1 kcal/mol in the DZP basis. In the absence of any other information we assume this would also occur in the larger basis. This is certainly preferable to not making any correction for the effects of selection. As far as basis set effects are concerned, we see an increase of about 1.7 kcal/mol at the CASSCF or valence CI level from increasing the basis, while at the MRCI or QCISD(T) level the increase is 2.5 kcal/mol. That is, the basis set effect on the dynamical correlation is about 0.8 kcal/mol, which is rather small. This reasoning suggests that the MRCI+Q(0.01) result in the [5*s* 3*p* 2*d* 1*f*] basis would be about 4.2 kcal/mol; the remaining error in this result should come dominantly from the one-particle basis set. Scuseria has recently studied the basis set effects on the separation using CCSD(T) wave functions [91]. From his results the effect of going to a complete one-particle basis could be as large as 2.5 kcal/mol, while it seems highly unlikely that any effect could decrease the separation below our MRCI+Q(0.01) [5*s* 3*p* 2*d* 1*f*] basis estimate of 4.2 kcal/mol. Combining these limits with our best computed value of 5.2 kcal/mol from Table 5.13, we can assert with some confidence that the true separation should be $5.2^{+1.5}_{-1.0}$ kcal/mol [89].

Our result thus demonstrates that C_3^+ has a cyclic structure. The estimated cyclic/linear separation is larger than the QCISD(T) results, mainly as a result of N-particle space effects. It is somewhat smaller than the best CCSD(T) results of Scuseria [91], although our uncertainty range includes his values. While these observations seem to suggest a much larger difference between QCISD(T) and CCSD(T) than might have been expected, it should be kept in mind that the QCI calculations are based on a UHF reference, while the CC calculations are based on RHF. Examination of CISD results suggests that half the difference can be attributed directly to reference treatments.

More details of the C_3^+ saga, including an answer to the question of whether the linear structure is a minimum or a saddle point can be found in the paper by Taylor and co-workers [89]. Symmetry breaking in the linear structure is also considered there. What we have described here illustrates the approaches that can be taken to calibrating calculations and estimating uncertainties in cases of severe nondynamical correlation effects, and for which calibration by full CI studies is not possible.

5.5 Other Examples

We have discussed here only a few different molecular properties and their calibration. We list here some other studies of interest by the Ames group that cover a wider range of properties. For benchmarking of dipole moments and dipole moment functions the extensive investigations of the ground-state dipole moment function of the OH radical by Langhoff and co-workers [92, 93] provide calibration of the N-particle and one-particle spaces. For excited states and valence/Rydberg mixing there are extensive calculations on AlH by Bauschlicher and Langhoff [94]. The same authors have also performed a detailed calibration study of covalent/ionic curve crossing in LiF [95]. The spectroscopy of FeO provides an excellent illustration of many of the difficulties encountered in accurate treatments of transition-metal systems, as well as problems obtaining adequate reference spaces for multireference ACPF calculations [32]. Finally, more recent work on the CH and CC binding energies in small hydrocarbon molecules extend some of the earlier accurate binding energy studies to polyatomic molecules [38, 38, 96].

Chapter 6

"Small" Effects

6.1 Other Sources of Error

The quoting of "small" in the title of this chapter is deliberate — we are about to discuss a number of effects that are generally regarded as small or even negligible. As we shall see, these terms may well be significant if we are attempting to obtain results of the highest accuracy. One approximation that we have made in most of the work discussed here is that only correlation effects in the valence shell are important, and we shall explore first the consequences of this approximation and how to correct for it. A second approximation is the neglect of relativistic effects, and we shall consider this briefly in the next section. Finally, we shall discuss basis set superposition error, a much-misunderstood effect that pervades all our calculations. Some other small sources of error, such as the Born-Oppenheimer approximation, will not concern us here. For the properties we have considered in this work the effects of this approximation are negligible.

6.2 Core-core and Core-valence Correlation

In most of the calculations reported in this work we have included correlation only for the valence electrons. In general, the neglect of correlation between core electrons, *core-core correlation* (CC), and between core and valence electrons, *core-valence correlation* (CV), is not motivated merely by a desire to reduce the number of electrons included in the correlation treatment. The first lesson about CC and CV correlation is that tossing the core electrons into a calculation designed to treat valence correlation is a totally inadequate approach. Indeed, as we shall mention in Sec. 6.4 it may lead to additional errors. This is because the basis set requirements for CC and CV correlation are given by the same reasoning developed in Sec. 4.1. That is, we need additional functions of the same spatial extent as the occupied orbitals being correlated, but with additional radial and angular nodes. Since the spatial extent of the core MOs is very small, the exponents of these correlating functions must be very large, and valence-correlation-optimized basis sets simply do not have exponents large enough to do the job. Hence basis sets for CC and CV calculations must be substantially increased over those used for valence correlation treatments.

The N-particle space and one-particle space requirements for describing CC and CV correlation have been explored by Bauschlicher and co-workers [97]. In full CI comparisons for several atoms, it was found that convergence of CC correlation with the N-particle space was relatively poor, even when large MRCI wave functions were used. In fact, agreement between MRCI and full CI was obtained only when the CASSCF calculation performed to obtain orbitals and reference configurations essentially accounted for all of the valence shell correlation! It is, of course, quite impractical in general to account for dynamical correlation at the CASSCF level in molecules. Convergence of CV correlation with the N-particle space was considerably better: the CV correlation was treated here by excluding from the wave function any double excitations from the core. Double excitations in which one electron is from the core and one from the valence orbitals are then responsible for most of the CV correlation contribution. Adding this class of excitation to valence MRCI treatments seems to give a satisfactory description of CV correlation effects, provided of course that the basis has been augmented with suitable high exponent functions. Since a number of investigations suggest that the effect of CV correlation on spectroscopic constants, for example, is much greater than is the effect of CC correlation, it seems a reasonable compromise to account for the former and to neglect the latter.

The increased basis set requirements when core electrons are correlated can make it difficult or impossible to carry out some calculations. One attractive possibility that is hard to realize in practice is to extend ANO contractions to include core correlation [48, 97]. The difficulty arises from the nature of core correlation and the behaviour of natural orbital occupation numbers. The correlation energy of the $1s^2$ pair is essentially constant across the first row. Viewed from a perturbation theory standpoint, the second-order energy numerators scale as Z^2, hence the denominators must also scale as Z^2. But the perturbation theory natural orbital occupation numbers then scale as Z^{-2}: in effect, we get the same CC correlation energy for exciting less of the electron density into correlating orbitals as Z increases. If we now perform an atomic CI calculation in which all the electrons are correlated, the NOs designed to correlate the core will become mixed with the very weakly occupied NOs for correlating the valence shell. Since the latter have many nodes the resulting orbitals are a mess, and certainly do not look promising for inclusion in an ANO basis set. Another approach would be to perform separate CC and valence calculations and combine the ANO sets (after orthogonalization, if desired). This leads to large basis sets that do not describe CV correlation especially well. And combining ANOs from separate CC,

CV and valence calculations leads to almost insuperable linear dependence problems in the resulting basis set [97]. We should note that these problems are perhaps at their worst for the first row. For heavier elements the separation between the valence shell and the "core" (meaning the next innermost shell here) is not as great, so the disparities in correlating orbital occupation number are less. Suitable ANO sets can often be obtained by correlating all the desired electrons [48, 98].

To give some idea of the quantitative contribution of core correlation, we can recall that CV correlation increases the binding energy of N_2 by less than 1 kcal/mol, although in recent calculations Werner and Knowles [70] have included both CC and CV correlation (in suitably augmented basis sets) and obtained almost 1.5 kcal/mol. In CH_4 CV correlation is responsible for a 0.0015 Å contraction in the bond length [99] — it was essential to account for this effect before trying to resolve the issue of a 0.0035 Å discrepancy between the then best theoretical results and experiment [100]. The effect of CV correlation on the singlet/triplet separation in CH_2 is to increase it by about 0.35 kcal/mol, as we noted in Sec. 4.5; since improvement in the valence correlation treatment decreases the separation it is vital to account for CV correlation in comparing the best computed results with experiment.

As we have seen, the basis set requirements for CC and CV correlation are very stringent. There is therefore considerable attraction in methods that treat these effects semiempirically. One approach is to treat CV correlation effects by an effective operator. The core polarization potential used by Müller *et al.* for CV correlation in the alkali atoms and alkali dimers is one such approach [101]. This method has been used successfully for other atoms, such as copper [98].

6.3 Relativistic Effects

Relativistic quantum chemistry is currently an active area of research (see, for example, the review volume edited by Wilson [102]), although most of the work is beyond the scope of this course. Much of the effort is based on Dirac's relativistic formulation of the Schrödinger equation: this results in wave functions that have four components rather than the single component we conventionally think of. As a consequence the mathematical and computational complications are substantial. Nevertheless, it is very useful to have programs for *Dirac-Fock* (the relativistic analogue of Hartree-Fock) calculations available, as they can provide calibration comparisons for more approximate treatments. We have developed such a program and used it for this purpose [103].

In general, the computational requirements for full four-component relativistic calculations on molecules are so severe that cheaper alternatives must be explored.

The next most sophisticated approach is probably the reformulation of the problem in terms of a single effective component wave function. One of the advantages of this approach is that it requires much less modification of existing methodology. In particular, electron correlation can be incorporated using basically the same programs as are currently used. Again, this sort of approach is really beyond the scope of the present course, but the interested reader can refer to the original literature [104].

An even simpler approach to relativistic effects that has been much exploited in calculations on fairly light elements (such as the first-row transition metals) is the use of perturbation theory. By treating relativity as a perturbation, Cowan and Griffin produced a rather simple correction, in first order, to be applied to the nonrelativistic result [105, 106]. The correction involves two one-electron integrals, over the mass-velocity and Darwin operators. These integrals are very straightforward and can be incorporated into a molecular property integral program with little effort. The first-order relativistic correction can then be evaluated as a property expectation value along with the dipole moment, etc. Obviously, correlated wave functions can be used as well as SCF wave functions. It should be noted that since the Cowan-Griffin correction is not bounded satisfactorily, some care is required if the operators are included in the Hamiltonian (for optimization or calculation of properties as energy derivatives). In general, the evaluation of the first-order relativistic correction is so cheap and simple that it should probably be done as a matter of course in calculations involving, say, transition metals. Even for heavy elements the effects of relativity on spectroscopic constants are remarkably well reproduced by first-order perturbation theory, compared to full Dirac-Fock results [103]. This was the approach taken for evaluation of the relativistic correction to the heavier rare-gas polarizabilities and hyperpolarizabilities described in Sec. 5.3 [82].

An even simpler approach to relativity, for heavy elements, is to use effective core potentials (ECPs) to represent the core electrons, taking the potentials from various compilations in the literature that explicitly include relativistic effects in the generation of the ECPs. References to such ECPs are given by Dyall *et al.* [103]. These *relativistic ECPs (RECPs)* allow the inclusion of some relativistic effects into a nonrelativistic calculation. Since ECPs will be treated in detail elsewhere, we will not pursue this approach further here. We may note, however, that recent comparisons with Dirac-Fock calculations suggest that the main weakness in the RECPs is *not* the treatment of relativity but the quality of the ECPs themselves [103]. Different RECPs gave spectroscopic constants with a noticeable scatter, compared to Dirac-Fock, but the relativistic corrections (difference between an RECP and the corresponding ECP value) were fairly consistent with one another.

One other consequence of including relativistic effects in a rigorous manner is worth noting. Solutions to the one-electron Dirac equation have singularities at the origin for a point nucleus — this can be avoided by employing a nucleus with a finite radius. Such a nuclear model is, of course, more physically plausible anyway. The shape of the nuclear charge distribution does not seem to be critical: both constant density "hard-sphere" and Gaussian models have been used. The latter have the advantage of yielding particularly simple formulas for molecular integrals when Gaussian basis functions are used. One of the most important consequences of using a finite nuclear size is that the wave function no longer has a cusp at the nuclear origin. Hence one often-quoted advantage of Slater-type functions, their nuclear cusp, becomes irrelevant.

6.4 Basis Set Superposition Error

Of all the problems that can beset quantum chemical calculations, *basis set superposition error (BSSE)* is probably the problem most cheerfully ignored. Sometimes this approach is justified, but this requires investigation that is seldom performed. Some understanding of BSSE is indispensable to anyone who hopes to perform accurate and reliable calculations. The problem of BSSE is a simple one: in a system comprising interacting fragments A and B, the fact that in practice the basis sets on A and B are incomplete means that the fragment energy of A will necessarily be improved by the basis functions on B, irrespective of whether there is any genuine binding interaction in the "molecule" or not. The improvement in the fragment energies will lower the molecular energy, giving a spurious increase in the binding energy.

This effect is probably best known from accurate calculations on the potential curve for He_2, where BSSE gives a well several times deeper than reality even in enormous basis sets [107]. It is probably for this reason that BSSE tends to be regarded as an effect that one need worry about only in calculations on very weakly interacting systems. This is simply (and unfortunately) not true. BSSE is an ever-present phenomenon and calculations with any pretension to real accuracy should include an investigation of BSSE. The approach most commonly taken to estimating the effect of BSSE is the *counterpoise correction* [108]: the separated fragment energies are computed not in the individual fragment basis sets, but in the total basis set for the system. Note that if multiple geometries of the molecule are to be calculated a counterpoise correction must be evaluated at each geometry. A comprehensive and lucid discussion of BSSE and counterpoise correction is provided by thr recent article by Liu and McLean [107]. These authors go so far as to consider not only BSSE from the one-particle basis, which is what most other authors mean by BSSE, but errors that arise from inconsistencies in the definition of the N-particle space between molecule and fragments. These "configurational' superposition errors are usually very small and can probably be neglected in all but the most exacting work on very weakly bound systems.

For an example of the size of BSSE in strongly bound systems, we can consider the large basis set N_2 dissocation energy study discussed in Sec. 5.1. Using the smaller primitive basis, (13*s* 8*p* 6*d* 4*f* 2*g* 1*h*), Almlöf and co-workers computed the BSSE contribution to D_e, which was defined to be twice the energy lowering obtained by performing a calculation on N atom with and without a ghost basis at the appropriate distance [68]. For the [6*s* 5*p* 4*d* 3*f*] contracted ANO set this BSSE contribution was 0.76 kcal/mol, for the [6*s* 5*p* 4*d* 3*f* 2*g*] set it was 0.36 kcal/mol, and for [6*s* 5*p* 4*d* 3*f* 2*g* 1*h*] 0.28 kcal/mol. It seems safe to assume that in the very largest calculations of Sec. 5.1 the BSSE does not exceed 0.3 kcal/mol, an assertion that is supported by extensive tests at the MP2 level. Nevertheless, in a study that attempts to achieve 1 kcal/mol accuracy in bond energies, a proper investigation of BSSE, and, if necessary, correction for it, is clearly mandatory.

This contention is certainly not without its opponents, and several workers are of the opinion that there is some sort of qualitative difference between BSSE in weakly bound and strongly bound systems. I do not see any justification for such a distinction myself. However, I would certainly agree that in attempts to estimate basis set limit values there may be alternative uses for the counterpoise correction that simply subtracting it from the computed bond energies. Since BSSE is in some sense a measure of basis set incompleteness, one can contemplate *increasing* the bond energy by some fraction of the counterpoise correction to correct for this residual incompleteness, rather than decreasing it to correct the computed result for BSSE. This is a completely empirical approach, but we have found (for strong interactions) that in large basis sets (up to *g* functions, say) increasing the computed values by 150% of the calculated BSSE gives a good approximation to the best extrapolations to the basis set limit that we can perform from very large basis set studies. The alternative to this sort of approach is to explicitly compute a range of results with progressively larger basis sets, corrected for BSSE, and extrapolate from that. This is undoubtedly more satisfactory from a formal point of view, but it is also much more expensive. Another difficulty is identifying which fragments to use in calculations on polyatomic molecules, as discussed by Parasuk and co-workers [109]. We may note finally here that inclusion of core correlation without appropriate augmentation of the basis set can lead to a substantial increase in BSSE.

One or two issues involving BSSE surface periodically in the literature. One is the question as to whether a counterpoise correction should be determined with a ghost basis that includes or excludes the occupied orbitals of the ghost fragment.

Arguments based on the Pauli principle have been used to support the exclusion of the occupied ghost orbitals, although in practical investigations such an approach seems to give more erratic counterpoise corrections than including them. This issue seems to have been laid to rest by detailed formal arguments by Gutowski *et al.* [110], as amplified by Liu and McLean [107]. They provide a proper foundation for the use of the full ghost basis, establishing, in the words of McLean [111], that "the Pauli principle can be left to take care of itself".

A second issue that arises in the context of BSSE is the use of basis sets that include *bond functions*, basis functions in the bonding region (commonly at the bond midpoint) that are included to try to accelerate convergence of results with the one-particle basis. Early studies that produced spectacularly good binding energies for diatomic molecules [112] were shown to be due almost entirely to BSSE [113]: the bond functions contribute in the molecule but not in the atoms. Accounting for BSSE gave much poorer (but much more plausible) bond energies. Some of the bond function enthusiasts returned to the fray with what seems to be an even less satisfactory approach — relying on a cancellation of errors (specifically, cancellation of BSSE with inadequate one-particle space errors) to produce acceptable results [114]. As with any uncontrolled approximation, relying on such a cancellation to occur in a wide range of situations seems to be ludicrously optimistic [115]. In more recent work [116, 117, 118], a much more careful approach has been taken to establishing how BSSE contributes to bond function results, and how to construct basis sets that suffer as little as possible from such errors, but considerable care is required to use bond functions well.

Chapter 7

Maxims

7.1 General Remarks

This chapter contains a number of sections covering different aspects of quantum chemical calculations. Within each section are items of advice or issues that should be borne in mind when designing a particular calculation.

7.2 One-particle Space

- *The ultimate accuracy of any calculation is determined by the one-particle basis set.*

This is one of the most obvious, yet one of the most ignored, observations about quantum chemical calculations. For example, it is in general just not possible to get the right answer for the right reason using a DZP basis. This is not an argument against using such sets, but their limitations must be kept constantly in mind.

- *A DZP or split-valence plus polarization basis is a minimal basis for correlated calculations.*

If the aim of the calculation is to describe dynamical correlation effects, there is seldom any point to using a basis that is smaller than double zeta in the valence shell and lacks polarization functions. Such a basis recovers so little of the correlation energy, and such an unrepresentative fraction (because of the poor treatment of angular correlation), that the results can be hopelessly unreliable. In this sense we should regard a DZP basis (or equivalent) as a minimal, in the sense of minimum acceptable, basis set for correlated calculations.

- *Most of the common basis sets do not form a well-defined sequence.*

In most calculations, improvement of the basis set involves an increase in both the primitive set and the contracted set. It is not generally obvious how one basis set is related to another (other than that larger basis sets are expected to perform better than smaller sets), and consequently it is not always clear how to extrapolate from a set of calculations to the basis set limit. It is preferable to use sequences of basis sets that form a well-defined chain of subspaces.

- *Spherical harmonics are preferable to Cartesian functions.*

At the very least it is desirable to project out the contaminant functions from the basis, even if the integrals are evaluated over Cartesians. Including the contaminants exacerbates linear dependence problems, and if their inclusion makes a big difference the basis set being used should be viewed with suspicion.

- *Bond functions are for bond-function experts, of whom there are disturbingly few.*

Adding bond functions is a good way to create major problems with superposition error. There are many subtleties to using bond functions properly (that is, getting the right answer for the right reason), and these are neither well known nor widely used. If you are not going to take the trouble to learn the pitfalls and how they (or some of them) can be avoided, steer clear of bond functions.

7.3 N-particle Space

- *Size-extensivity or size-consistency can rarely be ignored.*

Other than MRCI calculations with relatively few electrons (say six or fewer), methods that do not even try to achieve correct scaling are unlikely to prove satisfactory. Even a Davidson-type correction is preferable to nothing at all. CPF/MCPF and especially CCSD/QCISD should be strongly preferred over CISD. And unless an exactly size-consistent method is being used, fragment energies should be computed using a supermolecule approach.

- *Selection of reference configurations for MRCI must be done consistently.*

If an MRCI(CAS) calculation will be too large, reference selection must be employed. It is absolutely crucial that selection be performed by merging configuration lists from all geometries of interest. This is especially a problem in computing potential energy surfaces for reaction dynamics: a large part of the surface must be computed first at the CASSCF or MCSCF level in order to provide enough sampling to define the reference list. Finding out partway through an MRCI study that one or more important CSFs need to be added to the reference space at particular geometries is an excellent way to use up spare computer capacity.

- *Reference selection should be carried out at several thresholds, if possible.*

That is, some calibration of reference selection is desirable. There is little evidence for significant differences between MRCI(CAS) and MRCI(0.01) results, but a threshold of 0.01 may generate too many CSFs in a large basis. A threshold of 0.05 is about as large as is generally safe, and it can be too large in some cases, so some comparisons should be made at a threshold of, say, 0.025. Calibrations in a smaller basis may be helpful.

- *Special care is needed in selecting reference configurations for MRACPF calculations.*

The ACPF method, at least in its original formulation, is extremely sensitive to the choice of reference CSFs. Reference lists that were perfectly satisfactory for MRCI calculations (or, at least, that gave sensible results and identified no other important configurations) may produce variational collapse in an ACPF calculation with the same list. Even if the ACPF calculation converges, the ACPF wave function may look odd, with a rather small weight in the reference configurations and large weights for configurations outside it. Any such configuration should be added to the reference list (although this may make the calculation too large, in which case the method can simply not be used). It is quite possible that this refinement of the reference list will have to be carried out iteratively, several times. The most irritating aspect of this procedure is that when the reference space is finally stable, most of the reference configurations that were added will go back to having small coefficients in the wave function, just as they did in the MRCI case.

This is a common problem in selected reference ACPF calculations, not a rarity. As discussed elsewhere, modifications of the ACPF procedure have been tried in order to avoid the problem, but these have met with mixed success. In general, multireference ACPF calculations should be monitored very carefully. Generation of potential energy surfaces with the MRACPF method will require a great deal of preliminary exploration to identify all required reference configurations, even more so than in the MRCI case discussed above.

- *Negative ion character in wave functions requires special attention.*

As discussed in Sec. 3.3, our understanding of the N-particle space requirements for treating correlation effects in negative ions is incomplete. This also extends to molecules or parts of potential energy surfaces where there is significant negative ion character in the wave function. For instance, the transition state for the reaction

$$F + H_2 \rightarrow HF + H \tag{7.1}$$

contains some $F^-H_2^+$ character. At the very least, when electronegative species like F or O are involved the MCSCF active orbital space should be expanded to include the next atomic p shell, since radial correlation effects of the form $p^2 \rightarrow p'^2$ are crucial to obtaining reasonable electron affinities for these atoms. Here p' is an orbital with the same spatial extent as p but with one more radial node. Including such orbitals in the MCSCF procedure may lead to additional important reference configurations for an MRCI calculation, but this is a price that must be paid.

Another difficulty is that coupling between the one-particle and N-particle spaces seems to be larger for negative ions than for other species. For example, in small or moderate-sized basis sets the correlation contribution from the $2s$ electrons reduces the EA of O and F, but as the basis set is expanded this contribution decreases to zero and then changes sign, eventually *increasing* the EA. Clearly, choosing a dynamical correlation treatment based on small basis set results would give an entirely false picture here.

7.4 Molecular Orbitals

- *SCF orbitals can give a biased description when nondynamical correlation effects are important.*

In general, the SCF method overestimates ionic character in molecular wave functions, and where several atomic states with different occupations are similar in energy (e.g., transition metals) the SCF method may also overestimate the importance of one such occupation at the expense of the others. When the MOs from such a calculation are used in a correlated treatment, the results may still suffer from this bias. Some correlation treatments react to this bias, as reflected in a large $\mathcal{T}_1$ diagnostic for CCSD (see Sec. 2.8), or instability in the energy or excessively large coefficients for some correlating configurations. However, some, like CISD, may not. This is another argument in favour of the more elaborate single-reference treatments: they seem more sensitive to bias in the MOs. Note that if the MOs are biased enough CISD may yield a perfectly plausible wave function, showing no other important configurations, yet the results will be quite wrong.

In view of the insensitivity of CISD to bias built into the MOs by the SCF step, the use of SCF MOs in MRCI calculations would appear to be a risky business.

- *Attention should be paid to degeneracies and to the symmetry and equivalence properties of the MOs.*

It seems to make little sense to set out to describe phenomena that involve degenerate states without ensuring that these degeneracies can appear in our calculations. The use of state-averaging to obtain degeneracies required by point-group symmetry was described in Sec. 5.4, and the formal analysis of imposing symmetry and equivalence restrictions was treated elsewhere. Using broken-symmetry MOs makes it difficult or impossible to identify individual electronic states, and it is vital in studies of symmetry breaking to ensure that the description allows the system to become symmetric if it wants to.

- *Watch out for SCF/MCSCF rotations that mix MOs between classes.*

Special care is necessary when the method used to optimize MOs involves orbital rotations to which the dynamical correlation treatment is not invariant. For example, MCSCF active MOs with very small occupation numbers may mix in an uncontrolled way with the low-lying virtuals, and those with occupation numbers very close to two may mix with the core orbitals. (In principle, MOs with occupation numbers very close to two or zero should not be in the active space anyway, but this may be hard to avoid in practice.) Dynamical correlation treatments used are not invariant to mixing of core and valence orbitals, and erroneous results will be obtained if this mixing happens at one geometry and not others. N_2 provides a concrete example. With a six electron/six active orbital CASSCF treatment, the $1s$ and $2s$ orbitals are inactive, and rotations between them would be excluded as redundant. Although these orbitals may mix strongly during the CASSCF iterations, at convergence some inactive orbital "canonicalization" procedure, like diagonalizing a Fock-type operator, will usually be employed, and this will give well-defined core $1s$ orbitals, and valence $2s$ orbitals for use in a ten valence electron correlation treatment. With a ten electron/eight active orbital CASSCF treatment (that is, a full valence CASSCF), at long bond distances $1s/2s$ mixing (which is not redundant for this case, at least in principle) will occur. This happens because a large amount of core correlation energy can be recovered from a small amount of core excitation, as discussed in Sec. 6.2. Canonicalization cannot undo this, as the $1s$ and $2s$ orbitals are now inactive and active, respectively. Hence at long distances a ten valence electron correlation treatment will be contaminated by core correlation that is absent near equilibrium. This can give quite surprising MRCI bond energies!

7.5 Interaction with Experiment

- *Understand exactly what the experimentalist needs to know, and to what level of uncertainty.*

In general, experimentalists (and other theorists, such as dynamicists, whom we group under this banner here) have a tendency to approach quantum chemists not with the question they actually want answered, but with a related question *that they think we can answer*. Sometimes their understanding of quantum chemistry is good enough that it is reasonable for them to decide on a model system when the real system is too elaborate, but much of the time it is not. Do not accept the questions they pose at face value — quiz them to obtain a better understanding of the problem and what it is they need to know. It may well be possible to bypass model systems and attack the real problem directly. On one occasion I was asked to compute potential energy curves for rare-gas diatomic ions. The actual problem concerned rare-gas cluster ions, but the experimentalists concerned were convinced that anything larger than a diatomic would be too large for quantum chemistry, so they planned to take an accurate diatomic potential and then use it in a crude (and inappropriate, as it turned out) model to generate a potential for the cluster. In fact, it was perfectly possible to generate these cluster potentials quantum chemically, but if I had not questioned them at length about what they wanted the diatomic potentials for, I would never have discovered that this was appropriate. This is far from being my only experience of this type.

A related issue is how accurately certain quantities must be calculated. It is almost pointless to ask a dynamicist, for example, what is an acceptable uncertainty in a calculation of computed points for a potential energy surface. He/she will answer with some unrealistically small value (and then go off and fit the points with an RMS accuracy of about 2 kcal/mol). Similarly, most experimentalists will answer "as accurately as you can" when asked how accurately they need some quantity. The best way of discovering how accurately a number is needed is to understand as much as possible of what is being done experimentally (or theoretically, by dynamicists). It may be that simply establishing the sign of a few quantities, or their order of magnitude, may be all that is needed to allow experimental results to be consistently and successfully analyzed, as Pulay has pointed out in the context of computing molecular force fields [119].

- *Set out to compute what can be compared with experiment.*

Most "experimental" results are the end-product of a considerable amount of data processing, usually with a theoretical analysis built in. It is often an enormous help to avoid these manipulations and assumptions by computing quantities that are closer to what is actually measured in the instrument. For example, the dissociation energy of the ground state of BH was uncertain for some time, even though these was an extensive band system from an excited state to the ground state, going up to very high vibrational levels. The uncertainty arose because the excited state was known to have a potential maximum at large r. In order to extract the ground state D_e from the spectroscopic measurements, the barrier height was needed, and this could not be determined from experiment. The D_e value in Herzberg was actually obtained using a crude theoretical estimate (quoted as uncertain ±100%!) of the barrier. Instead of just computing the barrier with a better calculation, by computing vibrational levels and tunnelling lifetimes in the excited state it was possible to obtain an accurate estimate of the effective barrier, which is noticeably lower than the maximum in the potential curve. It was thus possible to establish D_e to high accuracy by comparing theoretically computed quantities with experiment.

7.6 Interaction with Other Theory

- *The only way to be certain of a theoretical result is to compute it yourself.*

Just because a calculation has found its way into the literature does not, in itself, mean that the result is correct. Over time one builds up a mental list of people whose calculations can be trusted, and the converse. Bear in mind that, commonly, the more distinguished is the senior author, the less likely it is that he/she has had much involvement in running the calculations. It is seldom a waste of computer time to rerun one or two calculations if there is any concern about whether they're right or not. And if this is not possible because, say, the discussion in the paper is inadequate to allow reconstruction of the basis set used, this is grounds for concern about the results anyway.

- *The cost of a calculation is not simply related to the quality.*

In particular, statements like "this calculation consumed three weeks of CPU time on a CONVEX" or "this calculation involved 17 000 000 CSFs in the CI expansion" do not, in themselves say anything about the quality of the results. An all valence electron CCSD(T) calculation on the cytosine-guanine base pair using a split-valence basis would involve about 23 000 000 T_1 and T_2 amplitudes, for example, but the results would undoubtedly be compromised by the inadequate basis set. The criteria that should be used to judge the quality of a calculation are those discussed in this course. Nevertheless, there is a variant of the pathetic fallacy that encourages people to believe that the most expensive calculation they can do must have *some* significance, simply because it was so expensive. Not so.

7.7 Small Effects

- *Basis set superposition error is a genuine problem. It cannot be wished away.*

A study with any pretensions to accuracy should involve some effort to explore BSSE. This does not necessarily mean including a counterpoise correction — the BSSE energy lowering may be small enough to be ignored or it may be useful as a guide to remaining basis set incompleteness. However, the only way to decide this is to compute the fragment energies in the full AO basis and examine the results.

- *Correlation effects involving core electrons may be indispensable for achieving high accuracy.*

For the very highest accuracy, the effect of at least core-valence correlation should be explored. This *must* be accompanied by some serious effort to extend the basis so that core-correlating functions are included. Using valence-optimized basis sets and including core correlation is not only a waste of computer time, but a potential source of problems, as it can substantially increase BSSE. This point is not well appreciated: the prevailing view appears to be that no harm can come of correlating the core when the basis set is inadequate. This is not so.

- *For heavier systems it is useful to make some simple estimate of relativistic effects.*

The first-order perturbation theory estimate of relativistic effects (inclusion of the mass-velocity and one-electron Darwin terms as suggested by Cowan and Griffin) is cheap and easy to compute as a property value at the end of a calculation. It is therefore very valuable as a check on the importance of relativistic effects, and should certainly be included in accurate calculations on, for example, transition-metal compounds. For even heavier elements relativistic effective core potentials should be used.

Bibliography

[1] G. H. Golub and C. F. Van Loan. *Matrix Computations.* Johns Hopkins University Press, Baltimore, 1989.

[2] G. Maroulis, M. Sana, and G. Leroy. *Int. J. Quantum Chem.*, **19**, 43, 1981.

[3] P.-O. Lõwdin. *Adv. Chem. Phys.*, **2**, 207, 1959.

[4] R. Shepard. *Adv. Chem. Phys.*, **69**, 63, 1987.

[5] B. O. Roos. *Adv. Chem. Phys.*, **69**, 399, 1987.

[6] F. W. Bobrowicz and W. A. Goddard. In H. F. Schaefer, editor, *Methods of Electronic Structure Theory.* Plenum, New York, 1977.

[7] W. J. Hehre, L. Radom, P. v. R. Schleyer, and J. A. Pople. *Ab Inition Molecular Orbital Theory.* Wiley-Interscience, New York, 1986.

[8] I. Shavitt. In H. F. Schaefer, editor, *Methods of Electronic Structure Theory.* Plenum, New York, 1977.

[9] N. C. Handy, P. J. Knowles, and K. Somasundram. *Theoret. Chim. Acta*, **68**, 87, 1985.

[10] S. J. Cole and R. J. Bartlett. *J. Chem. Phys.*, **86**, 873, 1987.

[11] R. J. Bartlett. *J. Phys. Chem.*, **90**, 4356, 1989.

[12] Y. S. Lee, S. A. Kucharski, and R. J. Bartlett. *J. Chem. Phys.*, **81**, 5906, 1984.

[13] M. Urban, J. Noga, S. J. Cole, and R. J. Bartlett. *J. Chem. Phys.*, **83**, 8041, 1985.

[14] K. Raghavachari, G. W. Trucks, J. A. Pople, and M. Head-Gordon. *Chem. Phys. Lett.*, **157**, 479, 1989.

[15] J. A. Pople, M. Head-Gordon, and K. Raghavachari. *J. Chem. Phys.*, **87**, 5968, 1987.

[16] T. J. Lee, A. P. Rendell, and P. R. Taylor. *J. Phys. Chem.*, **94**, 5463, 1990.

[17] R. J. Bartlett and G. D. Purvis. *Int. J. Quantum Chem.*, **14**, 561, 1978.

[18] J. A. Pople, J. S. Binkley, and R. Seeger. *Int. J. Quantum Chem. Symp.*, **10**, 1, 1976.

[19] S. R. Langhoff and E. R. Davidson. *Int. J. Quantum Chem.*, 8, 61, 1974.

[20] P. E. M. Siegbahn. *Chem. Phys. Lett.*, **55**, 386, 1978.

[21] E. R. Davidson and D. W. Silver. *Chem. Phys. Lett.*, **52**, 403, 1977.

[22] J. A. Pople, R. Seeger and R. Krishnan. *Int. J. Quantum Chem. Symp.*, **11**, 149, 1977.

[23] J. M. L. Martin, J. P. François, and R. Gijbels. *Chem. Phys. Lett.*, **172**, 346, 1990.

[24] W. D. Laidig and R. J. Bartlett. *Chem. Phys. Lett.*, **104**, 424, 1984.

[25] M. R. A. Blomberg and P. E. M. Siegbahn. *J. Chem. Phys.*, **78**, 5682, 1983.

[26] A. C. Hurley. *Electron Correlation in Small Molecules.* Academic Press, London, 1976.

[27] W. Kutzelnigg. In H. F. Schaefer, editor, *Methods of Electronic Structure Theory.* Plenum, New York, 1977.

[28] R. Ahlrichs, P. Scharf, and C. Ehrhardt. *J. Chem. Phys.*, **82**, 890, 1985.

[29] D. P. Chong and S. R. Langhoff. *J. Chem. Phys.*, **84**, 5606, 1986.

[30] R. J. Gdanitz and R. Ahlrichs. *Chem. Phys. Lett.*, **143**, 413, 1988.

[31] T. J. Lee and P. R. Taylor. *Int. J. Quantum Chem. Symp*, **23**, 199, 1989.

[32] C. W. Bauschlicher, S. R. Langhoff, and A. Komornicki. *Theoret. Chim. Acta*, **77**, 263, 1990.

[33] R. J. Buenker, S. D. Peyerimhoff, and W. Butscher. *Mol. Phys.*, **35**, 771, 1978.

[34] B. Huron, P. Rancurel, and J. P. Malrieu. *J. Chem. Phys.*, 58, 5745, 1973.

[35] R. K. Nesbet, T. L. Barr, and E. R. Davidson. *Chem. Phys. Lett.*, **4**, 203, 1969.

[36] C. W. Bauschlicher and P. R. Taylor. *J. Chem. Phys.*, **85**, 6510, 1986.

[37] C. W. Bauschlicher and P. R. Taylor. *J. Chem. Phys.*, **85**, 2779, 1986.

[38] C. W. Bauschlicher, S. R. Langhoff, and P. R. Taylor. *Adv. Chem. Phys.*, **77**, 103, 1990.

[39] T. J. Lee, R. B. Remington, Y. Yamaguchi, and H. F. Schaefer. *J. Chem. Phys.*, **89**, 408, 1988.

[40] T. Helgaker and J. Almlöf. *J. Chem. Phys.*, **89**, 4889, 1988.

[41] J. W. Richardson, W. C. Nieuwpoort, R. R. Powell, and W. F. Edgell. *J. Chem. Phys.*, **36**, 1057, 1962.

[42] T. H. Dunning. *J. Chem. Phys.*, **53**, 2823, 1970.

[43] S. Huzinaga. Approximate atomic functions. University of Alberta, Edmonton.

[44] B. Roos and P. Siegbahn. *Theoret. Chim. Acta*, **17**, 209, 1970.

[45] F. B. van Duijneveldt. IBM Technical Report RJ945, 1971.

[46] R. S. Grev and H. F. Schaefer. *J. Chem. Phys.*, **91**, 7305, 1989.

[47] R. Ahlrichs and P. R. Taylor. *J. Chim. Phys.*, **78**, 316, 1981.

[48] J. Almlöf and P. R. Taylor. *J. Chem. Phys.*, **86**, 4070, 1987.

[49] J. Almlöf and P. R. Taylor. *Adv. Quantum Chem.*, in press.

[50] J. Almlöf and P. R. Taylor. *J. Chem. Phys.*, **92**, 551, 1990.

[51] J. Almlöf, T. Helgaker, and P. R. Taylor. *J. Phys. Chem.*, **92**, 3029, 1988.

[52] P. O. Widmark, P.-Å. Malmqvist, and B. O. Roos. *Theoret. Chim. Acta*, **77**, 291, 1990.

[53] T. H. Dunning. *J. Chem. Phys.*, **90**, 1007, 1989.

[54] R. C. Raffenetti. *J. Chem. Phys.*, **58**, 4452, 1973.

[55] R. Moccia and M. Zandomeneghi. *Adv. Nucl. Quad. Res.*, **2**, 135, 1975.

[56] E. R. Davidson and D. Feller. *Chem. Rev.*, **86**, 681, 1986.

[57] A. D. McLean and G. S. Chandler. IBM Technical Report RJ2665, 1979.

[58] T. H. Dunning. *J. Chem. Phys.*, **66**, 1383, 1977.

[59] A. J. Wachters. *J. Chem. Phys.*, **52**, 1033, 1970.

[60] P. J. Hay. *J. Chem. Phys.*, **66**, 4377, 1977.

[61] H. Partridge. *J. Chem. Phys.*, **90**, 1043, 1989.

[62] H. Partridge. *J. Chem. Phys.*, **87**, 6643, 1987.

[63] H. Partridge and K. Fægri. to be published.

[64] K. Jankowski, R. Becherer, P. Scharf, H. Schiffer, and R. Ahlrichs. *J. Chem Phys.*, **82**, 1413, 1985.

[65] C. W. Bauschlicher. *Chem. Phys. Lett.*, **142**, 71, 1987.

[66] C. W. Bauschlicher, S. R. Langhoff, and P. R. Taylor. *J. Chem. Phys.*, **87**, 387 1987.

[67] C. W. Bauschlicher and S. R. Langhoff. *J. Chem. Phys.*, **86**, 5595, 1987.

[68] J. Almlöf, B. J. DeLeeuw, P. R. Taylor, C. W. Bauschlicher, and P. Siegbahn *Int. J. Quantum Chem. Symp*, **23**, 345, 1989.

[69] R. Ahlrichs, P. Scharf, and K. Jankowski. *Chem. Phys.*, **98**, 381, 1985.

[70] H.-J. Werner and P. J. Knowles. *J. Chem. Phys.*, **94**, 1264, 1991.

[71] P. Pyykkö. *Chem. Rev.*, **88**, 563, 1988.

[72] T. J. Lee and J. E. Rice. *J. Chem. Phys.*, **94**, 1215, 1991.

[73] G. E. Scuseria. private communication.

[74] T. J. Lee, A. P. Rendell, and P. R. Taylor. *J. Chem. Phys.*, **92**, 489, 1990.

[75] C. W. Bauschlicher and P. R. Taylor. *Theoret. Chim. Acta*, **74**, 63, 1988.

[76] A. P. Rendell, T. J. Lee, and P. R. Taylor. *J. Chem. Phys.*, **92**, 7050, 1990.

[77] J. D. Watts, I. Černušák, J. Noga, R. J. Bartlett, C. W. Bauschlicher, T. Lee, A. P. Rendell, and P. R. Taylor. *J. Chem. Phys.*, **93**, 8875, 1990.

[78] T. J. Lee, A. P. Rendell, and P. R. Taylor. *J. Chem. Phys.*, **93**, 6636, 1990.

[79] T. J. Lee, A. P. Rendell, and P. R. Taylor. *Theoret. Chim. Acta*, in press.

[80] I. Černušák, G. H. F. Diercksen, and A. J. Sadlej. *Phys. Rev. A*, **33**, 814, 1986.

[81] P. R. Taylor, T. J. Lee, J. E. Rice, and J. Almlöf. *Chem. Phys. Lett.*, **163**, 359, 1989.

[82] J. E. Rice, P. R. Taylor, T. J. Lee, and J. Almlöf. *J. Chem. Phys.*, **94**, 4972, 1991.

[83] D. Kolb, W. R. Johnson, and P. Shorer. *Phys. Rev. A*, **26**, 19, 1982.

[84] D. P. Shelton. *Phys. Rev. A*, **42**, 2578, 1990.

[85] D. P. Shelton. private communication.

[86] R. S. Grev, I. L. Alberts, and H. F. Schaefer. *J. Phys. Chem.*, **94**, 3379, 1990.

[87] K. Raghavachari. *Chem. Phys. Lett.*, **171**, 249, 1990.

[88] J. M. L. Martin, J. P. François, and R. Gijbels. *J. Chem. Phys.*, **93**, 5037, 1990.

[89] P. R. Taylor, J. M. L. Martin, J. P. François, and R. Gijbels. *J. Phys. Chem.*, in press.

[90] R. S. Grev, I. L. Alberts, and H. F. Schaefer. *J. Phys. Chem.*, **94**, 8744, 1990.

[91] G. E. Scuseria. *Chem. Phys. Lett.*, **176**, 27, 1991.

[92] S. R. Langhoff, C. W. Bauschlicher, and P. R. Taylor. *J. Chem. Phys.*, **86**, 6992, 1987.

[93] S. R. Langhoff, C. W. Bauschlicher, and P. R. Taylor. *J. Chem. Phys.*, **91**, 5953, 1989.

[94] C. W. Bauschlicher and S. R. Langhoff. *J. Chem. Phys.*, **89**, 2116, 1988.

[95] C. W. Bauschlicher and S. R. Langhoff. *J. Chem. Phys.*, **89**, 4246, 1988.

[96] S. R. Langhoff, C. W. Bauschlicher, and P. R. Taylor. *Chem. Phys. Lett.*, **180**, 88, 1991.

[97] C. W. Bauschlicher, S. R. Langhoff, and P. R. Taylor. *J. Chem. Phys.*, **88**, 2540, 1988.

[98] L. G. M. Pettersson and H. Åkeby. *J. Chem. Phys.*, **94**, 2968, 1991.

[99] P. E. M. Siegbahn. *Chem. Phys. Lett.*, **119**, 515, 1985.

[100] P. Bowen-Jenkins, L. G. M. Pettersson, P. Siegbahn, J. Almlöf, and P. R. Taylor. *J. Chem. Phys.*, **88**, 6977, 1988.

[101] W. Müller, J. Flesch, and W. Meyer. *J. Chem. Phys.*, **80**, 3297, 1984.

[102] S. Wilson, editor. *Methods in Computational Chemistry Vol. 2*. Plenum, New York, 1988.

[103] K. G. Dyall, P. R. Taylor, K. Fægri, and H. Partridge. *J. Chem. Phys.*, in press.

[104] B. Hess. *Phys. Rev. A*, **32**, 756, 1975.

[105] R. D. Cowan and D. C. Griffin. *J. Opt. Soc. Am.*, **66**, 1010, 1976.

[106] R. L. Martin. *J. Phys. Chem*, **87**, 750, 1983.

[107] B. Liu and A. D. McLean. *J. Chem. Phys.*, **91**, 2348, 1989.

[108] S. F. Boys and F. Bernardi. *Mol. Phys.*, **19**, 553, 1970.

[109] V. Parasuk, J. Almlöf, and B. J. DeLeeuw. *Chem. Phys. Lett.*, in press.

[110] M. Gutowski, J. H. van Lenthe, F. B. van Duijneveldt, and G. Chałąsinski. *Chem. Phys. Lett.*, **124**, 370, 1986.

[111] A. D. McLean. private communication.

[112] W. Butscher, S.-K. Shih, R. J. Buenker, and S. D. Peyerimhoff. *Chem. Phys. Lett.*, **52**, 457, 1977.

[113] C. W. Bauschlicher. *Chem. Phys. Lett.*, **74**, 277, 1980.

[114] J. S. Wright and R. J. Buenker. *Chem. Phys. Lett.*, **106**, 570, 1984.

[115] C. W. Bauschlicher. *Chem. Phys. Lett.*, **122**, 572, 1985.

[116] J. M. L. Martin, J. P. François, and R. Gijbels. *J. Comput. Chem.*, **10**, 152, 1989.

[117] J. M. L. Martin, J. P. François, and R. Gijbels. *J. Comput. Chem.*, **10**, 875, 1989.

[118] J. M. L. Martin, J. P. François, and R. Gijbels. *Theoret. Chim. Acta*, **76**, 195, 1989.

[119] P. Pulay. In H. F. Schaefer, editor, *Applications of Electronic Structure Theory*. Plenum, New York, 1977.

THE EFFECTIVE CORE POTENTIAL METHOD.

Ulf Wahlgren
Institute of Theoretical Physics, University of Stockholm.
Vanadisvägen 9, S-11346 Stockholm, Sweden.

I. Introduction.

The ECP method dates back to 1960, when Phillips and Kleinman suggested an approximation scheme for discarding core orbitals in band calculations [1]. They replaced the full Fock-operator with the following operator:

$$\hat{F} \rightarrow \hat{F} + \sum_{c} (\epsilon_v - \epsilon_c) \mid c >< c \mid$$

where ϵ_v is a valence orbital energy and ϵ_c are the core orbital energies. The c:s in the projection operators are the core orbitals, and F_{val} is the usual Fock-operator defined in the valence space only. It is easily realized that the effect of this operator is to make the valence orbital and the core orbitals degenerate. The idea was to rotate the solutions corresponding to the core and the valence orbitals such as to remove the nodes in the valence orbital while keeping its valence character in the outer atomic regions. This trick of course reduces the computational effort in the core region, since in the case of basis set expansions a much smaller basis set will be needed to describe the valence orbital

This method was originally called the Pseudopotential method. Phillips and Kleinman only considered the case of one single valence orbital. in which case the total energy of the valence electron is well defined (and equal to the orbital energy). This detail was of some importance for the generalization of the method to molecular applications. Szads pointed out that the total energy could not be rigorously separated into a core and a valence part when more than one valence orbital was considered[2]. This objection is quite formal, however, since other approximations also have to be made in order to obtain a useful method.

The generalization of the pseudopotential method to molecules was done by Bonifacic and Huzinaga[3] and by Goddard, Melius and Kahn[4] some ten years after Phillips and Kleinman's original proposal. In the molecular pseudopotential or Effective Core Potential (ECP) method all core-valence interactions are approximated with l dependent projection operators. and a totally symmetric screening type potential. The new operators, which are parametrized such that the ECP operator should reproduce atomic all electron results, are added to the Hamiltonian and the one electron ECP equations are obtained variationally in the same way as the usual Hartree Fock equations. Since the total energy is calculated with respect to this approximative Hamiltonian the separability problem becomes obsolete.

There are essentially two types of ECP:s in general use, one which follows Phillips and Kleinmans original suggestion and uses explicit core orbitals in the projection operators, and one which uses projection operators on the orbital angular momentum with a parametrized local radial part. Using the symbol $\hat{M}(r)$ for the totally symmetric screening potential the two types of ECP operators can be written:

$$\hat{F} = \hat{F}^{val} + \hat{M}(r) + \sum_{c}^{core} \mid \phi_c > B_c < \phi_c \mid$$

$$\hat{F} = \hat{F}^{val} + \hat{M}(r) + \sum_{l}^{valence} \mid l > f_l(r) < l \mid$$

where B_c are parameters and $f_l(r)$ are parametrized functions of the form

$$\sum_i A_{i,l} r^{-n_{i,l}} exp(-\alpha_{i,l} r^2)$$

The totally symmetric operators $\hat{M}(r)$ are of the same form as the operators $f_l(r)$. The ϕ_c:s are atomic core orbitals expressed in the full all-electron basis set, and $\hat{F}^{val}$ are normal Fock operators defined in the valence space only. The valence orbitals $\phi_v = \sum_p c_p \chi_v$ are expressed in an appropriate valence basis set, determined through some optimization procedure, which is considerably smaller than the original all electron basis set.

Bonifacic and Huzinaga[3] use explicit core orbital projection operators, while orbital angular momentum projection operators are used by Goddard, Kahn and Melius[4], by Barthelat and Durand[5] and others. Explicit core orbital projection operators can, in the full basis set, be viewed as shift operators which ensure that the first root in the Fock matrix really corresponds to a valence orbital. However, in applications the basis set is always modified and the role of the core orbital projection operators thus partly changes.

The use of Effective Core Potential operators reduces the computational problem in three ways: the primitive basis set can be reduced, the contracted basis set can be reduced and the occupied orbital space can be reduced. The reduction of the occupied orbital space is almost inconsequential in molecular calculations, since it neither affects the number of integrals nor the size of the matrices which has to be diagonalized. The reduction of the primitive basis set is of course more important, but since the integral evaluation time is in general not the bottleneck in molecular calculations, this reduction is still of limited importance. There are some cases where the size of the primitive basis set indeed is important, e.g. in direct SCF procedures. The size of the contracted basis set is very important, however. The bottleneck in normal SCF or CI calculations is the disc storage and/or the iteration time. Both the disc storage and the iteration time depend strongly on the number of contracted functions.

Nodeless valence orbitals are used with Goddard-Kahn-Melius type ECP:s, while the nodal structure in general is kept in conjunction with Huzinaga-type ECP:s. In both cases the valence basis set is determined by some fitting procedure. When the nodal structure of the valence orbitals is kept typically one primitive function is used to describe an inner node.

Experience shows that the two types of ECP:s are of a comparable accuracy. This is not so surprising since the projection operators are l dependent in both cases (the ϕ_c:s of course depend on l) and the operators are parametrized such as to reproduce selected atomic properties, usually valence orbital energies and shapes. There are, however, some practical differences between the methods. If only local r-dependent operators are used the ECP:s become reasonably basis set independent. On the other hand it is easier to optimize the ECP parameters if explicit core orbital projection operators are used (and the nodal structure of the valence orbitals is retained) because of the level shift character of the projection operator (more on this subject below). Of course there is a price to pay if the full nodal structure of the valence orbitals are kept, since a larger primitive basis set has to be used. However, as mentioned above the size of the primitive basis set is in general not crucial in a calculation, and there are some advantages in keeping the full valence nodal structure, for example in CI calculations.

The ECP method which will be discussed henceforth is derived from Huzinaga and Bonifacic's equations, and the full nodal structure of the valence orbitals is always kept. In the early ECP application on first row transition metals the only orbitals which were variationally determined were 3d and 4s[6]. However, experience showed that in certain cases it was important also to include the 3s and the 3p orbitals in the valence space[7-9], and ECP:s with these characteristics were accordingly developed[10].

The equations used in our ECP formalism are as follows:

$$\hat{F} = \hat{h}^{eff} + \sum_{c}^{val}(2\hat{J}_i - \hat{K}_i)$$

where

$$\hat{h}^{eff} = \hat{T} + \hat{V}^{eff} + \hat{P}$$

$$\hat{V}^{eff} = (\frac{-Z^{eff}}{r})(1 + M_1 + M_2)$$

$$M_1 = \sum_{p} A_p exp(-\alpha_p r^2)$$

$$M_2 = \sum_{q} rC_q exp(-\gamma_p r^2)$$

$$\hat{P} = \sum_{k} \mid \phi_k > B_k < \phi_k \mid$$

The parameters which appear in these equations are B_k, A_p, α_p, C_q and γ_q. The B_k values are usually chosen to be the absolute value of the corresponding core orbital energies. A_p, α_p, C_q and γ_q are calibrated to reproduce the orbital energies and shapes of the valence orbitals as obtained from all electron calculations on the atom. The outer core orbitals are in principle included in the valence space but these orbitals are often kept frozen during both the parameter fitting and the molecular calculations.

A full nodal structure of the valence orbitals and the particular form of the Bonifacic Huzinaga type ECP makes it relatively easy to determine the ECP parameters. The energetically most important effect to be described by the ECP is the Pauli repulsion between the core and the valence orbitals. If valence orbitals with nodes are used this effect is to a large extent accounted for by the core orbital projection operators, and the screening potential becomes a fairly small correction term which is reasonably simple to fit.

With the development of new, powerful integral programs and generalized contraction schemes the ECP formalism has become relatively less efficient for atoms lighter than the second row transition metals (TM). If we consider a first row transition row element, an ECP will have to describe the 3s, 3p, 3d, 4s and 4p AO:s explicitly. The size of the contracted basis will thus be at least 3 s-type, 3 p-type and 3 d-type basis functions, which amounts to 30 basis functions. A full basis, using a general contraction scheme will comprise 5 s-type, 4 p-type and 3 d-type basis functions, i.e. 35 basis functions in all. Of course ECP:s may still be important in cases where a substantial number of relatively light atoms are involved, or in metal cluster calculations. One example of the former could be metal complexes such as MCl_6. If ECP:s are used to describe the chlorines the calculation becomes no more time consuming than if the ligands were fluorines.

For heavy atoms the ECP:s quickly become essential. The basis set size at the ECP level does not change if we go to the second or third row TM, while a generally contracted basis set for a second row TM would comprise 45 basis functions. The difference increases even more drastically for the third row TM:s where in addition the 4f orbital can be included in the core in the ECP case.

If elements heavier than the first row transition elements are considered, relativistic effects must be accounted for. This can be done quite easily using the ECP formalism, since the relativistic effects (excluding the spin-orbit coupling) can be accommodated directly into the ECP operators (see also the next section).

Another quite different area where ECP:s have proven to be very useful for the development of transition metal cluster models. By using a very simplified description of the metal atoms, where all electrons including the d-electrons are considered as core, certain properties of the solid material such as chemisorption on metal surfaces or the reactivity of metal clusters has been studied theoretically with considerable success.

In the two following subsections we will describe relativistic spin free ECP:s and the one electron ECP:s used to model metal clusters and metal surfaces.

II. Relativistic ECP:s.

Relativity becomes important for elements heavier than the first row transition elements. Most methods applicable on molecules are derived from the Dirac equation. The Dirac equation itself is difficult to use, since it involves a description of the wave function as a four component spinor. The Dirac equation can be approximately brought to a two-component form using e.g. the Foldy-Wouthuysen (FW) transformation[11,12]. Unfortunately the FW transformation, as originally proposed, is both quite complicated and also divergent in the expansion in the momentum (for large momenta), and it can thus only be carried out approximately (usually to low orders). The resulting equations are not variationally stable, and they are used only in first order perturbation theory.

There are three terms which appears in the first order relativistic expression: the mass-velocity term, the Darwin term and the spin-orbit term[12]. Out of these terms the first two are comparatively easy to calculate, while the spin-orbit interaction term is more complicated. Fortunately, the spin-orbit interaction is often not too important for chemical properties, at least for the second row transition elements. It is therefore usual to neglect it in quantum chemical calculations.

First order perturbation theory is well known to yield good results for elements up to and including the second row transition elements. Perturbation theory can in principle also be used at the ECP level. However, it has been shown that in order for perturbation theory to work the 4s and the 4p shells must also be split[13], and larger basis sets must thus be used in this case. If ECP:s are used to describe the heavier atoms there is no need to rely on first order methods. A variationally stable procedure, which does not require any splitting of the outer core orbitals, is obtained simply by fitting the ECP parameters to relativistic atomic results. One possible weakness with this method, which does not appear to have any practical implications however, is that the operators used to accommodate the relativistic effects do not have the proper functional form. Clearly the Gaussian type operators used with the ECP do not have much resemblance to the Darwin and mass-velocity operators (of course the relativistic effects could in principle be exactly described using Gaussian expansions, but the expansions used in the ECP:s are quite short, usually 2-4 terms).

Another possibility to describe relativistic effects is to use modified convergent FW transformations. A promising procedure along these lines, the no-pair external field method, has been suggested by Douglas and Kroll[14] and further developed by Hess[15]. This method has yielded quite promising results for several second and third row TM compounds. SCF calculations on AuH using the no-pair operator gives a binding energy of 1.54 eV and a bond distance of 1.59 Å. An MCPF calculation gave a binding energy of 2.90 eV, while the experimental bond distance and binding energy is 1.52 Å and 3.35 eV. These results, which are typical, agree well with results obtained with relativistic ECP:s. It is in principle easy to combine the no-pair relativistic method and the ECP method, in which case the relativistic effects no longer have to be described by the ECP operators. However, calculations on AgH and AuH show that the difference between results obtained using explicit no-pair operators in conjunction with ordinary ECP operators, and results obtained with conventional relativistic ECP:s where the relativistic effects are included in the ECP operators, are practically the same.

In summary, conventional relativistic ECP:s provide an efficient mean to calculate molecular properties up to and including the third row transition elements in cases where the spin-orbit coupling is weak. ECP:s can also be used together with explicit relativistic no-pair operators. Such ECP:s are somewhat more precise at at the atomic level, but of essentially the same quality as conventional relativistic ECP:s in molecular applications. It should also be possible to combine the ECP formalism with full Fock-Dirac methods, but this has yet not been done.

III. One electron ECP:s used to model transition metal clusters.

With present day computers it is not realistic to carry out quantitative ab initio calculations on more than 5-10 metal atoms, neither with all electron methods nor with normal ECP:s. If the troublesome d shells can be approximated much larger cluster can be handled, however. Melius et.al.[16-18] were the first to suggest a computational scheme in which the d electrons are also included in an ECP. This type of ECP, in which only the 4s electrons were explicitly accounted for in the calculations, was used by Goddard et al[17] in calculations of oxygen and hydrogen chemisorption on nickel surfaces. These and similar one electron ECP:s have also been used successfully by Bagus et al, by Bauschlicher and by Siegbahn et al in application where one metal atom is described at the all electron level and the one electron ECP atoms are used to mimic a metallic surrounding[19,21].

There are several prerequisites which have to be fulfilled for the one electron ECP approach to be applicable. In the case of metal clusters the atomic configuration must be known, i.e. one must safely be able to assume a $d^n s^2$ or a $d^{n+1}s^1$ configurations on the atoms in the cluster. The d orbitals should not form covalent bonds neither within the cluster nor between the cluster and the adsorbate. Ferromagnetic metals and copper are likely to have these properties. For other metals this is not so clear. Indications are that e.g. the ground state of the Pt_5 cluster is low spin with developed covalent intra cluster d-d bonds[22].

Since the d orbitals are not allowed to relax in a one electron ECP it may appear that a third prerequisite is that the frozen d approximation should be valid, i.e. the relaxation of the d orbitals should not influence the bonding appreciably. In reality the effect of d-shell relaxation on various metal cluster properties is appreciable; e.g. the d-shell relaxation effect on chemisorption energy of oxygen on a Ni_5 cluster is about 40 kcal/mol[23]. However, a small d orbital relaxation is not a necessary prerequisite for the development of a one electron ECP provided that the relaxation is not dominated by covalency effects. The covalent contribution to the bonding of an oxygen atom to a Cu_5 cluster (where the d-shell relaxation contribution to the binding energy is 17 kcal/mol) is only a few kcal/mol[24].

The d orbital projection operator will not play the role of a simple level shift operator in the one electron ECP formalism. On the contrary, it will enter actively in the bonding by raising the energy of the system when an approaching orbital is starting to overlap with the d orbital, which of course is precisely what the "real" d orbital does as well. The only way to determine the parameters of the d projection operators at the *atomic* level would be to consider excited states of the *atom* with an occupied 4d orbital. The sensitivity of the 4d orbital to the parameters of the d projection operator is, however, very small. The only remaining alternative is to optimize the d orbital projection operator by comparing with molecular or cluster calculations at the all electron level. In our one electron ECP approach we have decided to make the parametrization of the d projection operator by comparing with all electron calculations on oxygen chemisorption at the hollow position on a five atom metal cluster[23]. In order to obtain a good description of both the binding energy and the bond distance we modified the form of the d projection operator to "soften" it in the outer regions by introducing a second diffuse and slightly attractive d projection operator. In order to obtain stable results at short adsorbate-surface distances we also included the 3s orbital (frozen) in the valence space.

In cluster models constructed to mimic adsorption at on top positions an all electron description must always used for the on top metal centre. The accuracy of this type of embedding model is very high compared with full all electron calculations [19-21]. Modelling the adsorption at bridge positions should ideally be done using two all electron centra. Such calculations do, however, become rather costly, and a simplified approach is to correct the ECP results close to the bridge site by comparing ECP and all electron results obtained for two metal atoms and the adsorbate.

Typically the basis set used in conjunction with the cluster model consists of 3-4 contracted s functions (with nodes) and one or two p functions. The exponents in the basis set and the expansion coefficients are determined by a least squares fit to the all electron valence orbitals. All the parameters entering the .screening potential are determined by fitting the shape and the energy of the 4s orbital to the corresponding all electron results, keeping the outer core orbitals (3s and sometimes 3p) frozen. The B_k values entering the s and p type. projection operators were chosen in the standard way, i.e. as the negative orbital energies of the corresponding core orbitals.

The B_{3d} parameters are determined on a small metal cluster by comparing computed all-electron and ECP binding energies and distances for one adsorbate (hitherto only oxygen adsorbed on an fourfold site on a five atom metal cluster has been used for this purpose). The binding energies and the bond distances were reproduced to within 1 kcal/mol and 0.05 bohr for Ni_5O[25] and Cu_5O[26]. The vibrational frequencies computed using the one electron ECP:s are somewhat too low, by 40 cm^{-1} for Ni_5O and by 10 cm^{-1} for Cu_5O, indicating a somewhat underestimated repulsion at shorter bond distances.However, the geometrical configuration, with an oxygen atom sort of sliding in between four nickel atoms, will make the vibrational frequencies very sensitive even to minor differences in the interatomic potentials. Calculations on Ni_5H and Cu_5H using ECP parameters determined for oxygen yielded satisfactory results. The difference between the ECP and the all electron binding energies were 3 kcal/mol for Ni_5H and 1 kcal/mol for Cu_5H. The distance above the surface was overestimated by .1 a.u. for Ni_5H. This is not much considering that the potential energy surface is very flat, and these results serves as an indication that the cluster model is reasonably adsorbate independent.

Adsorption of oxygen and hydrogen at hollow sites on Ni(100) and Cu(100) are thus described satisfactorily by one electron ECP clusters. When adsorption at on top positions are considered effects directly involving the metal d shells become important. An example of this is the dissociation of molecular hydrogen at the on top position on a Ni(100) surface where s-d hybridization is very important both for intermediate states and for the barrier to dissociation. In such cases one all electron atom is always used at the on top site. For reactions which take place at bridge sites it is desirable to use two all electron atoms, but in this case the results from all ECP calculations can be corrected by comparing with calculations with an adsorbate and only two all electron metal atoms.

IV. Summary.

In the present chapter we have described the Effective Core Potential method, the special case of relativistic ECP:s and the cluster model. The development of new integral programs and of generalized contractions has made traditional all-electron methods quite competitive in many cases, although the ECP method still leads to large savings for example for small metal clusters or for metal complexes with many heavy ligands. For heavy atoms the ECP method is important, not to say necessary, in particular since relativistic effects (in particular within the spin free formalism) are quite easy to incorporate. In modelling work, such as for cluster models of transition metals, the method has been used with considerable success, and there are as of yet no seriously competing methods within the ab-initio framework.

References

1. Phillips,J.C. and Kleinman,L. (1959), 'New method for calculating wave functions in crystals and molecules', Phys.Rev. **116**, 287
2. Szadz,L. and Brown,L. (1976), 'New formulation of the pseudopotential theory for atoms with two valence electrons', J.Chem.Phys. **65**, 1393
3. Bonifacic,V. and Huzinaga,S. (1974), J.Chem.Phys.,'Atomic and molecular calculations with the model potential method', J.Chem.Phys.**60**, 2779
4. Melius,C.F. and Goddard,W.A. (1974), '*ab initio* effective potentials for use in molecular quantum mechanics', Phys.Rev. **A10**, 1528
5. Barthelat,J.C., Durand,Ph. and Serafini,A. (1977), 'Non-empirical pseudopotentials for molecular calculations I. The PSIBMOL algorithm and test applications', Mol.Phys. **33**, 159
6. Wahlgren,U. (1978), 'Pseudopotential Calculations on some first, second and third row molecules. A comparative study', Chem.Phys. **32**, 215
7. Gropen,O., Wahlgren,U. and Pettersson,L.G.M. (1982), 'Effective core potential calculations on small molecules containing transition metal ions', Chem.Phys. **66**, 459
8. Gropen,O., Wahlgren,U. and Pettersson,L.G.M. (1982), 'Effective core potential calculations on the NiH_4^{2-} ion as a test case for studying rotational barriers' Chem.Phys. **66**, 453
9. Pettersson,L.G.M. and Strömberg,A. (1983), 'A study of the value of the interaction integrals in effective core potential applications', Chem.Phys. Lett. **99**, 122
10. Pettersson,L.G.M., Wahlgren,U. and Gropen,O. (1983), 'Effective core potential calculations using frozen orbitals. Applications to transition metals' Chem.Phys. **80**, 7
11. Foldy,L.L. and Wouthuysen (1950), 'On the Dirac theory of spin $\frac{1}{2}$ particles and its non-relativistic limit', Phys.Rev **78**, 29
12. Moss,R.E. (1973), in: 'Advanced molecular quantum mechanics', Chapman and Hall, London
13. Blomberg,M.R.A. and Wahlgren,U., (1988), 'On the effect of core orbital

relaxation in first-order relativistic calculations', Chem.Phys.Lett. **145**, 393

14. M.Douglas and N.M.Kroll (1974) 'Quantum electrodynamical corrections to the fine structure of helium', Ann.Phys. (N.Y.) **82**, 89
15. Hess,B.A. (1989), 'Relativistic electronic-structure calculations employing a two-component no-pair formalism with external-field projection operators' Phys.Rew.A **33** 3742
16. Melius,C.F., Upton,T.H. and Goddard,W.A. (1978), 'Electronic properties of metal clusters (Ni_{13} to Ni_{87}) and implications for chemisorption', Solid State Comm. **28**, 501
17. Upton,T.H. and Goddard,W.A. (1981),'Chemisorption of H, Cl, Na,O and S on Ni(100) surfaces: A theoretical study using Ni_{20} clusters', in CRC critical reviews, solid state and materials sciences, CRC press, Boca Raton, 1981, 261
18. Upton,T.H. and Goddard,W.A. (1979), 'Chemisorption of atomic hydrogen on large scale nickel cluster surfaces', Phys.Rev.Lett, **42**, 472
19. Bauschlicher,C.W., Walch,S.P., Bagus,P.S. and Brundle,C.R. (1983), 'comment on "Evidence for two states of chemisorbed oxygen on Ni(100)"', Phys.Rev.Lett. **50**, 864
20. Bagus,P.S., Bauschlicher,C.W., Nelin,C.J. and Laskowski,B.C. (1984), 'A proposal for the proper use of pseudopotentials in molecular orbital cluster model studies of chemisorption', J.Chem.Phys. **81**, 3594
21. Bauschlicher,C.W. (1986),'Coverage dependent effects on metal surfaces: O, S,F and Cl on Ni', J.Chem.Phys. **84**, 250
22. Gropen,O. ,Almlof,J. and Wahlgren,U. (1991) 'Model studies of chemisorption on Pt surfaces', in *Cluster Models for Surface and Bulk Phenomena*, NATO ASI Series, (edited by G. Pacchioni and P.S. Bagus, Plenum,New York)
23. I. Panas, P. Siegbahn and U. Wahlgren (1987), 'Model Studies of the Chemisorption of Hydrogen and Oxygen on Nickel Surfaces. I. The Design of a One Electron Effective Core Potential Which Includes 3d Relaxation Effects', Chem.Phys. **112**, 325
24. U. Wahlgren, L.G.M. Pettersson and P. Siegbahn (1989), 'Cu 3d Covalency in Chemisorption ?', J.Chem.Phys. **90**, 4613
25. I. Panas, P. Siegbahn and U. Wahlgren (1988), 'Model Studies of the Chemisorption of Hydrogen and Oxygen on Nickel Surfaces. II. Atomic Chemisorption on Ni(100)', Theor.Chim.Acta **74**, 167
26. A. Mattsson, I. Panas, P. Siegbahn, U. Wahlgren and H. Åkeby (1987), 'Model studies of the chemisorption of hydrogen and oxygen on Cu(100)', Phys.Rev. B **36** (1987) 7389.

Lecture Notes in Chemistry

For information about Vols. 1–18
please contact your bookseller or Springer-Verlag

Vol. 19: E. Clementi, Computational Aspects for Large Chemical Systems. V, 184 pages. 1980.

Vol. 20: B. Fain, Theory of Rate Processes in Condensed Media. VI, 166 pages. 1980.

Vol. 21: K. Varmuza, Pattern Recognition in Chemistry. XI, 217 pages. 1980. (out of print)

Vol. 22: The Unitary Group for the Evaluation of Electronic Energy Matrix Elements. Edited by J. Hinze. VI, 371 pages. 1981.

Vol. 23: D. Britz, Digital Simulation in Electrochemistry. X, 120 pages. 1981. (out of print) 2nd edition. ISBN 3-540-18979-3

Vol. 24: H. Primas, Chemistry, Quantum Mechanics and Reductionism. XII, 451 pages. 1981. (out of print) 2nd edition, ISBN 3-540- 12838-7

Vol. 26: S. Califano, V. Schettino and N. Neto, Lattice Dynamics of Molecular Crystals. VI, 309 pages. 1981.

Vol. 27: W. Bruns, I. Motoc, and K.F. O'Driscoll, Monte Carlo Applications in Polymer Science. V, 179 pages. 1982.

Vol. 28: G.S. Ezra, Symmetry Properties of Molecules. VIII, 202 pages. 1982.

Vol. 29: N.D. Epiotis, Unified Valence Bond Theory of Electronic Structure VIII, 305 pages. 1982.

Vol. 30: R.D. Harcourt, Qualitative Valence-Bond Descriptions of Electron-Rich Molecules: Pauling "3-Electron Bonds" and "Increased-Valence" Theory. X, 260 pages. 1982.

Vol. 31: H. Hartmann, K.-P. Wanczek, Ion Cyclotron Resonance Spectrometry II. XV, 538 pages. 1982.

Vol. 32: H.F. Franzen Second-Order Phase Transitions and the Irreducible Representation of Space Groups. VI, 98 pages. 1982. (out of print)

Vol. 33: G.A. Martynov, R.R. Salem, Electrical Double Layer at a Metal-dilute Electrolyte Solution Interface. VI, 170 pages. 1983.

Vol. 34: N.D. Epiotis, Unified Valence Bond Theory of Electronic Structure · Applications. VIII, 585 pages. 1983.

Vol. 35: Wavefunctions and Mechanisms from Electron Scattering Processes. Edited by F.A. Gianturco and G. Stefani. IX, 279 pages. 1984.

Vol. 36: J. Ugi, J. Dugundji, R. Kopp and D. Marquarding, Perspectives in Theoretical Stereochemistry. XVII, 247 pages. 1984.

Vol. 37: K. Rasmussen, Potential Energy Functions in Conformational Analysis. XIII. 231 pages. 1985.

Vol. 38: E. Lindholm, L. Åsbrink, Molecular Orbitals and their Energies, Studied by the Semiempirical HAM Method. X, 288 pages. 1985.

Vol. 39: P. Vany´sek, Electrochemistry on Liquid/Liquid Interfaces. 2, 3–108 pages. 1985.

Vol. 40: A. Plonka, Time-Dependent Reactivity of Species in Condensed Media. V, 151 pages. 1986.

Vol. 41: P. Pyykkö, Relativistic Theory of Atoms and Molecules. IX, 389 pages. 1986.

Vol. 42: W. Duch, GRMS or Graphical Representation of Model Spaces. V, 189 pages. 1986.

Vol. 43: F.M. Fernández, E.A. Castro, Hypervirial Theorems. VIII, 373 pages. 1987.

Vol. 44: Supercomputer Simulations in Chemistry. Edited by M. Dupuis. V, 312 pages. 1986.

Vol. 45: M.C. Böhm, One-Dimensional Organometallic Materials. V, 181 pages. 1987.

Vol. 46: S.J. Cyvin, I. Gutman, Kekulé Structures in Benzenoid Hydrocarbons. XV, 348 pages. 1988.

Vol. 47: C.A. Morrison, Angular Momentum Theory Applied to Interactions in Solids. 8,9–159 pages. 1988.

Vol. 48: C. Pisani, R. Dovesi, C. Roetti, Hartree-Fock Ab Initio Treatment of Crystalline Systems. V, 193 pages. 1988.

Vol. 49: E. Roduner, The Positive Muon as a Probe in Free Radical Chemistry. VII, 104 pages. 1988.

Vol. 50: D. Mukherjee (Ed.), Aspects of Many-Body Effects in Molecules and Extended Systems. VIII, 565 pages. 1989.

Vol. 51: J. Koca, M. Kratochvíl, V. Kvasnicka, L. Matyska, J. Pospíchal. Synthon Model of Organic Chemistry and Synthesis Design. VI, 207 pages. 1989.

Vol. 52: U. Kaldor (Ed.), Many-Body Methods in Quantum Chemistry. V, 349 pages. 1989.

Vol. 53: G.A. Arteca, F.M. Fernández, E.A. Castro, Large Order Perturbation Theory and Summation Methods in Quantum Mechanics. XI, 644 pages. 1990.

Vol. 54: S.J. Cyvin, J. Brunvoll, B.N. Cyvin, Theory of Coronoid Hydrocarbons. IX, 172 pages. 1991.

Vol. 55: L.T. Fan, D. Neogi, M. Yashima. Elementary Introduction to Spatial and Temporal Fractals. IX, 168 pages. 1991.

Vol. 56: D. Heidrich, W. Kliesch, W. Quapp, Properties of Chemically Interesting Potential Energy Surfaces. VIII, 183 pages. 1991.

Vol. 57: P. Turq, J. Barthel, M. Chemla, Transport, Relaxation, and Kinetic Processes in Electrolyte Solutions. XIV, 206 pages. 1992.

Vol. 58: B. O. Roos (Ed.), Lecture Notes in Quantum Chemistry. VII, 421 pages. 1992.

Editorial Policy

This series aims to report new developments in chemical research and teaching - quickly, informally and at a high level. The type of material considered for publication includes:

1. Preliminary drafts of original papers and monographs
2. Lectures on a new field, or presenting a new angle on a classical field
3. Seminar work-outs
4. Reports of meetings, provided they are
   a) of exceptional interest and
   b) devoted to a single topic.

Texts which are out of print but still in demand may also be considered if they fall within these categories.

The timeliness of a manuscript is more important than its form, which may be unfinished or tentative. Thus, in some instances, proofs may be merely outlined and results presented which have been or will later be published elsewhere. If possible, a subject index should be included. Publication of Lecture Notes is intended as a service to the international chemical community, in that a commercial publisher, Springer-Verlag, can offer a wider distribution to documents which would otherwise have a restricted readership. Once published and copyrighted, they can be documented in the scientific literature.

Manuscripts

Manuscripts should comprise not less than 100 and preferably not more than 500 pages. They are reproduced by a photographic process and therefore must be typed with extreme care. Symbols not on the typewriter should be inserted by hand in indelible black ink. Corrections to the typescript should be made by pasting the amended text over the old one, or by obliterating errors with white correcting fluid. Authors receive 50 free copies and are free to use the material in other publications. The typescript is reduced slightly in size during reproduction; best results will not be obtained unless the text on any one page is kept within the overall limit of 18 x 26.5 cm (7 x $10^1/_2$ inches). The publishers will be pleased to supply on request special stationary with the typing area outlined.

Manuscripts should be sent to one of the editors or directly to Springer-Verlag, Heidelberg.